Basics of Environmental Science

*Basics of Environ_____ iental study. The book
offers everyone s_____ nderstanding of natural
environments an_____ of the environmental
sciences, providin_____ sses and systems and
the effects of hun_____

In this second edit_____ ies is clearly explained.
These include gl_____ n, acid rain, deforestation, human popt_____ generation. There are
also descriptions

Michael Allaby is_____ on various aspects of
environmental sc_____ scientific dictionaries
and edited an ant

Basics of Environmental Science
2nd Edition

Michael Allaby

London and New York

First published 1996
by Routledge
11 New Fetter Lane, London EC4P 4EE

Simultaneously published in the USA and Canada
by Routledge
29 West 35th Street, New York, NY 10001

Second edition 2000
Routledge is an imprint of the Taylor & Francis Group

Typeset by ⚡ Tek-Art, Croydon, Surrey
Printed and bound in Great Britain by
TJ International Ltd, Padstow, Cornwall

British Library Cataloguing in Publication Data
A catalogue record for this book is available from the British Library

Library of Congress Cataloguing in Publication Data
A catalog record for this book is available from the Library of Congress

ISBN 0-415-21175-1 (hbk)
0-415-21176-X (pbk)

Contents

Figures

Tables

 # Preface to the Second Edition

Three years have passed since the first edition of *Basics of Environmental Science* appeared. During this time new concerns have arisen, the controversy in Britain over the safety and desirability of genetically modified foods being the most spectacular example. At the same time, our understanding of other issues has improved as more information about them has been gathered.

Revising the book for its new edition has given me the opportunity to add more information where it is now available and to outline some of the new controversies, including that over genetically modified food. At the same time I have been able to study the whole of the text and to bring it up to date where necessary.

At intervals throughout the book I have added links to sites on the World Wide Web. This has now become an invaluable educational resource and I am delighted to have been able to weave this book into its fabric.

Revised, updated, and modernized, I hope that the new edition will be of value and interest to everyone seeking to broaden their understanding of the science behind environmental issues.

Michael Allaby
Wadebridge, Cornwall
November 1999

How to Use This Book

Basics of Environmental Science will introduce you to most of the topics included under the general heading of 'environmental science'. In this text, these topics are arranged in six chapters: Introduction; Earth Sciences; Physical Resources; Biosphere; Biological Resources; and Environmental Management. Within these chapters, each individual topic is described in a short section. There are 62 of these sections in all, numbered in sequence. All are listed on the contents pages.

You can dip into the book anywhere to read a chapter that interests you. Each is self-contained. It is not quite possible to avoid some overlap, however. This means you may find in one section a technical term that is not fully explained. In the section 'Energy from the Sun' (section 11), for example, you will come across a mention of the 'greenhouse effect', but without a detailed explanation of what that is. When you encounter a difficulty of this kind, refer to the contents pages. In this example you will find a section, number 13, devoted to the 'greenhouse effect', in which the phenomenon is explained fully. If there is no section specifically devoted to the term you find troublesome, look in the index. Almost certainly the term will be explained somewhere, and the index will tell you where to look. Some of the terms that you may find less familiar are defined in the glossary.

At the end of each chapter you will find a list of sections that contain explanations of terms you have just encountered.

This procedure may seem cumbersome, but it would be impractical to provide a full explanation of terms each time they occur.

 Introduction

When you have read this chapter you will have been introduced to:

- a definition of the disciplines that comprise the environmental sciences
- cycles of elements and environmental interactions
- the difference between ecology and environmentalism
- the history of environmental science
- attitudes to the natural world and the way they change over time

1 What is environmental science?

There was a time when, as an educated person, you would have been expected to converse confi-
dently about any intellectual or cultural topic. You would have read the latest novel, been familiar
with the work of the better-known poets, have had an opinion about the current state of art, musical composition and both musical and theatrical performance. Should the subject of the conversation
have changed, you would have felt equally relaxed discussing philosophical ideas. These might
well have included the results of recent scientific research, for until quite recently the word 'philosophy' was used to describe theories derived from the investigation of natural phenomena as well
as those we associate with philosophy today. The word 'science' is simply an anglicized version
of the Latin *scientia*, which means 'knowledge'. In German, which borrowed much less from Latin,
what we call 'science' is known as *Wissenschaft*, literally 'knowledge'. 'Science' did not begin to be
used in its restricted modern sense until the middle of the last century.

As scientific discoveries accumulated it became increasingly difficult, and eventually impossible,
for any one person to keep fully abreast of developments across the entire field. A point came
when there was just too much information for a single brain to hold. Scientists themselves could
no longer switch back and forth between disciplines as they used to do. They became specialists
and during this century their specialisms have divided repeatedly. As a broadly educated person
today, you may still have a general grasp of the basic principles of most of the specialisms, but
not of the detail in which the research workers themselves are immersed. This is not your fault
and you are not alone. Trapped inside their own specialisms, most research scientists find it difficult to communicate with those engaged in other research areas, even those bordering their own.
No doubt you have heard the cliché defining a specialist as someone who knows more and more
about less and less. We are in the middle of what journalists call an 'information explosion' and
most of that information is being generated by scientists.

Clearly, the situation is unsatisfactory and there is a need to draw the specialisms into groups
that will provide overarching views of broad topics. It should be possible, for example, to fit the
work of the molecular biologist, extracting, cloning, and sequencing DNA, into some context that
would relate it to the work of the taxonomist, and the work of both to that of the biochemist. What
these disciplines share is their subject matter. All of them deal with living or once-living organisms. They deal with life and so these, as well as a whole range of related specialisms, have come
to be grouped together as the life sciences. Similarly, geophysics, geochemistry, geomorphology,
hydrology, mineralogy, pedology, oceanography, climatology, meteorology, and other disciplines

are now grouped as the earth sciences, because all of them deal with the physical and chemical nature of the planet Earth.

The third, and possibly broadest, of these groupings comprises the environmental sciences, sometimes known simply as 'environmental science'. It embraces all those disciplines which are concerned with the physical, chemical, and biological surroundings in which organisms live. Obviously, environmental science draws heavily on aspects of the life and earth sciences, but there is some unavoidable overlap in all these groupings. Should palaeontology, for example, the study of past life, be regarded as a life science or, because its material is fossilized and derived from rocks, an earth science? It is both, but not necessarily at the same time. The palaeontologist may date a fossil and determine the conditions under which it was fossilized as an earth scientist, and as a life scientist reconstruct the organism as it appeared when it was alive and classify it. It is the direction of interest that defines the grouping.

Any study of the Earth and the life it supports must deal with process and change. The earth and life sciences also deal with process and change, but environmental science is especially concerned with changes wrought by human activities, and their immediate and long-term implications for the welfare of living organisms, including humans.

At this point, environmental science acquires political overtones and leads to controversy. If it suggests that a particular activity is harmful, then modification of that activity may require national legislation or an international treaty and, almost certainly, there will be an economic price that not everyone will have to pay or pay equally. We may all be environmental winners in the long term, but in the short term there will be financial losers and, not surprisingly, they will complain.

Over the last thirty years or so we have grown anxious about the condition of the natural environment and increasingly determined to minimize avoidable damage to it. In most countries, including the United States and European Union, there is now a legal requirement for those who propose any major development project to calculate its environmental consequences, and the resulting environmental impact assessment is taken into account when deciding whether to permit work to proceed. Certain activities are forbidden on environmental grounds, by granting protection to particular areas, although such protection is rarely absolute. It follows that people engaged in the construction, extractive, manufacturing, power-generating or power-distributing, agricultural, forestry, or distributive industries are increasingly expected to predict and take responsibility for the environmental effects of their activities. They should have at least a general understanding of environmental science and its application. For this reason, many courses in planning and industrial management now include an environmental science component.

This book provides an overview of the environmental sciences. As with all the broad scientific groupings, opinions differ as to which disciplines the term covers, but here the net is cast widely. All the topics it includes are generally accepted as environmental sciences. That said, the approach adopted in *Basics of Environmental Science* is not the only one feasible. In this rapidly developing field there is a variety of ideas about what should be included and emphasized and what constitutes an environmental scientist.

This opening chapter provides a general introduction to environmental science, its history, and its relationship to environmental campaigning. It is here that an important point is made about the overall subject and the content of the book: environmental science and 'environmentalism' are not at all the same thing. Environmental science deals with the way the natural world functions; environmentalism with such modifications of human behaviour as reformers think appropriate in the light of scientific findings. Environmentalists, therefore, are concerned with more than just science. As its title implies, *Basics of Environmental Science* is concerned mainly with the science.

The introduction is followed by four chapters, each of which deals with an aspect of the fundamental earth and life sciences on which environmental science is based, in each case emphasizing the importance of process and change and, where appropriate, relating the scientific description of what happens to its environmental implications and the possible consequences of perturbations to the system. The fifth and final chapter deals with environmental management, covering such matters as wildlife conservation, pest control, and the control of pollution.

You do not have to be a scientist to understand *Basics of Environmental Science*. Its language is simple, non-technical, and non-mathematical, but there are suggestions for further reading to guide those who wish to learn more. Nor do you have to read the book in order, from cover to cover. Dip into it in search of the information that interests you and you will find that each short block is quite self-contained.

It is the grouping of a range of disciplines into a general topic, such as environmental science, which makes it possible to provide a broad, non-technical introduction. The grouping is natural, in that the subjects it encompasses can be related to one another and clearly belong together, but it does not resolve the difficulty of scientific specialization. Indeed, it cannot, for the great volume of specialized information that made the grouping desirable still exists. Except in a rather vague sense, you cannot become an 'environmental scientist', any more than you could become a 'life scientist' or an 'earth scientist'. Such imprecise labels have very little meaning. Were you to pursue a career in the environmental sciences you might become an ecologist, perhaps, or a geomorphologist, or a palaeoclimatologist. As a specialist you would contribute to our understanding of the environment, but by adding detailed information derived from your highly specialized research.

Environmental science exists most obviously as a body of knowledge in its own right when a team of specialists assembles to address a particular issue. The comprehensive study of an important estuary, for example, involves mapping the solid geology of the underlying rock, identifying the overlying sediment, measuring the flow and movement of water and the sediment it carries, tracing coastal currents and tidal flows, analysing the chemical composition of the water and monitoring changes in its distribution and temperature at different times and in different parts of the estuary, sampling and recording the species living in and adjacent to the estuary and measuring their productivity.[1] The task engages scientists from a wide range of disciplines, but their collaboration and final product identifies them all as 'environmental scientists', since their study supplies the factual basis against which future decisions can be made regarding the environmental desirability of industrial or other activities in or beside the estuary. Each is a specialist; together they are environmental scientists, and the bigger the scale of the issue they address the more disciplines that are likely to be involved. Studies of global climate change currently engage the attention of climatologists, palaeoclimatologists, glaciologists, atmospheric chemists, oceanographers, botanists, marine biologists, computer scientists, and many others, working in institutions all over the world.

You cannot hope to master the concepts and techniques of all these disciplines. No one could, and to that extent the old definition of an 'educated person' has had to be revised. Allowing that in the modern world no one ignorant of scientific concepts can lay serious claim to be well educated, today we might take it to mean someone possessing a general understanding of the scientific concepts from which the opinions they express are logically derived. In environmental matters these are the concepts underlying the environmental sciences. *Basics of Environmental Science* will introduce you to those concepts. If, then, you decide to become an environmental scientist the book may help you choose what kind of environmental scientist to be.

2 Environmental interactions, cycles, and systems

Inquisitive children sometimes ask whether the air they breathe was once breathed by a dinosaur. It may have been. The oxygen that provides the energy to power your body has been used many times by many different organisms, and the carbon, hydrogen, and other elements from which your body is made have passed through many other bodies during the almost four billion years that life has existed on our planet. All the materials found at the surface of the Earth, from the deepest ocean trenches to the top of the atmosphere, are engaged in cycles that move them from place to place. Even the solid rock beneath your feet moves, as mountains erode, sedimentary rocks are subducted into the Earth's mantle, and volcanic activity releases new igneous rock. There is nothing new or original in the idea of recycling!

The cycles proceed at widely differing rates and rates that vary from one part of the cycle to another. Cycling rates are usually measured as the time a molecule or particle remains in a particular part of the cycle. This is called its 'residence time' or 'removal time'. On average, a dust or smoke particle in the lower atmosphere (the troposphere) remains airborne for a matter of a few weeks at most before rain washes it to the surface, and a water molecule remains in the air for around 9 or 10 days. Material reaching the upper atmosphere (the stratosphere) resides there for much longer, sometimes for several years, and water that drains from the surface into ground water may remain there for up to 400 years, depending on the location.

Water that sinks to the bottom of the deep oceans eventually returns to the surface, but this takes very much longer than the removal of water molecules from the air. In the Pacific Ocean, for example, it takes 1000 to 1600 years for deep water to return to the surface and in the Atlantic and Indian Oceans it takes around 500 to 800 years (MARSHALL, 1979). This is relevant to concerns about the consequences of disposing industrial and low-level radioactive waste by sealing it in containers and dumping them in the deep oceans.

Those monitoring the movement of materials through the environment often make use of labelling, different labels being appropriate for different circumstances. In water, chemically inert dyes are often used. Certain chemicals will bond to particular substances. When samples are recovered, analysis reveals the presence or absence of the chemical label. Radioisotopes are also used. These consist of atoms chemically identical to all other atoms of the same element, but with a different mass, because of a difference in the number of neutrons in the atomic nucleus. Neutrons carry no charge and so take no part in chemical reactions, the chemical characteristics of an element being determined by the number of protons, with a positive charge, in its atomic nucleus.

You can work out the atmospheric residence time of solid particles by releasing particles labelled chemically or with radioisotopes and counting the time it takes for them to be washed back to the ground, although the resulting values are very approximate. Factory smoke belching forth on a rainy day may reach the ground within an hour or even less; the exhaust gases from an aircraft flying at high altitude will take much longer, because they are further from the ground to start with and in much drier air. It is worth remarking, however, that most of the gases and particles which pollute the air and can be harmful to health have very short atmospheric residence times. Sulphur dioxide, for example, which is corrosive and contributes to acid rain, is unlikely to remain in the air for longer than one month and may be washed to the surface within one minute of being released. The atmospheric residence time for water molecules is calculated from the rate at which surface water evaporates and returns as precipitation.

The deep oceans are much less accessible than the atmosphere, but water carries a natural label in the form of carbon-14 (^{14}C). This forms in the atmosphere through the bombardment of nitrogen (^{14}N) by cosmic radiation, but it is unstable and decays to the commoner ^{12}C at a steady rate. While water is exposed to the air, both ^{12}C and ^{14}C dissolve into it, but once isolated from the air the decay of ^{14}C means that the ratio of the two changes, ^{12}C increasing at the expense of ^{14}C. It is assumed that ^{14}C forms in the air at a constant rate, so the ratio of ^{12}C to ^{14}C is always the same and certain assumptions are made about the rate at which atmospheric carbon dioxide dissolves into sea water and the rate at which water rising from the depths mixes with surface water. Whether or not the initial assumptions are true, the older water is the less ^{14}C it will contain, and if the assumptions are true the age of the water can be calculated from its ^{14}C content in much the same way as organic materials are ^{14}C-dated.

Carbon, oxygen, and sulphur are among the elements living organisms use and they are being cycled constantly through air, water, and living cells. The other elements required as nutrients are also engaged in similar biogeochemical cycles. Taken together, all these cycles can be regarded as components of a very complex system functioning on a global scale. Used in this sense, the concept of a 'system' is derived from information theory and describes a set of components which interact to form a coherent, and often self-regulating, whole. Your body can be considered as a system in which each organ performs a particular function and the operation of all the organs is coordinated so that you exist as an individual who is more than the sum of the organs from which your body is made.

Biochemical cycles

The surface of the Earth can be considered as four distinct regions and because the planet is spherical each of them is also a sphere. The rocks forming the solid surface comprise the lithosphere, the oceans, lakes, rivers, and icecaps form the hydrosphere, the air constitutes the atmosphere, and the biosphere contains the entire community of living organisms.

Materials move cyclically among these spheres. They originate in the rocks (lithosphere) and are released by weathering or by volcanism. They enter water (hydrosphere) from where those serving as nutrients are taken up by plants and from there enter animals and other organisms (biosphere). From living organisms they may enter the air (atmosphere) or water (hydrosphere). Eventually they enter the oceans (hydrosphere), where they are taken up by marine organisms (biosphere). These return them to the air (atmosphere), from where they are washed to the ground by rain, thus returning to the land.

The idea that biogeochemical cycles are components of an overall system raises an obvious question: what drives this system? It used to be thought that the global system is purely mechanical, driven by physical forces, and, indeed, this is the way it can seem. Volcanoes, from which atmospheric gases and igneous rocks erupt, are purely physical phenomena. The movement of crustal plates, weathering of rocks, condensation of water vapour in cooling air to form clouds leading to precipitation – all these can be explained in purely physical terms and they carry with them the substances needed to sustain life. Organisms simply grab what they need as it passes, modifying their requirements and strategies for satisfying them as best they can when conditions change.

Yet this picture is not entirely satisfactory. Consider, for example, the way limestone and chalk rocks form. Carbon dioxide dissolves into raindrops, so rain is very weakly acid. As the rain water washes across rocks it reacts with calcium and silicon in them to form silicic acid and calcium bicarbonate, as separate calcium and bicarbonate ions. These are carried to the sea, where they react to form calcium carbonate, which is insoluble and slowly settles to the sea bed as a sediment that, in time, may be compressed until it becomes the carbonate rock we call limestone. It is an entirely inanimate process. Or is it? If you examine limestone closely you will see it contains vast numbers of shells, many of them minute and, of course, often crushed and deformed. These are of biological origin. Marine organisms 'capture' dissolved calcium and bicarbonate to 'manufacture' shells of calcium carbonate. When they die the soft parts of their bodies decompose, but their insoluble shells sink to the sea bed. This appears to be the principal mechanism in the formation of carbonate rocks and it has occurred on a truly vast scale, for limestones and chalks are among the commonest of all sedimentary rocks. The famous White Cliffs of Dover are made from the shells of once-living marine organisms, now crushed, most of them beyond individual recognition.

Here, then, is one major cycle in which the biological phase is of such importance that we may well conclude that the cycle is biologically driven, and its role extends further than the production of rock. The conversion of soluble bicarbonate into insoluble calcium carbonate removes carbon, as carbon dioxide, from the atmosphere and isolates it. Eventually crustal movements may return the rock to the surface, from where weathering returns it to the sea, but its carbon is in a chemically stable form. Other sedimentary rock on the ocean floor is subducted into the mantle. From there its carbon is returned to the air, being released volcanically, but the cycle must be measured in many millions of years. For all practical purposes, most of the carbon is stored fairly permanently. As the newspapers constantly remind us, carbon dioxide is a 'greenhouse gas', one of a number of gases present in the atmosphere that are transparent to incoming, short-wave solar radiation, but partially opaque to long-wave radiation emitted from the Earth's surface when the Sun has warmed it. These gases trap outgoing heat and so maintain a temperature at the surface markedly higher than it would be were they absent. Since the Earth formed, some 4.6 billion years ago, the Sun has grown hotter by an estimated 25 to 30 per cent, and the removal of carbon dioxide from the air, to a significant extent as a result of biological activity, has helped prevent surface temperatures rising to intolerable levels.

Gaia

A hypothesis, proposed principally by James Lovelock, that all the Earth's biogeochemical cycles are biologically driven and that on any planet which supports life conditions favourable to life are maintained biologically. Lovelock came to this conclusion as a result of his participation in the preparations for the explorations of the Moon and Mars. One object of the Mars programme was to seek signs of life on the planet. Martian organisms, should they exist, might well be so different from organisms on Earth as to make them difficult to recognize as being alive at all. Lovelock reasoned that the one trait all living organisms share is their modification of the environment. This occurs when they take materials from the environment to provide them with energy and structural materials, and discharge their wastes into the environment. He argued that it should be possible to detect the presence of life by an environment, especially an atmosphere, that was far from chemical equilibrium. Earth has such an atmosphere, with anomalously large amounts of nitrogen and oxygen, as well as methane, which cannot survive for long in the presence of oxygen. It then occurred to him that the environmental modifications made and sustained by living organisms actually produced and maintained chemical and physical conditions optimum for those organisms themselves. In other words, the organisms produce an environment which suits them and then 'manage' the planet in ways that maintain those conditions.

Does this suggest that our climate is moderated, or even controlled, by biological manipulation? Certainly this is the view of James Lovelock, whose Gaia hypothesis takes the idea much further, suggesting that the Earth may be regarded as, or perhaps really *is*, a single living organism. It was this idea of a 'living planet' that he came to call 'Gaia' (LOVELOCK, 1979).

His hypothesis has aroused considerable interest, but Gaia remains controversial and there are serious objections to it. Expressed in its most extreme form, which is that almost all surface processes are biologically driven, it appears circular, with an explanation for everything, as when the existence of Gaia is introduced to explain the hospitable environment and the hospitable environment proves the existence of Gaia (JOSEPH, 1990). On the other hand, the more moderate version, which emphasizes the biological component of biogeochemical cycles more strongly than most traditional accounts, commands respect and promises to be useful in interpreting environmental phenomena, although not all scientists would associate this with the name 'Gaia' (WESTBROEK, 1992). It has been found, for example, that the growth of marine plankton can be stimulated by augmenting the supply of iron, an essential and, for them, limiting nutrient, with implications for the rate at which carbon dioxide is transferred from the atmosphere to the oceans and, therefore, for possible climate change (DE BAAR *ET AL.*, 1995).

Authorities differ in the importance they allot to the role of the biota (the total of all living organisms in the world or some defined part of it) in driving the biogeochemical cycles, but all agree that it is great, and it is self-evident that the constituents of the biota shape their environment to a considerable extent. Grasslands are maintained by grazing herbivores, which destroy seedlings by eating or trampling them, so preventing the establishment of trees, and over-grazing can reduce semi-arid land to desert. The presence of gaseous oxygen in the atmosphere is believed to result from photosynthesis.

We alter the environment by the mere fact of our existence. By eating, excreting, and breathing we interact chemically with our surroundings and thereby change them. We take and use materials, moving them from place to place and altering their form. Thus we subtly modify environmental conditions in ways that favour some species above others. In our concern that our environmental modifications are now proceeding on such a scale as to be unduly harmful to other species and possibly ourselves, we should not forget that in this respect we differ from other species only in degree. All living things alter their surroundings, through their participation in the cycles that together comprise the system which is the dynamic Earth.

3 Ecology and environmentalism

Our concern over the condition of the natural environment has led to the introduction of a new concept, of 'environmental quality', which can be measured against defined parameters. To give one example, if the air contains more than 0.1 parts per million (ppm) of nitrogen dioxide (NO_2)

or sulphur dioxide (SO_2) persons with respiratory complaints may experience breathing difficulties, and if it contains more than about 2.5 ppm of NO_2 or 5.0 ppm of SO_2 healthy persons may also be affected (KUPCHELLA AND HYLAND, 1986). These are quantities that can be monitored, and there are many more. It is also possible, though much more difficult, to determine the quality of a natural habitat in terms of the species it supports and to measure any deterioration as the loss of species.

These are matters that can be evaluated scientifically, in so far as they can be measured, but not everything can be measured so easily. We know, for example, that in many parts of the tropics primary forests are being cleared, but although satellites monitor the affected areas it is difficult to form accurate estimates of the rate at which clearance is proceeding, mainly because different people classify forests in different ways and draw different boundaries to them. The United Nations Environment Programme (UNEP) has pointed out that between 1923 and 1985 there were at least 23 separate estimates of the total area of closed forest in the world, ranging from 23.9 to 60.5 million km^2. The estimate UNEP prefers suggests that in pre-agricultural times there was a total of 12.77 million km^2 of tropical closed forest and that by 1970 this had been reduced by 0.48 per cent, to 12.29 million km^2, and that the total area of forests of all kinds declined by 7.01 per cent, from 46.28 to 39.27 million km^2, over the same period (TOLBA ET AL., 1992). Edward O. Wilson, on the other hand, has written that in 1989 the total area of *rain* forests was decreasing by 1.8 per cent a year (WILSON, 1992). (A rain forest is one in which the annual rainfall exceeds 2540 mm; most occur in the tropics, but there are also temperate rain forests.) Similar differences occur in estimates of the extent of land degradation through erosion and the spread of deserts (called 'desertification'). Before we can devise appropriate responses to these examples of environmental deterioration we have to find some way of reconciling the varying estimates of their extent. After all, it is impossible to address a problem unless we can agree on its extent.

Even when quantities can be measured with reasonable precision controversy may attend interpretations of the measurements. We can know the concentration of each substance present in air, water, soil, or food in a particular place at a particular time. If certain of those substances are not ordinarily present and could be harmful to living organisms we can call them 'pollutants', and if they have been introduced as a consequence of human activities, rather than as a result of a natural process such as volcanism, we can seek to prevent further introduction of them in the future. This may seem simple enough, but remember that someone has to pay for the measurement: workers need wages, and equipment and materials must be bought. Reducing pollution is usually inconvenient and costly, so before taking action, again we need to determine the seriousness of the problem. The mere presence of a pollutant does not imply harm, even when the pollutant is known to be toxic. Injury will occur only if susceptible organisms are exposed to more than a threshold dose, and where large numbers of very different species of plants, animals, and microorganisms are present this threshold is not easily calculated.

Nor is it easy to calculate thresholds for human exposure, because only large populations can be used for the epidemiological studies that will demonstrate effects, and small changes cannot always be separated statistically from natural fluctuations. (Epidemiology is the study of the incidence, distribution, and control of illness in a human population.) It has been estimated that over several decades the 1986 accident at the Chernobyl nuclear reactor may lead to a 0.03 per cent increase in radiation-induced cancer deaths in the former Soviet Union and a 0.01 per cent increase in the world as a whole, increases that will not be detectable against the natural variations in the incidence of cancer from year to year (ALLABY, 1995).

Where there is doubt, prudence may suggest we set thresholds very low, and in practice this is what happens. With certain pesticide residues in food, for example, the EU operates a standard of 'surrogate zero' by setting limits lower than the minimum quantity that can be detected.

Where the statistical evaluation of risk is unavoidably imprecise yet remedial action seems intuitively desirable, decisions cannot be based solely on scientific evidence and are bound to be more or less controversial. Since decisions of any kind are necessarily political, and will be argued this way and that, people will take sides and issues will tend to become polarized.

At this point, environmental science gives way to environmental campaigning, or environmentalism, and political campaigns are managed by those activists best able to publicize their opinion. In their efforts to attract public attention and support, spokespersons are likely to be drawn into oversimplifying complex, technical issues which, indeed, they may not fully understand, and to exaggerate hazards for the sake of dramatic effect.

Environmental science has a long history and concern with the condition of the environment has been expressed at intervals over many centuries, but the modern environmental movement emerged during the 1960s, first in the United States and Britain. The publication of *Silent Spring* in 1962 in the United States and 1963 in Britain provided a powerful stimulus to popular environmental concern and may have marked the origin of the modern movement. This was the book in which Rachel Carson mounted a strong attack on the way agricultural insecticides were being used in North America. The dire consequences of which she warned were essentially ecological: she maintained that the indiscriminate poisoning of insects by non-selective compounds was capable of disrupting food chains, the sequences of animals feeding on one another as, for example, insects → blackbirds → sparrowhawks. The 'silent spring' of her title referred to the absence of birds, killed by poisons accumulated through feeding on poisoned insects, but the 'fable' with which the book begins also describes the deaths of farm livestock and humans. The catastrophe was ecological and so the word 'ecology' acquired a political connotation. A magazine devoted to environmental campaigning, founded in 1970, was (and still is) called *The Ecologist*.

Ecology is a scientific discipline devoted to the study of relationships among members of living communities and between those communities and their abiotic environment. Intrinsically it has little to do with campaigning for the preservation of environmental quality, although individual ecologists often contribute their professional expertise to such campaigns and, of course, their services are sought whenever the environmental consequences of a proposed change in land use are assessed.

To some non-scientists, however, 'ecology' suggests a kind of stability, a so-called 'balance of nature' that may have existed in the past but that we have perturbed. This essentially metaphysical concept is often manifested as an advocacy for ways of life that are held to be more harmonious or, in the sense in which the word is now being used, 'ecological'. The idea is clearly romantic and supported by a somewhat selective view of history, but it has proved powerfully attractive. In her very detailed study of it Meredith Veldman, a historian at Louisiana State University, locates the development of environmentalism in Britain firmly in a long tradition of romantic protest that also includes the fiction of J.R.R. Tolkien and the Campaign for Nuclear Disarmament (VELDMANN, 1994).

'Ecology', then, is at one and the same time a scientific discipline and a political, at times almost religious, philosophy which inspires a popular movement and 'green' political parties in many countries. As a philosophy, it no longer demands piecemeal reform to achieve environmental amelioration, but calls for the radical restructuring of society and its economic base. The two meanings attached to the word are now quite distinct and it is important not to confuse them. When people say a particular activity or way of life is 'ecologically sound' they are making a political statement, not a scientific one, even though they may be correct in supposing the behaviour they approve to have less adverse effect on human health or the welfare of other species than its alternatives. 'Ecologically sound' implies a moral judgement that has no place in scientific argument; to a scientist the phrase is meaningless.

This is not to denigrate those who use the word 'ecology' in one sense or the other, simply to point out that the meanings are distinct and our attitudes to the environment are shaped by historical, social, and economic forces. They are not derived wholly from a scientific description of the environment or understanding of how it works. The nuclear power industry, for example, is opposed on ecological grounds, but there is no evidence that it has ever caused the slightest injury to non-humans, apart from vegetation around the Chernobyl complex following the accident there, and its adverse effects on human health are extremely small, especially when compared with those resulting from other methods of power generation; indeed, it is extremely unlikely that the correct routine operation of a nuclear power plant has any harmful effect at all, on humans or non-humans.[2] The anti-nuclear wing of the environmental movement is highly influential and has done much to erode public confidence in the industry, but whether this is environmentally beneficial is open to debate, to say the least. In contrast, on those occasions when scientists and campaigners collaborate, say in devising (scientifically) the best way to manage an area in order to maximize its value as natural habitat then campaigning (politically) to have the area protected from inappropriate development, they can achieve their useful and practicable goal. While it is certainly true that some ecological (i.e. environmentalist) campaigns owe little to ecology (the science), others, though not necessarily the most populist, are scientifically well informed. It is also true that if we confine our interest to the acquisition of an abstract understanding of the way the world is, that understanding will be of limited practical value. If damage to the environment is to be avoided or past damage remedied, scientific understanding must be applied and this is possible only through political processes.

This book will introduce you to the environmental sciences, of which ecology is one and, therefore, the word 'ecology' will henceforth be used only in its scientific sense. When issues of concern to environmentalists are discussed, as obviously they must be, they will be evaluated scientifically rather than politically. If your knowledge of environmental matters until now has been derived principally from campaigning literature, you may find the scientific accounts describe a world that is far more complex than you may have supposed and about which rather less is known than the campaigners sometimes imply. You should not be disheartened, for that is the way it is, and much remains to be discovered – perhaps one day, by you.

4 History of environmental science

By the time their civilization reached its peak in the Fifth Dynasty (after about 2480 BC) the ancient Egyptians seem to have become happy people. According to accounts described by the late Joseph Campbell (CAMPBELL, 1962), a leading authority on the ways people have seen themselves and the world around them, they had a joyful, outward-looking view of the world around them. True, they were somewhat preoccupied with the after-life, but that was seen pretty much as a continuation of their present lives and was celebrated in some of the most beautiful art and magnificent architecture the world has ever seen. Their pharaoh was described as 'good' rather than 'great' and the land he ruled was paradise, mythologically and to some extent literally. Life was very predictable and secure. Each year, the appearance of Sirius, the star of Isis, on the horizon at dawn heralded the flooding of the Nile. The reliable flood brought water and silt to enrich the cultivated land and guarantee the bountiful harvest that would follow. No doubt the work was hard, as it always is, but there was ample time for festivals and celebrations.

The Egyptians did not develop what we would recognize as science. Their view of the world was mythological and magical. Nevertheless, they did have a view of the world and a practical knowledge of those aspects of it that mattered to them. They knew much about agriculture, plants and animals, water and how to use it effectively, they knew how to make bricks and were expert in the use of stone. People have always constructed mental frameworks to describe and explain the world around them. Not all were as positive as that of the Egyptians, but humans have an inherent need to understand, to make sense of their surroundings and locate themselves in them.

If we are to understand the world about us we must discover an order underlying phenomena or, failing that, impose one. Only then can we categorize things and so bring coherence to what would otherwise be chaotic. Most early attempts at classification were based on a mythological world-view. The anthropologist Mary Douglas has suggested, for example, that the biblical distinction between 'clean' and 'unclean' animals arose because Hebrew priests believed that sheep and goats, both ruminant animals with cloven hoofs, fitted into what they supposed to be the divine scheme, but pigs did not, because they have cloven hoofs but are not ruminants (BOWLER, 1992, pp. 11–12).

Science, in those days called 'philosophy' ('love of wisdom'), began with Thales (c. 640–546 BC), who lived in the Greek trading town of Miletus on the Aegean coast of what is now Turkey. He and his followers became known as the Ionian or Milesian school and the radical idea they introduced was that phenomena could be discussed rationally. That is to say, they suggested the mythical accounts of creation could be tested and rational explanations proposed for the order underlying the constant change we see everywhere. It is this critical attitude, allowing all ideas to be challenged by rational argument based on evidence and weaker theories to be replaced by stronger ones, which distinguishes science from non-science and pseudoscience. It originated only once; other civilizations developed considerable technological skills, but it was only among the Greeks living on the shores of Asia Minor that the modern concept of a 'scientific approach' emerged. All our science is descended from that beginning, and it began with environmental science. The Greek development reached its peak with the Academy, founded by Plato (429–347 BC), a student of Socrates, and the Lyceum, founded by Plato's disciple Aristotle (384–322 BC). Aristotle wrote extensively on natural history. His studies of more than 500 species of animals included accurate descriptions, clearly based on personal observation, that were not confirmed until many centuries later. He recorded, for example, the reproduction of dogfish and the mating of squid and octopus. He also wrote about the weather in a book called *Meteorologica* ('discourse on atmospheric phenomena'), from which we derive our word 'meteorology'.

Roman thinkers continued the Greek tradition, Pliny the Elder (c. AD 23–79) being the best-known Roman naturalist. His *Natural History*, covering what are now recognized as botany, zoology, agriculture, geography, geology, and a range of other topics, was based on fact, although he mingled records of his own observations with myths and fantastic travellers' tales. Muslim scholars translated the Greek and Latin texts into Arabic, but it was not until the thirteenth century that they became generally available in Europe, as Latin translations from the Arabic.

Throughout this long history the central purpose of the enterprise has survived. There have been digressions, confusions, theories that led into blind alleys, but always the principal aim has been to replace mythical explanations with rational ones. Since myth is very often enshrined in religious texts, it may seem that the scientific agenda is essentially atheistic. Indeed, it has been so at times and in respect of some religions, and to this day scientists are often accused of atheism, but most modern thinkers regard the conflict as much more apparent than real. The writings of the Arabian physician Avicenna (979–1037) and philosopher Averroës (1126–98) kept classical ideas current in the Muslim world, where they were accommodated quite comfortably by Islam, and St Thomas Aquinas (c. 1224–74) used the natural order revealed by Aristotle as a proof of the existence of God, thus permitting science and religion to coexist in Christendom.

This is not to say that the dividing line between mythical and rational explanations was always clearly drawn, nor to deny that interpretations which undermined traditional beliefs sometimes generated fierce arguments. Scientists were often engaged in attempts to reconcile the two views and, then as now, scientific ideas could be attacked on essentially political grounds. In the years following the French Revolution, for example, conservatives in Britain used scriptural authority to justify the preservation of the social order. This led scientists supporting them to adapt the Neptunian theory of Abraham Gottlob Werner (1749–1817) so that it appeared to substantiate the story of Noah's flood. Werner proposed that the Earth was once covered entirely by an ocean, from which some rocks had crystallized and beneath which others had been deposited as sediment, the rocks being exposed through the gradual and continuing retreat of the waters. This obsession with the biblical flood continues in some English-speaking countries to the present day, from time to time with 'discoveries' of the remains of the Ark, although scientists elsewhere in Europe had ceased to take it seriously by the eighteenth century (BOWLER, 1992, pp. 129–130).

Much of the history of the environmental sciences revolves about the reconstruction of the history of the planet since it first formed. To a considerable extent, this reconstruction was based on interpretations of fossils, which were by no means always seen as the obvious remains of once-living organisms.[3] Even when it became possible to use the fossils entrapped within them to arrange rock strata in a chronological sequence controversy continued over the assignment of dates to those strata, the mechanisms by which the rocks had assumed their present forms and distribution, and over the total age of the Earth itself. It was in his effort to solve this puzzle that in 1650 James Ussher (1581–1656), an Irish scholar and archbishop of Armagh, constructed what may have been the first theoretical model. Basing his chronology on the Old Testament, he concluded the Earth had been created in 4004 BC!

If the development of environmental science seems to have been dominated by the study of rocks and fossils, it is perhaps because elucidating the history of the planet was a necessary first step toward an understanding of its present condition and, in any case, the classification and distribution of plants and animals played a major role in it. The theory of evolution by natural selection was derived from Earth history, and Charles Darwin (1809–82) began his career as a geologist.

A unifying theme was supplied by Alexander von Humboldt (1769–1859). Mining engineer, geologist, geophysicist, meteorologist, and geographer, Humboldt spent the years from 1799 to 1804 exploring in tropical South America with his friend, the botanist Aimé Bonpland (1773–1858). His subsequent accounts greatly advanced knowledge of plant geography and his five-volume *Kosmos*, completed after his death, sought to demonstrate how physical, biological, and human activities combined to regulate the environment (BOWLER, 1992, pp. 204–211). This helped establish biogeography as a scientific discipline and applied a range of disciplines to the study of environments. Humboldt is also credited with having shifted science generally from its rather abstract preoccupations in the eighteenth century to its much greater reliance on observation and experiment characteristic of the nineteenth and twentieth.

Biogeography also fed back into the earth sciences. Plotting the distribution of present and extinct plants and animals played a major part in the development of the theory of continental drift by the German climatologist Alfred Wegener (1880–1930), who sought to explain the apparent fit between the coasts of widely separated continents, such as the west coast of Africa and east coast of South America, by postulating that the continents were once joined and have since drifted apart. He published this in 1915 as *Die Entstehung der Kontinente und Ozeane* (it did not appear in English until 1924, as *The Origin of Continents and Oceans*), which led in turn to the theory of sea-floor spreading, proposing that continental drift is driven by the expansion and contraction of the crust beneath the ocean floor, and then, in the 1960s and 1970s, to the unifying concept of plate tectonics.

Ecology grew partly out of theories of evolution that were being discussed during the eighteenth and nineteenth centuries. Darwinism is an ecological theory, after all, but this line of development branched, the other strand leading into German Romanticism. This was a very influential intellectual movement based on the idea that individual freedom and self-expression would bring people into close touch with a sublime reality surrounding us all and of which we long to become part. The discipline of ecology also originated in a quite different concept, that of the 'economy of nature'. This led to an idyllic view of nature as the harmonious product of all the countless interactions among living organisms and well able to supply human needs. Indeed, the view had strong links to natural theology, according to which God had so endowed all plants and animals with needs and the means to satisfy them as to guarantee that harmony among them would be preserved. This is the origin of the idea of a 'balance of nature' and, sentimental though it sounds, it taught that the interactions among organisms relate them in complex ways, and by early in the eighteenth century, long before the word 'ecology' was coined (by Ernst Haeckel (1834–1919) in 1866), it had generated some ideas with a startlingly modern ring. The writer Richard Bradley (1688–1732), for example, noted that insect species tend to specialize in the plants on which they feed and he advised farmers not to kill birds in their fields, because the birds feed on insects that would otherwise damage crops.

Environmental science ranges so widely that much of the history of science is relevant to its own development. Even such apparently unrelated discoveries as the gas laws relate very directly to meteorology, climatology and, through them, to weather forecasting and considerations of possible climate change. Today, many disciplines contribute to environmental science and its practitioners are equipped with instruments and techniques that enable them to begin compiling an overall, coherent picture of the way the world functions. The picture remains far from complete, however, and we must be patient while we wait to discover whether some of what are popularly perceived as environmental problems are really so and, if they are, how best to address them.

5 Changing attitudes to the natural world

When Julius Caesar (100–44 BC) became emperor of Rome, in 47 BC, traffic congestion was one of the pressing domestic problems he faced. He solved it by banning wheeled traffic from the centre of Rome during daytime, with the predictable result that Romans were kept awake at night by the incessant rumbling of iron-shod wheels over cobblestones. Nevertheless, Claudius (10 BC–AD 54, reigned from 41) later extended the law to all the important towns of Italy, Marcus Aurelius (AD 121–80, reigned from 161) made it apply to every town in the empire, and Hadrian (AD 76–138, reigned from 117) tightened it by restricting the number of vehicles allowed to enter Rome even at night (MUMFORD, 1961). The problem then, as now, was that a high population density generates a high volume of traffic and no one considered the possibility of designing towns with lower population and housing densities, as an alternative to building more and bigger roads.

If environmental science has a long history, so do the environmental problems that concern us today. We tend to imagine that urban air pollution is a recent phenomenon, dating mainly from the period of rapid industrialization in Europe and North America that began in the late eighteenth century. Yet in 1306 a London manufacturer was tried and executed for disobeying a law forbidding the burning of coal in the city, and the first legislation aimed at reducing air pollution by curbing smoke emissions was enacted by Edward I in 1273. The early efforts were not particularly successful and they dealt only with smoke from the high-sulphur coal Londoners were importing by ship from north-east England and which was, therefore, known as 'sea coal'. A wide variety of industries contributed to the smells and dust and poured their effluents into the nearest river. The first attempts to reduce pollution of the Thames date from the reign of Richard II (1367–1400, reigned from 1377). It was because of the smoke, however, that Elizabeth I refused to enter the city in 1578, and by 1700 the pollution was causing serious damage by killing vegetation, corroding buildings, and ruining clothes and soft furnishings in every town of any size (THOMAS, 1983). Indeed, the pall of smoke hanging over them was often the first indication approaching travellers had of towns.

Filthy it may have been, but 'sea coal' was convenient. It was a substitute for charcoal rather than wood, because of the high temperature at which it burned, and it was probably easier to obtain. If its use were to be curtailed, either manufacturing would suffer, with a consequent reduction in employment and prosperity, or charcoal would be used instead, in which case pollution might have been little reduced overall. Environmental protection always involves compromise between conflicting needs.

Much of the primary forest that once covered most of lowland Britain, which Oliver Rackham, possibly the leading authority on the history of British woodland, has called the 'wildwood', had been cleared by the time of the Norman invasion, in 1066, mainly to provide land on which to grow crops. It did not disappear, as some have suggested, to provide fuel for eighteenth-century iron foundries, or to supply timber to build ships. Paradoxically, the iron foundries probably increased the area of woodland, by relying for fuel on managed coppice from sources close at hand, and reports of a shortage of timber for shipbuilding had less to do with a lack of suitable trees than with the low prices the British Admiralty was prepared to pay (ALLABY, 1986, p. 110).

As early as the seventh century there were laws restricting the felling of trees and in royal forests a fence was erected around the stump of a felled tree to allow regeneration (ALLABY, 1986, p. 198). By the thirteenth century there were laws forbidding the felling of trees, clearing of woodland, and even the taking of dead wood, although they were seldom enforced, except as a means to raise revenue by fining an offender the value of the trees felled (RACKHAM, 1976).

For most of history, however, the conflict between farms and forests was resolved in favour of farms, although in England there is a possibility of confusion over the use of the word 'forest'. Today, the word describes an extensive tract of land covered with trees growing closely together, sometimes intermingled with smaller areas of pasture. Under Norman law, however, it had a different meaning, derived from the Latin *foris*, meaning 'outdoors', and applied to land beyond the boundaries of the enclosed farmland or parklands and set aside for hunting. Much of this 'forest' belonged to the sovereign. Special laws applied to it and were administered by officers appointed for the purpose. It might or might not be tree-covered.

Forests were regarded as dark, forbidding places, the abode of dangerous wild animals and brigands.[4] When Elizabethan writers used the word 'wilderness' they meant unmanaged forest, and in North America the earliest European settlers contrasted the vast forests they saw unfavourably with the cultivated fields they hoped to establish. Until modern times, famine was a real possibility and the neater the fields, the fewer the weeds in them, and the healthier the crops, the more reassuring the countryside appeared.

Mountains, upland moors, and wetlands were wastelands that could not be cultivated and they were no less alarming. In 1808, Arthur Young (1741–1820), an agricultural writer appointed secretary to the Board of Agriculture established by Prime Minister, William Pitt, in 1793, submitted a report on the enclosure of 'waste' land, arguing strongly in favour of their improvement by cultivation (YOUNG, 1808).

What we would understand today as the conservation of forest habitats and wildlife began quite early in the tropics, where it was a curious by-product of colonial expansion. This led government agencies and private companies to employ scientists or, in the case of the British East India Company, surgeons, many of whom had time to spare and wide scientific interests. One of the earliest conservation experiments was begun in Mauritius in the middle of the eighteenth century by French reformers seeking to prevent further deforestation as part of their efforts to build a just society. Interestingly, they had perceived a relationship between deforestation and local climate change. In the British territories, scientists also noted this relationship. Forest reserves were established in Tobago in 1764 and St Vincent in 1791, and a law passed in French Mauritius in 1769 was designed to protect or restore forests, especially on hill slopes and near to open water. Plans for the planting and management of Indian forests began in 1847 (GROVE, 1992), the foresters being known as 'conservators', a title still used in Britain by the Forestry Commission.

At about the same time, Americans were also becoming aware of the need for conservation. George Perkins Marsh (1801–82), US ambassador to Italy from 1862 until his death, wrote *Man and Nature* while in Italy. Published in 1864, this book led to the establishment of forest reserves in the United States and other countries, but it also challenged the then accepted relationship between humans and the natural environment.[5] In 1892, Warren Olney, John Muir, and William Keith founded the Sierra Club (*www.sierraclub.org/index_right.htm*), and the National Audubon Society (*www.audubon.org/*), named after John James Audubon (1785–1851),[6] the renowned wildlife painter and conservationist, was founded shortly afterwards.

While 'wilderness' has always implied hostility (and nowadays the word is often applied to certain urban areas), to these early conservationists it also had another, quite different meaning. To them, and those who thought like them, the word suggested purity, freedom from human interference, and the place where humans may find spiritual renewal, although this idea was often combined, as it is still, with that of economic resources held in reserve until a use can be found for them. It is tempting to associate the spiritual view exclusively with European Romanticism, but it also occurs in non-European cultures and, even in Europe, there were a few writers who saw wilderness in this way prior to the eighteenth century.

Today, the love of wilderness and desire to protect it probably represents the majority view, at least in most industrialized societies. Similarly, most people recognize pollution as harmful and will support measures to reduce it, provided they are not too expensive or disruptive. As we have seen, however, these are far from being new ideas or new attitudes. They have emerged at various times in the past, then concern has waned. It may seem that public attitudes reflect some cyclical change, and this may be not far from the truth.

When the possibility of famine was real, the most beautiful landscape was one that was well and intensively farmed. When factory jobs were scarce and insecure, but for large numbers of people the only jobs available, smoking chimneys symbolized prosperity. No one could afford to care that the fumes were harmful, even that they were harmful to human health, for hunger and cold were still more harmful and more immediately so. When the first European colonists reached North America, they could make no living from the forest. They had to clear it to provide cultivable land, and they had to do so quickly. It was only the wealthy who had the leisure to contemplate wilderness and could afford to point out the dangers of pollution, with the risk that were their warnings heeded, factories might close.

Modern concerns continue to follow the cycle. The present wave of environmental concern began in Britain and the United States in the 1960s, at a time of rising prosperity. It continued into the 1970s and then, as economies began to falter and unemployment began to creep upwards, interest faded. It re-emerged in the 1980s, as economies seemed to revive, then waned again as recession began to bite hard. The fluctuations in public concern are recorded in the numbers of books on environmental topics published year by year. In the early 1970s vast numbers appeared, but far fewer books were being published by the middle 1970s. More 'green' titles were issued in the early 1980s, but by the end of the decade large numbers of copies were being returned to publishers unsold and by the early 1990s most publishers would not accept books with titles suggesting anything remotely 'ecological', 'environmental', or 'green'.

This should surprise no one. When times are hard people worry most about their jobs, their homes, and whether they will be able to feed their families. It is only when they have economic security that they feel able to relax sufficiently to turn their attention to other matters. The preservation of species or of a tranquil, attractive countryside in which to walk means little to the homeless teenager begging for food or the single mother whose child needs shoes.

It should surprise no one, but there is an important lesson to be learned from it. All governments now accept the need for environmental reform, but a perceptual gulf exists between rich and poor which parallels that between rich and poor within nations. In poor countries facing high levels of infant mortality and chronic shortages of supplies necessary for the provision of health care, housing, and education, the most pressing needs relate to the provision of employment and industrialization based, so far as possible, on the exploitation of indigenous resources. Environmental hazards seem less urgent, and efforts by the rich to persuade the poor to move them higher up the international agenda can be seen as attempts to increase development costs and so perpetuate economic inequality. It is well to remember that environmental issues that seem self-evidently urgent to Europeans and North Americans may not seem so to everyone.

End of chapter summary

Like the life and earth sciences, the environmental sciences comprise elements taken from many other disciplines. The all-round environmental scientist must be part biologist, part ecologist, part toxicologist, part pedologist (soil scientist), part geomorphologist, part limnologist (student of freshwater systems), and part meterologist, as well as being familiar with ideas taken from many other disciplines. In helping to resolve the disputes that frequently arise over conservation and land use, the environmental scientist must also possess tact and political skill.

It is important that the environmental scientist remain a scientist. Environmentalism consists in campaigning and proselytising in pursuit of essentially ideological objectives. It is not necessarily based on scientific assessment, nor should it be, because its appeal is primarily moral. This means there is a clear difference between environmental science and environmentalism and the two should not be confused.

End of chapter points for discussion

- To what extent is our attitude to the natural world linked to our level of economic prosperity?
- To what extent are environmentalist campaigns informed by science?
- What does Gaia hypothesis assert?

See also

Formation of the Earth (section 6)
Weathering (section 8)
Coasts, estuaries, sea levels (section 10)
Greenhouse effect (section 13)
Evolution and structure of the atmosphere (section 14)
Dating methods (section 19)
Climate change (section 20)
Fresh water (section 22)
Biogeography (section 32)

Nutrient cycles (section 33)
Ecology (sections 35–44)
Limits of tolerance (section 44)
Evolution (section 45)
Transnational pollution (section 62)

Further reading

The Ages of Gaia. James Lovelock. 1988. Oxford University Press, Oxford. Provides a general, non-technical introduction to and description of the Gaia theory, written by the scientist who first proposed it.

Fantasy, the Bomb, and the Greening of Britain. Meredith Veldman. 1994. Cambridge University Press, New York. The rise of the environmental movement in Britain, seen in the context of a long history of romantic protest; written by a historian.

The Fontana History of the Environmental Sciences. Peter J. Bowler. 1992. Fontana Press (HarperCollins), London. Probably the most authoritative account of the subject; written in non-technical language.

Gaia: The Growth of an Idea. Lawrence E. Joseph. 1990. Arkana (Penguin Books), London. Also provides an account of Gaia theory, simply written by a journalist, but includes objections to the idea and difficulties it raises.

Man and the Natural World. Keith Thomas. 1983. Penguin Books, London. A comprehensive account of changing attitudes in Britain between 1500 and 1800.

Thinking Green: An Anthology of Essential Ecological Writing. Michael Allaby (ed.). 1989. Barrie and Jenkins, London. A selection of excerpts from some of the most influential writing on environmental topics.

Notes

1 See, for example, *The Clyde Estuary and Firth; an assessment of present knowledge*, compiled by members of the Clyde Study Group (1974), NERC Publications Series C No. 11.
2 For a useful discussion of this topic, see KUPCHELLA AND HYLAND, pp. 160–162.
3 See Rudwick, Martin J.S. 1976. *The Meaning of Fossils: Episodes in the history of palaeontology*. Univ. of Chicago Press, Chicago.
4 This theme is explored in Allaby, Michael, 1999, *Ecosystems: Temperate Forests*. Fitzroy Dearborn, London, pp. 146–149.
5 For an excerpt from this book see Allaby, Michael (ed.). 1989. *Thinking Green: An anthology of essential ecological writing*. Barrie and Jenkins, London, pp. 61–68.
6 For a brief biography of Audubon, see *www.audubon.org/nas/jja.html*. There is also an excellent essay about his relationship with Darwin in Weissmann, Gerald, 1998, *Darwin's Audubon*. Plenum Press, New York, pp. 9–24.

References

Allaby, Michael. 1986. *The Woodland Trust Book of British Woodlands*. David and Charles, Newton Abbot. 1995. *Facing the Future*. Bloomsbury, London. p. 197.

Bowler, Peter J. 1992. *The Fontana History of the Environmental Sciences*. FontanaPress (HarperCollins), London.

Campbell, Joseph. 1962. *The Masks of God: Oriental Mythology*. Arkana (Penguin Books), London. pp. 95–98.

de Baar, Hein J.W., de Jong, Jeroen T.M., Bakker, Dorothee C.E., Loscher, Bettina M., Veth, Cornelius, Bathmann, Uli, and Smetacek, Victor. 1995. 'Importance of iron for plankton blooms and carbon dioxide drawdown in the Southern Ocean'. *Nature*, 373, 412–415.

Grove, Richard H. 1992. 'Origins of Western environmentalism'. *Scientific American*, July 1992. pp. 22–27.

Joseph, Lawrence E. 1990. *Gaia: The growth of an idea*. Arkana (Penguin Books), London. Ch. IV.

Kupchella, Charles E. and Hyland, Margaret C. 1986. *Environmental Science*. Allyn and Bacon, Needham Heights, Mass. 2nd edition, pp. 316–317.

Lovelock, James. 1979. Gaia: *A new look at life on Earth*. OUP, Oxford. 1988. *The Ages of Gaia*. OUP, Oxford.

Marshall, N.B. 1979. *Developments in Deep-Sea Biology*. Blandford, Poole, Dorset. p.32–33.

Mumford, Lewis. 1961. *The City in History.* Pelican Books, London. p. 254.

Rackham, Oliver. 1976. *Trees and Woodland in the British Landscape.* J.M. Dent, London. p. 154. (A second, 1990, edition is now available.)

Thomas, Keith. 1983. *Man and the Natural World.* Penguin Books, London. pp. 244–247.

Tolba, Mostafa K. and El-Kholy, Osama A. 1992. *The World Environment 1972–1992.* Chapman and Hall, London, on behalf of UNEP. pp. 160–162.

Veldman, Meredith. 1994. *Fantasy, the Bomb, and the Greening of Britain.* Cambridge University Press, New York.

Westbroek, Peter. 1992. *Life as a Geological Force.* W.W. Norton, New York.

Wilson, Edward O. 1992. *The Diversity of Life.* Penguin Books, London. p. 264.

Young, Arthur. 1808. *General Report on Enclosures.* Republished, 1971, by Augustus M. Kelley, New York.

2 Earth Sciences

When you have read this chapter you will have been introduced to:

- the formation and structure of the Earth
- rocks, minerals, and geologic structures
- weathering
- how landforms evolve
- coasts, estuaries, and changing sea levels
- solar energy
- albedo and heat capacity
- the greenhouse effect
- evolution, composition, and structure of the atmosphere
- general circulation of the atmosphere
- ocean currents and gyres
- weather and climate
- ice ages and interglacials
- climate change
- climatic regions and plants

6 Formation and structure of the Earth

Among the nine planets in the solar system, Earth is the only one which is known to support life. All the materials we use are taken from the Earth and it supplies us with everything we eat and drink. It receives energy from the Sun, which drives its climates and biological systems, but materially it is self-contained, apart from the dust particles and occasional meteorites that reach it from space (ADAMS, 1977, pp. 35–36). These may amount to 10 000 tonnes a year, but most are vaporized by the heat of friction as they enter the upper atmosphere and we see them as 'shooting stars'. At the most fundamental level, the Earth is our environment.

The oldest rocks, found on the Moon, are about 4.6 billion years old and this is generally accepted to be the approximate age of the Earth and the solar system generally. There are several rival theories describing the process by which the solar system may have formed.[1] The most widely accepted theory, first proposed in 1644 by René Descartes (1596–1650), proposes that the system formed from the condensation of a cloud of gas and dust, called the 'primitive solar nebula' (PSN). It is now thought this cloud may have been perturbed by material from a supernova explosion. Fusion processes within stars convert hydrogen to helium and in larger stars go on to form all the heavier elements up to iron. Elements heavier than iron can be produced only under the extreme conditions of the supernova explosion of a very massive star, and the presence of such elements (including zinc, gold, mercury, and uranium) on Earth indicates a supernova source.

As the cloud condensed, its mass was greatest near the centre. This concentration of matter comprised the Sun, the planets forming from the remaining material in a disc surrounding the star, and the whole system rotated. The inner planets formed by accretion. Small particles moved close to one another, were drawn together by their mutual gravitational attraction, and as their masses increased they gathered more particles and continued to grow. At some point it is believed that a collision between the proto-Earth and a very large body disrupted the planet, the material re-forming as two bodies rather than one: the Earth–Moon system. This explains why the Earth and Moon are considered to be of the same age and, therefore, why lunar rocks 4.6 billion years old are held to be of about the age of the Earth and Moon.

The material of Earth became arranged in discrete layers, like the skins of an onion. If accretion was a slow process compared to the rate at which the PSN cooled, the densest material may have arrived first, followed by progressively less dense material, in which case the layered structure has existed from the start and would not have been altered by melting due to the gravitational energy released as heat by successive impacts. This model is called 'heterogeneous accretion'. If material arrived quickly in relation to the rate of PSN cooling, then it would have comprised the whole range of densities. As the planet cooled from the subsequent melting, denser material would have gravitated to the centre and progressively less dense material settled in layers above it. This model is called 'homogeneous accretion' (ALLABY AND ALLABY, 1999).

As it exists today, the Earth has a mean radius of 6371 km, equatorial circumference of 40 077 km, polar circumference of 40 009 km, total mass of 5976×10^{24}g, and mean density of 5.517 g. cm^{-3}. Of its surface area, 149×10^6 km^2 (29.22 per cent) is land, 15.6×10^6 km^2 glaciers and ice sheets, and 361×10^6 km^2 oceans and seas (HOLMES, 1965, ch. II). Land and oceans are not distributed evenly. There is much more land in the northern hemisphere than in the southern, but at the poles the positions are reversed: Antarctica is a large continent, but there is little land within the Arctic Circle.

At its centre, the Earth has a solid inner core, 1370 km in radius, made from iron with some nickel (see Figure 2.1). This is surrounded by an outer core, about 2000 km thick, also of iron with nickel, but liquid, although of very high density. Movement in the outer core acts like a self-exciting dynamo and generates the Earth's magnetic field, which deflects charged particles reaching the Earth from space. Outside the outer core, the mantle, made from dense but somewhat plastic rock, is about 2900 km thick, and at the surface there is a thin crust of solid rock, about 6 km thick beneath the oceans and 35 km thick (but less dense) beneath the continents.

Miners observed long ago that the deeper their galleries the warmer they found it to work in them. Surface rocks are cool, but below the surface the temperature increases with depth. This is called the 'geothermal gradient'. A little of the Earth's internal heat remains from the time of the planet's formation, but almost all of it is due to the decay of the radioactive elements that are distributed widely throughout the mantle and crustal rocks. The value of the geothermal gradient varies widely from place to place, mostly between 20 and 40 °C for every kilometre of depth, but in some places, such as Ontario, Canada, and the Transvaal, South Africa, it is no more than

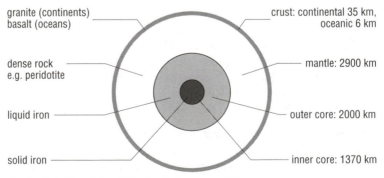

Figure 2.1 **Structure of the Earth (not to scale)**

9 or 10 °C per kilometre (HOLMES, 1965, ch. XXVIII, p. 995). Because of the low thermal conductivity of rock, very little of this heat reaches the surface and it has no effect on the present climate.

Where the gradient is anomalously high, however, it can be exploited as a source of geothermal energy. In volcanic regions, such as New Zealand, Japan, Iceland, and Italy, water heated below ground may erupt at the surface as geysers, hot springs, or boiling mud. More often it fails to reach the surface and is trapped at depth, heated by the surrounding rock. A borehole drilled into such a reservoir may bring hot water to the surface where it can be used. In some places a body of dry subsurface rock is much hotter than its surroundings. In principle this can also be exploited, although experimental drilling, for example some years ago in Cornwall, Britain, has found the resulting energy rather costly. The technique is to drill two boreholes and detonate explosive charges at the bottom, to fracture the rock between them and so open channels through it. Cold water is then pumped at pressure down one borehole; it passes through the hot rock and returns to the surface through the other borehole as hot water.

This exploitation of geothermal energy is not necessarily clean. Substances from the rock dissolve into the water as it passes, so it returns to the surface enriched with compounds some of which are toxic. The solution is often corrosive and must be kept isolated from the environment and its heat transferred by heat exchangers. Nor is the energy renewable. Removal of heat from the rock cools it faster than it is warmed by radioactive decay, so eventually its temperature is too low for it to be of further use. Similarly, the abstraction of subsurface hot water depletes, and eventually empties, the reservoir.

Although subsurface heat has no direct climatic effect, there is a sense in which it does have an indirect one. Material in the mantle is somewhat plastic. Slow-moving convection currents within the mantle carry sections of the crustal rocks above them, so that over very long time-scales the crustal material is constantly being rearranged.[2] On Earth, but possibly on no other solar-system planet, the crust consists of blocks, called 'plates', which move in relation to one another. The theory describing the process is known as 'plate tectonics' (GRAHAM, 1981). At present there are seven large plates, a number of smaller ones, and a still larger number of 'microplates'. The boundaries (called 'margins') between plates can be constructive, destructive, or conservative. At constructive margins two plates are moving apart and new material emerges from the mantle and cools as crustal rock to fill the gap, marked by a ridge. There are ridges near the centres of all the world's oceans. Where plates move towards one another there is a destructive margin, marked by a trench where one plate sinks (is subducted) beneath the other. At conservative margins two plates move past one another in opposite directions (see Figure 2.2). There are also collision zones, where continents or island arcs have collided. In these, all the oceanic crust is believed to have been subducted into the mantle, leaving only continental crust. Such zones may be marked in various ways, one of which is the presence of mountains made from folded crustal rocks. An island arc is a series of volcanoes lying on the side of an ocean trench nearest to a continent. The volcanoes are due to the subduction of material.

Slowly but constantly the movement of plates redistributes the continents carried on them. A glance at a map shows the apparent fit between South America and Africa, but for 40 million years or more prior to the end of the Triassic Period, about 213 million years ago, all the continents were joined in a supercontinent, Pangaea, surrounded by a single world ocean, Panthalassa. Pangaea then broke into two continents, Laurasia in the north and Gondwana in the south, separated by the Tethys Sea, of which the present Mediterranean is the last remaining trace. The drift of continents in even earlier times has now been reconstructed, with the proposing of a supercontinent called Rodinia that existed about 750 million years ago (DALZIEL, 1995). The Atlantic Ocean opened about 200 million years ago and it is still growing wider by about 3–5 cm a year. A little more than 100 million years ago India

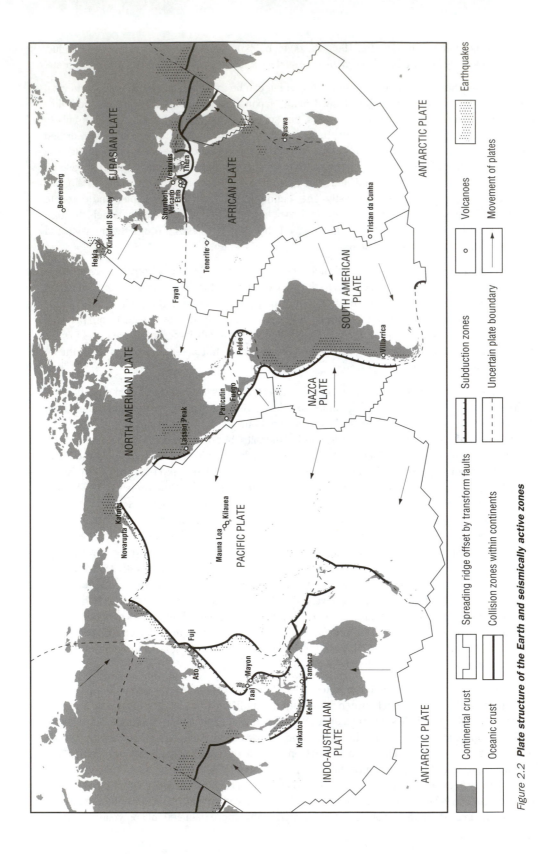

Figure 2.2 *Plate structure of the Earth and seismically active zones*

separated from Antarctica. The Indian plate began subducting beneath the Eurasian plate and as India moved north the collision, about 50 million years ago, raised the Himalayan mountain range. India is still moving into Asia at about 5 cm a year and the mountains are still growing higher (WINDLEY, 1984, pp. 161 and 310), although the situation is rather complicated. Rocks exposed at the surface are eroded by ice, wind, and rain, so mountains are gradually flattened. At the same time, the crumpling that produces mountains of this type increases the mass of rock, causing it to sink into the underlying mantle. This also reduces the height of large mountain ranges. It is possible, however, for the eroded material to lighten the mountains sufficiently to reduce the depression of the mantle, causing them to rise, and there is reason to suppose this is the case for the Himalayas (BURBANK, 1992). The Red Sea is opening and in time will become a new ocean between Africa and Arabia.

The distribution of land has a strong influence on climates. If there is land at one or other pole, ice sheets are more likely to form. The relative positions of continents modify ocean currents, which convey heat away from the equator, and the size of continents affects the climates of their interiors, because maritime air loses its moisture as it moves inland. The Asian monsoon is caused by pressure differences to the north and south of the Himalayas. In winter, subsiding air produces high pressure over the continent and offshore winds, with very dry conditions inland. The word 'monsoon' simply means 'season' (from the Arabic word for 'season', *mausim*) and this is the winter, or dry, monsoon. In summer, pressure falls as the land warms, the wind direction reverses, and warm, moist air flows across the ocean toward the continent, bringing heavy rain. This is the summer, wet monsoon. Plate tectonics exerts a very long-term influence, of course, and other factors modify climates in the shorter term, but the distribution of land and sea determines the overall types of climate the world is likely to have (HAMBREY AND HARLAND, 1981).

Plate tectonics affects the environment more immediately and more dramatically. The movement of plates causes earthquakes, because it tends to happen jerkily as accumulated stress is released, and is associated with volcanism due to weakening of the crust at plate margins. Earthquakes cause damage to physical structures, which is the direct cause of most injuries, and those which occur beneath the sea produce tsunami (*www.geophys.washington.edu/ tsunami/general/physics/physics.html*). These are shock waves affecting the whole water column. No more than a metre high and with a wavelength of hundreds of kilometres, but travelling at more than 700 km h^{-1}, on reaching shallow water they rise to great height and destructive power (ALLABY, 1998, pp. 54–60).

If volcanic ash reaches the stratosphere it can cause climatic cooling, but volcanic eruptions are more usually associated with damage to human farms and dwellings. This arises partly because of the beneficial effect volcanoes can have. Volcanic ash and dust are often rich in minerals and rejuvenate depleted soils. Farmers can grow good crops on them, which is why there tend to be cultivated fields at the foot and even on the lower slopes of active volcanoes.

7 The formation of rocks, minerals, and geologic structures

Volcanoes create environments. This was demonstrated very dramatically, and shown on television, in 1963, when a new submarine volcano called Surtsey (*volcano.und.nodak.edu/vwdocs/ volc_images/europe_west_asia/surtsey.html*) erupted to the south of Iceland. The eruption was extremely violent, because sea water entered the open volcanic vent, and steam, gas, pieces of rock, and ash were hurled many kilometres into the air. Since then eruptions of this type have been called 'Surtseyan'. The lava cone was high enough to rise above the surface, where it formed what is now the island of Surtsey. As it cooled, sea birds began to settle on it.[3] They carried plant seeds and slowly plants and animals began to colonize the new land.

Even the damage caused by destructive eruptions is repaired, although this can take a long time. The 1883 eruption of Krakatau, in the Sunda Strait between Java and Sumatra, Indonesia, destroyed almost every living thing on Krakatau itself and on two adjacent islands. Three years later the lava was covered in places by a thin layer of cyanobacteria, and a few mosses, ferns, and about 15 species of flowering plants, including four grasses, had established themselves. By 1906 there was some woodland, which is now thick forest. The only animal found in 1884 was a spider, but by 1889 there were many arthropods and some lizards. In 1908, 202 species of animals were living on Krakatau and 29 on one of the islands nearby, although bats were the only mammals. Rats were apparently introduced in 1918. Species continued to arrive and 1100 were recorded in 1933 (KENDEIGH, 1974, pp. 24–25).

Rock that forms from the cooling and crystallization of molten magma is called 'igneous', from the Latin *igneus*, 'of fire', and all rock is either igneous or derived from igneous rock. This must be so, since the molten material in the mantle is the only source for entirely new surface rock. If the magma reached the surface before cooling the rock is known as 'extrusive'; if it cooled beneath the surface surrounded by older rock into which it had been forced, it is said to be 'intrusive'. Intrusive rock may be exposed later as a result of weathering. It is not only igneous rocks that can form intrusions. Rock salt (NaCl) can accumulate in large amounts beneath much denser rocks and rise through them very slowly to form a salt dome. Salt domes are deliberately sought by geologists prospecting for oil but occasionally they can break through the surface. When this happens the salt may flow downhill like a glacier.

The character of the rock depends first on its chemical composition. If it is rich in compounds of iron and magnesium it will be dark (melanocratic); if it is rich in silica, as quartz and feldspars, it will be light in colour (leucocratic). Rock between the two extremes is called 'mesocratic'. The rock comprises minerals, each with a particular chemical composition, and minerals crystallize as they cool. Whole rock is quarried for building and other uses; many minerals are mined for the chemical substances they contain, especially metals, and some are valued as gemstones. Crystallization proceeds as atoms bond to particular sites on the surface of a seed crystal, forming a three-dimensional lattice. It can occur only where atoms have freedom to move and so the more slowly a molten rock cools the larger the crystals it is likely to contain. The crystal size gives the rock a grain structure, which also contributes to its overall character. The type of rock is also determined by the circumstances of its formation. Lava that flows as sheets across the land surface or sea bed often forms basalt, a dark, fine-grained, hard rock. Basalt covers about 70 per cent of the Earth's upper crust, making it the commonest of all rocks; most of the ocean floor is of basalt overlain by sediments and on land it produces vast plateaux, such as the Deccan Traps in India. Intrusive igneous rocks are usually of the light-coloured granite type. Beyond this, however, the identification and classification of igneous rocks are rather complicated.[4]

Rocks formed on the ocean floor may be thrust upward to become dry land or exposed when the sea level falls. Tectonic plate movements are now believed to be the principal mechanism by which this occurs. Where two plates collide the crumpling of rocks can raise a mountain chain, as is happening now between the Indian and Eurasian plates, raising the Himalayan chain. The Himalayas, which began to form some 52–49 million years ago following the closure of the Tethys Sea, are linked to the Alps, which began forming about 200 million years ago owing to very complex movements of a number of plates (WINDLEY, 1984, pp. 202–308). The formation of a mountain chain by the compression of crustal rocks is known as an 'orogeny' (or 'orogenesis').

The British landscape was formed by a series of orogenies. The first, at a time when Scotland was still joined to North America, began about 500 million years ago and produced the Caledonian–Appalachian mountain chain (WINDLEY, 1984, pp. 181–208) as well as the mountains of northern Norway. The Appalachians were later affected by the Acadian orogeny, about 360 million years ago, and the Alleghanian orogeny, about 290 million years ago. Europe was affected by the Hercynian and Uralian orogenies, both of which occurred at about the same time as the Alleghanian. Figure 2.3 shows the area of Europe affected by several orogenies.[5]

Igneous intrusions can be exposed through the weathering away of softer rocks surrounding them. Such an exposed intrusion, roughly circular in shape and with approximately vertical sides, is called a 'boss' if its surface area is less than 25 km^2 and a 'batholith' if it is larger (and they are often much larger). Dartmoor and Bodmin Moor, in Devon and Cornwall, Britain, lie on the surface of granite batholiths.

Mountains are not always formed from igneous rocks, however. There are fossil shells of marine organisms at high altitudes in the Alps and Himalayas, showing that these mountains were formed by the crumpling of rocks which had formed from sea-bed sediments.

Many sedimentary rocks are composed of mineral grains eroded from igneous or other rocks and transported by wind or more commonly water to a place where they settle. Others, said to be of 'biogenic' origin, are derived from the insoluble remains of once-living organisms. Limestones, for example, are widely distributed. Most sediments settle in layers on the sea bed, to which rivers have carried them. Periodic changes in the environmental conditions in which they are deposited may cause sedimentation to cease and then resume later, and chemical changes in the water or the sediment itself will be recorded in the sediments themselves and in the rocks into which they may be converted.

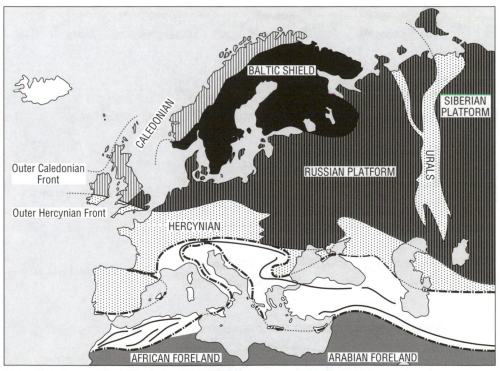

Figure 2.3 **The mountain-forming events in Europe**
Note: **The thick lines (– · – · –) mark the Alpine orogeny**

Sandstones are perhaps the most familiar sedimentary rocks, consisting mainly of sand grains, made from quartz (silica, SiO_2) which crystallized originally into igneous rock. Clay particles, much smaller than sand grains, can pack together to make mudstones. Sediments rich in calcium carbonate, often consisting mainly of the remains of shells and containing many fossils, form limestone and dolomite (sometimes known as 'dolostone' to distinguish it from the mineral called dolomite) (HOLMES, 1965, ch. VI, pp. 118–141). Particles deposited as sediments are changed into rock by the pressure of later deposits lying above them and the action of cementing compounds subsequently introduced into them. The process, occurring at low temperature, is called 'diagenesis'. Some sedimentary rocks are very hard and many, especially sandstones and limestones, make excellent and durable building stone. Once formed, a sedimentary rock is subject to renewed weathering, especially if it is exposed at the surface, so sedimentary rocks continually form and re-form.

Sediments are deposited in horizontal layers, called 'beds', but subsequent movements of the crust often fold or fracture them. It is not unusual for beds to be folded until they are upside down, and the reconstruction of the environmental conditions under which sediments were deposited from the study of rock strata often begins by seeking to determine which way up they were when they formed. All in all, the interpretation of sedimentary structures can be difficult.[6] Figure 2.4 shows the sequence of events by which sedimentary structures may be folded, sculptured, and then subside to be buried beneath later beds producing an unconformity.

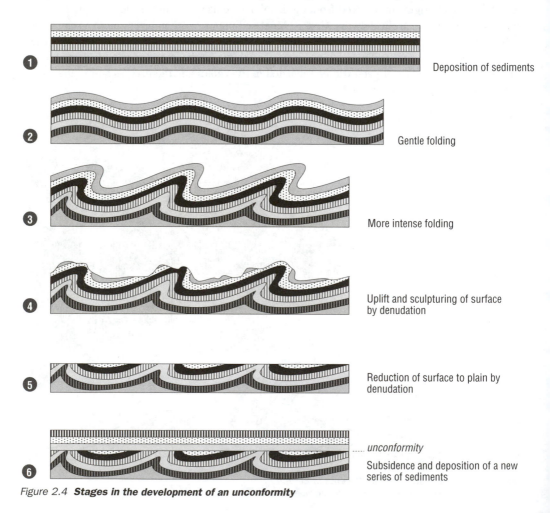

Figure 2.4 **Stages in the development of an unconformity**

The extreme conditions produced by the folding and shearing of rock can alter its basic structure by causing some of its minerals to recrystallize, sometimes in new ways. This process, called 'metamorphism', also happens when rock of any type comes into contact with molten rock, during the intrusion of magmatic material for example. Marble is limestone or dolomite (dolostone) that has been subjected to metamorphism at high temperature. Such shells as it contained are completely destroyed as the calcium carbonate recrystallizes as the mineral calcite. If quartz or clay particles are present, new minerals may form, such as garnet and serpentine. Hard limestone containing fossils is often called marble, but there are no fossils in true marble.

Slate is also a metamorphic rock, derived from mudstone or shale, in which the parallel alignment of the grains, due to the way the rock formed, allows the rock to cleave along flat planes (HOLMES, 1965, pp. 168–170). It may contain fossils, although they are uncommon and usually greatly deformed, because slate forms when the parent sedimentary rock is squeezed tightly between two bodies of harder rock that are moving in parallel but opposite directions, so its particles, and fossils, are dragged out. It is this that gives slate its property of 'slaty cleavage' which, with the impermeable surface imparted at the same time, makes it an ideal roofing and weatherproofing material. Metamorphic rocks are widely distributed and with practice you can learn to recognize at least some of them.[7]

All the landscapes we see about us and the mineral grains that are the starting material for the soils which form over their surfaces are produced by these processes. The intrusion or extrusion of igneous rock supplies raw material. This weathers to provide the mineral grains which become soil when they are mixed with organic matter, or is transported to a place where it is deposited as sediment. Pressure converts sediments into sedimentary rocks, which may then be exposed by crustal movements, so that erosion can recommence. Metamorphic rocks, produced when other rocks are subjected to high pressures and/or temperatures, are similarly subject to weathering. It is the cycling of rocks, from the mantle and eventually back to it through subduction, that produces the physical and chemical substrate from which living organisms can find subsistence.

8 Weathering

No sooner has a rock formed than it becomes vulnerable to attack by weathering. The word 'weathering' is slightly misleading. We associate it with wind, water, freezing, and thawing. These are important agents of weathering, but they are not the only ones. Weathering can be chemical as well as physical and it often begins below ground, completely isolated from the weather.

Beneath the surface, natural pores and fissures in rocks are penetrated by air, containing oxygen and carbon dioxide, and by water into which a wide variety of compounds have dissolved to make an acid solution. Depending on their chemical composition, rock minerals may dissolve or be affected by oxidation, hydration, or hydrolysis (HOLMES, 1965, pp. 393–400). Oxidation is a reaction in which atoms bond with oxygen or lose electrons (and other atoms gain them, and are said to be 'reduced'). Hydration is the bonding of water to another molecule to produce a hydrated compound; for example, the mineral gypsum ($CaSO_4.2H_2O$) results from the hydration of anhydrite ($CaSO_4$). Hydrolysis (lysis, from the Greek *lusis*, 'loosening') is a reaction in which some parts of a molecule react with hydrogen ions and other parts with hydroxyl (OH) ions, both derived from water, and this splits the molecule into two or more parts.

The result of chemical weathering can be seen in the limestone pavements found in several parts of England, Wales, and Ireland.[8] South Devon, England, is famous for its red sandstones, well exposed in the coastal cliffs of the Torbay area. These date from the Devonian Period, some 400 million years ago, when what is now Devon was a hot, arid desert. The desert sand contained some iron, which was oxidized to its insoluble red oxide, giving the sandstone its present colour.

Limestone pavement

A distinctive feature, sometimes covering a large area, that occurs in many parts of the world. It forms when horizontal limestone beds are exposed by the erosion of any material that may once have covered them and joints within them are penetrated by rain water carrying dissolved CO_2. This weak carbonic acid ($CO_2 + H_2O \rightarrow H^+ + (HCO_3)^-$) reacts with calcium carbonate to produce calcium bicarbonate, which is soluble in water and is carried away. This widens the joints to form deep crevices (called 'grikes' in England) separated by raised 'clints'. Small amounts of soil accumulating in the sheltered grikes provide a habitat for lime-loving plants, making limestone pavements valuable botanically. At a deeper level, the grikes may join to form caves. Particular areas of limestone pavement are protected in Britain by Limestone Pavement Orders issued under the Wildlife and Countryside Act 1981, mainly to prevent the stone being taken to build garden rockeries and for other ornamental uses.

Iron oxidizes readily and this form of weathering has produced hematite (Fe_2O_3), one of the most important iron ore minerals, some of which occurs in banded ironstone formations, 2–3 billion years old, composed of alternating bands of hematite and chert (SiO_2). Iron and other metals can also be concentrated by hydrothermal, or metasomatic, processes. Near mid-ocean ridges, where new basalt is being erupted on to the sea bed, iron, manganese, and some other metals tend to separate from the molten rock and are then oxidized and precipitated, where particles grow to form nodules, sometimes called 'manganese nodules' because this is often the most abundant metal in them. Vast fields of nodules, containing zinc, lead, copper, nickel, cobalt, silver, gold, and other metals as well as manganese and iron, have been found on the floor of all the oceans (KEMPE, 1981). A few years ago serious consideration was given to the possibility of dredging for them, but at present metals can be obtained more cheaply by conventional mining on land.

Hydrothermal weathering, in which hot solutions rise from beneath and react with the rocks they encounter, produces a range of commercially valuable minerals, perhaps the best known of which is kaolin, or 'china clay'. This material was first discovered in China in 500 BC and was used to make fine porcelain, hence the names 'china clay' and 'kaolin', from *kao ling*, meaning 'high ridge', the type of landscape in which it occurred. Today it is still used in white ceramics, but most is used as a filler and whitener, especially in paper. The paper in this book contains it. Kaolin deposits (*www.wbb.co.uk/)Welcome.htm*) occur in several countries, but the most extensively mined ones are in Cornwall and Devon, Britain.

Kaolin is a hydrated aluminium silicate, $Al_2O_32SiO_2.2H_2O$, obtained from the mineral kaolinite. The British deposits occur in association with the granite batholiths and bosses intruded during the Hercynian orogeny. Granites consist of quartz crystals, mica, and feldspars. Feldspars are variable in composition. All are aluminium silicates, those associated with the kaolinite deposits being plagioclase feldspars, relatively rich in sodium. As the intruded granite was cooling, it was successively exposed to steam, boron, fluorine, and vaporized tin. The feldspar reacted with these, converting it into kaolinite (the process is known as kaolinization), a substance consisting of minute white hexagonal plates which are separated from the rock industrially by washing and precipitation, leaving a residue of quartz grains (a white sand) and mica. About 15 per cent of the material is recovered as kaolin (*www.kaolin.com/ccpmin.htm*), 10 per cent is mica waste, and 75 per cent is sand, which is also waste although it has found some use for building and landscaping. In some places the kaolinization process has been completed from above, possibly by humic or other acids from overlying organic material, but most of the kaolinite formed at depth is overlain by unaffected granite, probably because the upward movement of acidic fluids was halted by the absence of veins or joints it could attack. The resulting deposits are funnel-shaped, extending in places to depths of more than 300 m.

Bauxite, the most important ore of aluminium, is also produced by the chemical weathering of feldspars, in this case by hydration. Bauxite is a mixture of hydrous aluminium oxides and hydroxides with various metals as impurities; to be suitable for mining it should contain 25–30 per cent of aluminium oxide.

Bauxite is a variety of laterite, one product of the kind of extreme weathering of soil called 'laterization'. The word 'laterite' is from the Latin *later*, meaning 'brick', and laterite is brick-hard. Laterization occurs only in some parts of the seasonal tropics, where soils are derived from granite parent material, but it is possible that removing the forest or other natural vegetation in such areas may trigger the formation of laterites. These can be broken by ploughing.

Except on steep slopes, tropical soils overlying granite can be up to 30 m deep. Naturally acidic water from the surface percolates through them, steadily eating away at the parent rock beneath, and plants draw the water up again through their roots. Water is also drawn upward by capillary attraction through tiny spaces between soil particles and evaporates from the surface. If the rainfall is fairly constant through the year, the movement of water is also constant, but if it is strongly seasonal, evaporation exceeds precipitation during the dry season and mineral compounds dissolved in the soil water are precipitated, the least soluble being precipitated first. Provided vegetation cover is adequate, with roots penetrating deep into the soil, the minerals will not accumulate in particular places and when the rains return they will be washed away. If there is little plant cover, however, they may accumulate near the surface. The most insoluble minerals are hydroxides of iron and aluminium (kaolinite) and they are what give many tropical soils their typically red or yellow colour (HOLMES, 1965, pp. 400–401). Soil developed over granite will contain sand, or quartz grains, and clays derived from feldspars in varying amounts. Figure 2.5 shows how these can grade almost imperceptibly from one to the other and from both into laterite. Laterite layers or nodules are hard, but not usually thick, because, being impermeable, they prevent further percolation of water downwards into the soil and thus bring the laterization process to an end. Erosion of the surface layer may then expose the laterite.

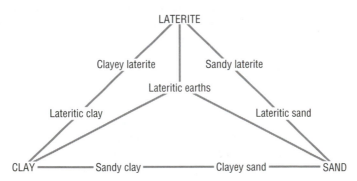

Figure 2.5 **Gradation of clay and sand to laterite**
Source: **Holmes, Arthur. 1965. Principles of Physical Geology. Nelson, Walton-on-Thames**

Laterization does not necessarily render a soil useless and many relatively laterized soils are cultivated, although some soils resembling lateritic soils, for example in parts of the eastern United States, are not truly laterized. Indeed, there are doubts about the extent to which laterization is occurring at present. Lateritic soils in the West Indies, Indonesia, Australia, India, and China may well be of ancient origin (HUNT, 1972, p. 193).

Living organisms contribute to weathering. By moving through soil they assist the penetration of air and water, and the decomposition of organic material releases acids and carbon dioxide, some of which dissolves into the soil water. Biological activity contributes greatly to the formation of soil.

Physical weathering is also important in soil formation, especially in its initial stages, but it can also degrade soils through erosion. Thermal weathering, which is the expansion and contraction due to repeated heating and cooling, causes rocks to flake, especially if water is held within small crevices. Small particles detached from the rock may then be carried by the wind and if they strike other rocks more particles may be chipped from them. Depending on their size, the particles may be carried well clear of the ground or may roll and bounce along the surface; the process is called 'saltation'. Most serious erosion is due to water, however. All water flowing across the land surface carries soil particles with it. This can lead to the formation of rills and gullies into which more particles are washed and then transported, or where water flows as sheets whole surface layers can be removed. In addition to this, all rivers erode their banks, and waves erode the shores of lakes and the sea (HUDSON, 1971, pp. 33–46).

These processes are entirely natural and part of the cycles by which originally igneous rocks are converted into sediments and landforms are made and age, but human activities can accelerate them. The UN estimates that in the world as a whole, some 1.093 billion hectares (ha) of land have been degraded by water erosion, 920 million ha by sheet and surface erosion and a further 173 million ha by the development of rills and gullies. Of the total area subject to serious degradation by water, 43 per cent is attributed to the removal of natural vegetation and deforestation, 29 per cent to over-grazing, 24 per cent to poor farming practices, such as the use of machinery that is too heavy for the soil structure to support and the cultivation of steep slopes, and 4 per cent to the over-exploitation of vegetation (TOLBA AND EL-KHODY, 1992, pp. 149–150). There is, however, some evidence that modern farming techniques can reduce soil erosion substantially. A study of a site in Wisconsin found that erosion in the period 1975–93 was only 6 per cent of the rate in the 1930s (TRIMBLE, 1999). This may be due to higher yields from the best land, combined with methods of tillage designed to minimize erosion (AVERY, 1995).

Weathering is the general name given to a variety of natural processes by which rock is recycled and soil and landscapes created. It creates and alters environments, but human activities can accelerate it on vulnerable land, degrading natural habitats and reducing agricultural productivity.

9 The evolution of landforms

The weathering of exposed rocks and the erosion and transport of loose particles create the landscapes we see and change them constantly. Change is usually slow, but not always. The 1952 Lynmouth flood was very sudden (see box), but not far away there are landscapes which record conditions long ago. During the most recent glaciation the ice sheets did not extend as far south as Devon, but on the high granite batholith of Dartmoor the climate was severe, with permanently frozen ground (permafrost), and to this day parts of Dartmoor are periglacial landscapes. Rock masses were shattered by the repeated freezing and thawing of water that penetrated crevices.

In winter the water expanded as it froze, widening the crevices, and in summer the water shrank as it melted, releasing flakes of rock and also large boulders. For those few weeks in summer when the weather was warm enough to thaw the surface layers of the permafrost, turning soil locked solid by ice into wet mud, the mud, together with large boulders embedded in it, slid down-hill, only to be brought to a halt when the temperature dropped and the mud froze again. Today, although there is no permafrost, the scattering of boulders around the tors remains as a record of the climate more than 10 000 years ago. Similar periglacial processes acting on the weak, jointed chalk of southern England caused slopes to retreat through the loss of material from their faces and produced large deposits of the angular debris comprising fragments of varying sizes called 'coombe rock' or sometimes 'head' (other definitions confine 'head' to deposits other than chalk). There are similar periglacial relics in North America and elsewhere in Europe.

The Lynmouth flood

On Exmoor, in south-west England, in the summer of 1952, almost 230 mm of rain fell in 24 hours on to land that was already waterlogged. The water drained northward, carried in two rivers, the East and West Lyn, which enter the sea together at the small village of Lynmouth, falling some 300 m in rather more than 1 km. Unable to carry the volume of water, during the night of 15 and 16 August both rivers flooded and the over-flow from the West Lyn cut a new channel that took it through Lynmouth, rejoining the original course at the mouth. Houses, roads, and bridges were destroyed, an estimated 40 000 tons of trees, soil, boulders, and rubble and masonry from collapsed structures piled up in the village, and 31 people were killed. The disaster was caused by nothing more than rain. Lynmouth was subsequently restored and is now a popular and attractive holiday resort.

Present permafrost regions occur in latitudes much higher than Britain. In Canada and Alaska within the Arctic Circle in places the permafrost is 400 m thick and in parts of Siberia it is 700 m thick. In Resolute Bay, in the Canadian Arctic, it extends to a depth of about 1000 m. Overall, nearly 20 per cent of the land area within the Arctic Circle is permafrost, and has remained in this state since the retreat of the ice sheets that once covered it.

Ice sheets are major sculptors of landscapes. As they move, they scour away all soil and other loose material, pushing it ahead and to the sides of them, where it may form moraines. They smooth angular rocks and the weight of the ice depresses the ground beneath. During a major glaciation ice sheets may grow to a thickness of more than 2500 m and depress the underlying surface by 600 m, which may take it to below sea level. As the ice retreats, the surface rises again, but it is a slow process, at least as measured on a human scale. Northern Canada, where shore-lines rose several tens of metres in less than 1000 years, and Scandinavia are still rising to compensate for the loss of their ice sheets around 10 000 years ago; in Scandinavia the surface was depressed by about 1000 m and has subsequently risen by 520 m. This 'glacioisostasy' demonstrates the slight flexibility of the Earth's crust.

Because there is a lag between the disappearance of the ice sheets and the recovery of the original surface elevation, bowls may remain where the ice was thickest. Depending on their location, these may be flooded by the sea or fill with fresh water. The North American Great Lakes and the Baltic Sea were made in this way. On a much smaller scale, so were the lakes of the English

Lake District. Ice accumulating in a pre-existing hollow will erode the sides to the open-sided, approximately circular shape of a cirque (also known as a 'corrie' or 'cwm'). Where a relatively narrow glacier flows into the sea the trough it excavates may later form a fjord, known in Scotland as a 'sea loch'. Some fjords are more than 1200 m deep. In latitudes higher than about 50°, ice has been the major geomorphological ('landscape-forming') agency.

Soil will tend to move slowly downslope by 'soil creep', caused by the expansion and contraction of material due to repeated wetting and drying, or 'solifluction', where the soil is lubricated by rain water (formerly the term 'solifluction' was applied only to periglacial environments where the ground is frozen for part of the year, but it is now used more widely and is recognized as an important process in some tropical areas). The rate of soil creep has been measured in the English Pennines as between 0.5 and 2.0 mm at the surface and 0.25 to 1.0 mm in the uppermost 10 cm (SMALL, 1970, p. 224). If the soil is deep and the underlying rock extensively weathered, large masses may slip suddenly and move rapidly as 'earth flows'. The collapse of coal tips at the Welsh village of Aberfan in 1966 was of this type (in this case known strictly as a 'flowslide'). The tips had been built over springs. Tip material absorbed the water, greatly increasing its weight but simultaneously lubricating it until it lost its inertia catastrophically (SMALL, 1970, p. 29–34). Earthquakes can break the bonds holding soil particles together, resulting in earth flows of dry material.

There are several ways in which masses of rock and earth can move downslope (HOLMES, 1965, p. 481). All such movements alter the shape of slopes, generally smoothing and reducing them. Figure 2.6 shows the stages by which this happens: (1) material from the free face is detached and falls to form a scree which buries a convex lower slope; (2) further falls cause the free face to retreat until it disappears altogether, leaving a slope that grades smoothly to the level of the higher ground; (3) the slope itself then erodes further. It can also happen that accumulated water increases the weight of a mass of weathered material until it shears away and slides down a concavely curved shear plane between it and the adjacent material. Because of the curved slope, the sliding layers are tilted backwards as they descend, so the toe of the slide is tipped upwards, forming a barrier behind which further debris will be held. This is a 'rotational slide', examples of which can be seen in several places along the south coast of England and in the Isle of Wight. Most failures are quite complex and involve more than one mechanism.

Geomorphology, the study of landforms and the processes by which they are produced and change, began with the work of an American geologist, William Morris Davis (1850–1934), of Harvard University. He proposed that landscapes evolve through a 'cycle of erosion' (the 'Davisian cycle'). This

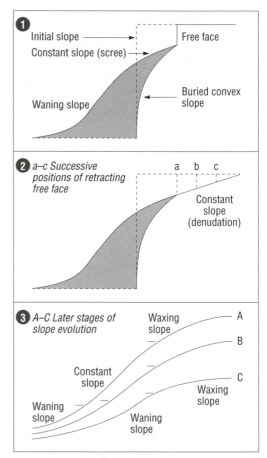

Figure 2.6 **Slope development**

begins when land is raised by tectonic movements. In a young landscape hills slope steeply and the slope of river beds is irregular. As the landscape matures, hill slopes become gentle and river beds slope smoothly. Eventually, the old landscape has eroded to a gently rolling peneplain ('almost a plain'), a word Davis coined. His idea was (posthumously) challenged, in 1924, by the German geologist Walther Penck (1888–1923), who argued that once a slope has settled at an angle which is mechanically stable for the material of which it is composed, it will maintain that angle. Erosion will wear away its face, but will not make it more shallow, so the face will retreat but the angle will remain fairly constant, and the steeper the slope the faster it will erode, because the slow-moving weathered material on a shallow slope will protect the underlying surface. Thus, if a slope is steeper near the bottom than it is higher up, the lower slope will erode faster than the upper slope and the structure will collapse. As the argument developed, geomorphologists came to realize that a true understanding of 'the slope problem' can best be gained from studies of low-latitude landscapes that have not been formed mainly by glacial action, as were those on which the theories of Davis and Penck were largely based (SMALL, 1970, pp. 194–224). Interest in the topic is not purely academic, for an understanding of how rock and soil behaves on sloping ground is necessary for engineers calculating the risks of landslides, erosion, and flooding, and devising schemes to minimize them. The matter is of major environmental importance.

Rivers provide the principal means by which particles eroded from surface rocks are transported from the uplands to the lowlands and eventually to the sea. Rivers are also major landscape features in their own right and, by cutting channels across the surface, important agents in the evolution of landscapes. It is not only mineral particles they transport, of course. Water draining into a river from adjacent land also contains organic matter and dissolved plant nutrients, and rivers also carry those substances we discharge into them as an apparently convenient method of waste disposal. They are also a major source of water supplied for domestic and industrial use.

Water drains from higher to lower ground, moving slowly as ground water between the freely draining soil and an impermeable layer of rock or clay, eventually emerging at the surface as a spring, seeping from the ground, or feeding directly into a river. The 'water table' is the upper limit of the ground water, below which the soil is fully saturated. These terms have the same meanings in British and North American usage, but confusion can arise over 'watershed', which has two different meanings. A drainage system removes water from a particular area, and one such area is separated from adjacent areas. In Britain, the area from which water is removed by a particular drainage system is called a 'catchment' and in North America it is called a 'watershed'. One catchment is separated from another by a 'divide', which in Britain is sometimes known as a 'watershed', and within a catchment the drainage system forms a pattern. Figure 2.7

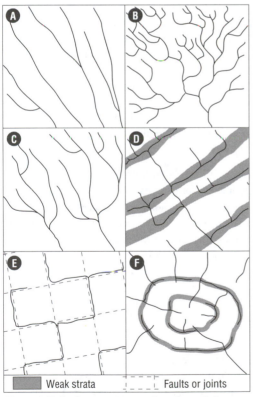

| Weak strata | Faults or joints |

Figure 2.7 **Drainage patterns. A, subparallel; B, dendritic; C, semi-denritic; D, trellised; E, rectangular; F, radial**

illustrates six of the commonest patterns, but others are possible and real patterns are seldom so clearly defined as the pictures may suggest. Climate, the type of rock, and the extent of erosion all play a part in determining the type of pattern that will develop. Dendritic patterns, for example, usually form on gently sloping land of fairly uniform geologic character. Radial patterns occur around domed hills and batholiths, and trellis patterns where rivers cross, more or less at right angles, alternating bands of relatively hard and soft rocks.

As they flow, rivers can be approximately classified into zones, mainly on a biological basis. The headstream, or highland brook, is small, usually torrential (which means its water flows at more than 90 cms^{-1}), and the water temperature varies widely. Few aquatic animals can survive in it. A little lower, trout can survive in what is still a fast-flowing stream, the troutbeck. Silt and mud begin to collect at the bottom of the minnow reach or grayling zone, some plants can survive, and the animal life becomes a little more diverse. In the lowland reach or bream zone the water flows slowly, the river is often meandering, and animal life is diverse. In this final zone the river flows across the coastal plain into the estuary.

10 Coasts, estuaries, sea levels

It seems natural to think of an estuary in terms of the river flowing into it, to see it as the end of the river, with a boundary somewhere offshore where the river meets and merges with the sea. Stand on a headland overlooking an estuary and this is how it looks, but the picture is misleading. An estuary is more accurately described as an arm of the sea that extends inland and into which a river flows. An estuary is dominated by the sea rather than its river, and many estuaries are in fact 'rias', or 'drowned river valleys', old river valleys which were flooded at some time in the past when the sea level rose. The estuaries of south-west England are good examples of rias. In several cases, such as the Camel in north Cornwall, before the marine transgression that began about 10 300 years ago the sea was 36 m below its present level (the sea is still rising at about 25 cm per century), and gently undulating land, with hills formed from igneous intrusions through Devonian slate which survive now as offshore islands, extended up to 5 km from the present coast. This land was blanketed with mixed deciduous forest. Remnants of the forest have been found on the sea bed at several points along the coast and its botanical and faunal composition determined (JOHNSON AND DAVID, 1982).

Sea levels change and at various times in the past they have been both higher and lower than they are today, and they are changing still. During glacial periods (ice ages), sea levels fall, because the volume of the oceans decreases as water evaporated from them accumulates in ice sheets. As the weight of ice depresses the land beneath it sea levels rise; as the ice sheets melt they also rise; and as land depressed by the weight of ice rises again when the ice has melted they fall. There is clear evidence in many places that sea levels were much lower at some time in the past. Raised beaches can be found that are several metres above the present high-tide level. These are areas of approximately level ground, nowadays usually vegetated, containing large numbers of shells of marine organisms. They can have been produced only by the movement of waves and tides over them, at a time when they formed the shore; they are ancient beaches now some distance from the sea.

The sea bed at the mouth of the Camel estuary is mainly sandy, with sand bars, and there are many sandy beaches along the adjacent coast. Sand consists primarily of quartz grains weathered and eroded from igneous rocks inland and transported by the river. They are deposited at the mouth of the estuary, then transported further by tides and sea currents. As they move they become mixed with varying amounts of sea shells, most of which are crushed to tiny fragments

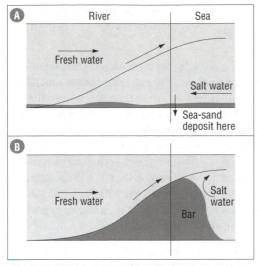

Figure 2.8 **Deposition of sand and formation of an estuarine sand bar**

through being battered by harder stones, producing a beach material with a relatively high calcium carbonate content; this was formerly used by farmers as 'lime', to raise the pH of their soils.

Sand that has been transported many miles by river is deposited where fresh water and sea water meet. Because sea water is denser than fresh water the two do not mix readily and tend to flow in separate channels. The configuration of these channels is determined by the topography of the estuary itself; they may flow side by side or form a wedge, in which fresh water rises over the sea water. On an incoming tide, freshwater and seawater currents often flow in opposite directions and marine fish can move considerable distances inland by keeping to the salt-water channel.

As the fresh water is forced to rise, it loses energy, and the quantity of material a river can transport, known as its 'traction load' or 'bed load', is directly proportional to the energy with which it flows. This depends in turn on such factors as the potential gravitational energy causing the water to flow (essentially the height of the river source above sea level), the gradient of the channel, and the amount of friction caused by contact with the banks and bed (SMALL, 1970, pp. 34–41). As the river water rises and loses energy, the sand grains sink, falling through the underlying salt water and on to the bed. Figure 2.8 shows what happens and how over time the deposition of sand can lead to the formation of a bar at the estuary mouth.

Sand does not remain static on the sea bed and material for the growth of a bar is also provided by sand being carried landward or along the shore by tides and sea currents. The sea water also loses energy as it pushes against the fresh water, and the sand it carries is again deposited.

Sand grains are much larger and heavier than the particles of silt rivers also carry. Silt particles are 2–60 micrometres (μm) in diameter, sand grains 60–2000 μm (in the British standard classification; in the widely used Udden–Wentworth classification they are 4–62.5 μm and 62.5–2000 μm respectively). Ordinarily, large particles would be expected to settle first and small ones later, but in an estuary the opposite occurs. Mudbanks, composed of silt, still smaller clay particles and, mixed with them, organic molecules from the decomposition of the waste products and dead bodies of biological organisms, form inland of the sand banks. Flocculation is the process responsible for this phenomenon. Many of the very small particles carry an electrical charge owing to the presence of bicarbonate (HCO_3^-), calcium (Ca^{2+}), sulphate (SO_4^{2-}), and chlorine (Cl^-) ions. In the boundary zone where fresh and salt water meet, these particles encounter chlorine, sodium (Na^+), sulphate, and magnesium (Mg^{2+}) ions, which bond to them and attract more silt particles, so the material forms clumps larger and heavier than sand grains, and these settle. The organic material mixed with them provides rich sustenance for bacteria and, closer to the surface, burrowing invertebrate animals, which provide food for wading birds. The environment is harsh because the salinity of the water varies widely, so although only a restricted number of species can regulate their osmosis well enough to survive in the mud, those which succeed do so in vast numbers. Estuarine waters may also be enriched by a 'nutrient trap', where the current pattern causes dissolved plant nutrients to be retained (CLARK, 1977, p. 6). In sheltered areas, plants rooted in the mud trap further sediment. In this way mangroves extend some tropical coastlines seaward and, in temperate regions, sediment trapped by saltmarsh vegetation raises the surface until it is beyond the reach of the tides and becomes dry land.

Beach material is transported by currents produced mainly by waves in directions determined by the angle between the shoreline and prevailing movements of the water. Tides have much less effect. If the angle between the approaching waves and the shore is less than 90°, water will flow parallel to the shore, carrying loose beach material with it. This is longshore drift. Its mechanics are complex, but its effect is to shift material to one end of the beach, where it may be trapped against resistant material, such as a cliff or groynes built to prevent beach erosion, swept away entirely (and commonly deposited elsewhere), or deposited where the shoreline turns into the mouth of a bay or estuary and the current becomes turbulent. Sand or shingle accumulating in this way may eventually rise above the water as a spit (HOLMES, 1965, pp. 820–822). Longshore drift moves material in such a way as to rearrange the shoreline so it is at right angles to prevailing waves. This is why waves often appear to meet the shore at right angles.

Waves are produced mainly by wind. This is obvious enough in the case of the huge waves thrown against the coast during storms, but less so in the case of the gentle swell that moves against ocean shores on even the calmest days and in fact is caused by strong winds 1500 km or more from land. The size of sea waves depends on wind force and the length of time during which it blows, the distance over which the water is affected (called the 'fetch'), and the influence of waves that were generated elsewhere. Some waves are also driven gravitationally. Low atmospheric pressure allows the sea level to rise beneath it and water flows downhill towards areas where higher pressure produces a lower sea level.

Beaches are built by waves, and especially by 'spilling breakers' which move a long way up the beach before breaking, spilling a little water from their crests. 'Plunging breakers', which collapse earlier, produce a strong backwash that tends to erode the beach. Where the usual wave pattern is of spilling breakers arriving at about six to eight a minute, the uppermost part of the beach is likely to end in a raised mound of coarse sand or pebbles (a 'berm').

Where waves crash against a vertical rock face, their force is considerable. It has been measured at up to 25 t m^{-2} (SMALL, 1970, p. 438). If the rock is soft, or has many joints or fissures, this is enough to erode it. In harder rock, it is enough to violently compress air held in crevices and allow the air to expand again as the water recedes. This weakens the rock and may produce further cracks into which air enters, until eventually sections are detached as boulders.

Sea cliffs are the result of the wave erosion of hills and as what used to be a hill is cut back, the base becomes a gently sloping wave-cut platform and the eroded material accumulates just below the low-tide limit as a wave-built terrace. Figure 2.9 illustrates the process and shows that it is the ultimate fate of sea cliffs to be eroded completely, until the land slopes gently from the upper limit of wave action to the low-tide line.

How long this takes depends on the resistance offered by the rock, the degree to which it is exposed to the full force of the waves, and the topography of the original high ground. In north Cornwall, Britain, the very impressive sea cliffs have taken around 10 000 years to reach their present condition.

The process may never be completed, because the sea level may reverse its present trend and fall again. This could happen were the ice sheets to advance in a new glaciation. Alternatively, cliff erosion may accelerate as the sea level rises. In some places the present sea-level rise is due to erosion, so it is really a sinking of the land rather than a rising of the sea. It is also believed to be due, in some places exclusively, to the expansion of sea water due to a warming of the sea as

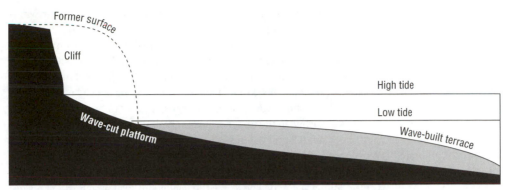

Figure 2.9 **The development of a sea cliff, wave-cut platform, and wave-build terrace**

a result of what may be a general climatic warming. Many climatologists predict that such a warming is likely to continue, but the consequences for sea levels are difficult to estimate and predictions vary widely.

11 Energy from the Sun

Tides are driven by gravitational energy and plate tectonics by the heat generated by the radioactive decay of elements in the Earth's mantle, but the energy driving the atmosphere, oceans, and living organisms is supplied by the Sun. To a limited extent this energy can also be harnessed directly to perform useful work for humans. Solar heat can be used directly to warm buildings and water, desalinate water, and cook food. Sunlight can be converted into electricity. Electrical power can also be generated from wind and sea waves and, because the atmospheric circulation responsible for wind and wind-driven waves is driven by solar heat, these are also forms of solar energy.

The outer layer of the Sun, which is what we see and the region from which the Sun radiates, is at a temperature of about 6000 K and it radiates energy at 73.5×10^6 W from every square metre of its visible surface (the photosphere; being entirely gaseous, the Sun has no solid surface). The figure can be calculated because the Sun behaves as a 'black body'. This is a body that absorbs all the energy falling on it and radiates energy at the maximum rate possible; the rate is calculated by using Stefan's law[9] and is proportional to the absolute temperature raised to the fourth power.

The Sun radiates in all directions and the Earth, being a very small target at a distance of 150 million km, intercepts 0.0005 per cent of the total. At the top of the Earth's atmosphere this amounts to about 1360 W m^{-2}, a value known as the 'solar constant'.

Solar output is not as constant as this name suggests. Between 1981 and 1984, it decreased by 0.07 per cent (HIDORE AND OLIVER, 1993, p. 166). This is a small deviation, but a decrease of about 0.1 per cent sustained over a decade would be sufficient to produce major climatic effects and a 5 per cent decrease might trigger a major glaciation. Cyclical variations in the Earth's rotation and orbit also alter the solar constant. These are believed to be the major cause of large-scale climatic change, and variations in solar output, marked by changes in sunspot activity, are linked to less dramatic changes, such as the Little Ice Age, a period when average temperatures were lower than at present which lasted from about 1450 to 1880. Some scientists believe that the recent climatic warming and rise in atmospheric carbon dioxide concentration are both wholly due to the marked increase in energy output of the Sun since about 1966 (CALDER, 1999).

Radiant heat and light are both forms of electromagnetic radiation, varying only in their wavelengths, and the Sun radiates across the whole electromagnetic spectrum. According to Wien's law[10], the wavelength at which a body radiates most intensely is inversely proportional to its temperature, so the hotter the body the shorter the wavelength at which it radiates most intensely. This is not surprising, because electromagnetic radiation travels only at the speed of light (beyond the Earth's atmosphere, in space, about 300 000 km s^{-1}) and the only way its energy can increase is by reducing the wavelength. Very short-wave (high-energy) gamma (10^{-4}–10^{-8} μm) and X (10^{-3}–10^{-5} μm) solar radiation is absorbed in the upper atmosphere and none reaches the surface. Radiation with a wavelength between 0.2 and 0.4 μm is called 'ultraviolet' (UV); at wavelengths below 0.29 μm, most UV is absorbed by stratospheric oxygen (O_2) and ozone (O_3). The wavelengths between 0.4 and 0.7 μm are what we see as visible light, with violet at the short-wave end of the spectrum and red at the long-wave end. These are the wavelengths at which the Sun radiates most intensely, with an intensity peak at around 0.5 μm in the green part of the spectrum. It is the part of the spectrum to which our eyes are sensitive, for the obvious reason that the most intense radiation is also the most useful, although some animals have eyes receptive to slightly shorter or longer wavelengths. Beyond the red end of the visible spectrum lie the infra-red wavelengths (0.7 μm to 1 mm) and, with increasing wavelengths, microwaves and radio waves, the longest of which have wavelengths up to about 100 km.

The atmosphere is transparent to wavelengths longer than 0.29 μm, although water vapour absorbs energy in several narrow bands between 0.9 and 2.1 μm (BARRY AND CHORLEY, 1982, pp. 10 and 15). When radiant energy, as light or heat, strikes the surface of land or water its energy is absorbed and the surface is warmed. The Earth is not warmed evenly and Figure 2.10 shows how the energy is distributed. The equator faces the Sun, which is always directly overhead at noon. Consequently, its radiation is most intense at the equator. With increasing distance from the equator, the Sun is lower in the sky at all seasons and its radiation covers a larger area less intensely.

Although latitude is obviously important, and places in high latitudes tend to receive less solar energy than those in low latitudes, cloudiness modifies the general distribution quite strongly. The equatorial region does not receive the most intense insolation, because for much of the time clouds shade the surface, reflecting incoming sunlight. Tropical and subtropical deserts, where skies are mainly clear, receive 50 to 100 per cent more insolation than the equator and the dry interiors of North America and Eurasia are much sunnier than maritime regions.

Rather less than half of the solar radiation reaching the top of the atmosphere penetrates all the way to the surface. As Figure 2.11 shows, most of the 'lost' incoming radiation is reflected directly back into space, and about 10 per cent is absorbed or scattered by ozone, water vapour, and particulate matter in the troposphere.

It is scattering that gives the sky its colour. Radiation bounces off particles (mainly molecules) of a particular size in relation to its wavelength. All that changes is the direction of the radiation. There is no loss of energy, but shorter wavelengths scatter more than longer ones. This is called Rayleigh scattering, after Lord Rayleigh (1842–1919) who discovered it, and it reflects radiation in all directions. When the Sun is high in a clear sky, violet light is scattered and absorbed very high in the atmosphere and blue below it. Scattering diffuses the blue light evenly and so the sky appears blue. If the sky is hazy, dust particles scatter light of all wavelengths and the sky appears white. When the Sun is low, dust particles scatter light in the orange and red wavelengths, but shorter wavelengths are absorbed during the much longer passage of the light through the air, and the sky appears orange or red. Spherical particles larger than those responsible for Rayleigh scattering (more than about 0.1 μm) scatter light of all wavelengths, mainly without changing its direction. This is Mie scattering, discovered by Gustav Mie in 1908, and it tends to darken the

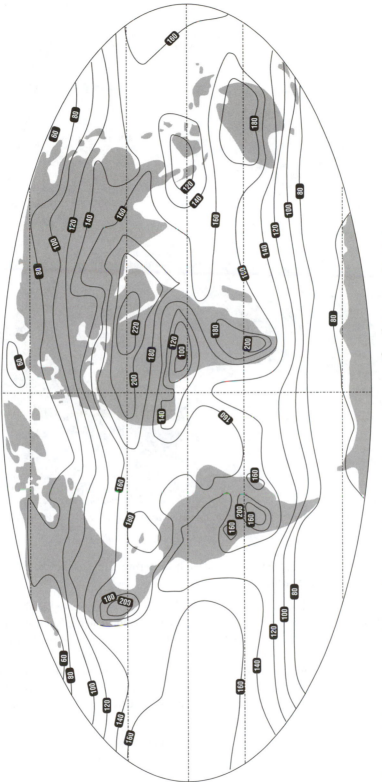

Figure 2.10 *Average amount of solar radiation reaching the ground surface, in kcal cm^{-2} yr^{-1} (1 kcal = 4186.8 J)*

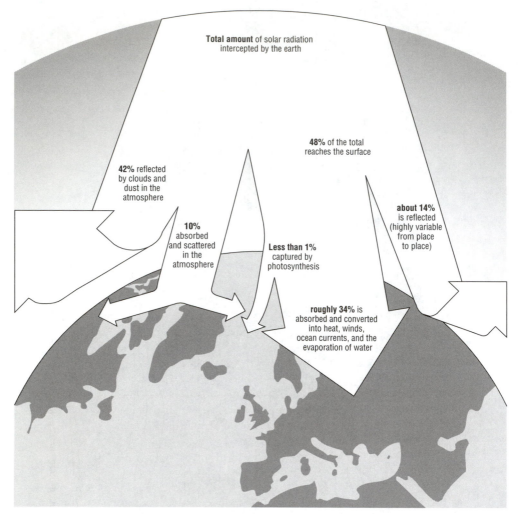

Total amount of solar radiation
intercepted by the earth

48% of the total
reaches the surface

42% reflected
by clouds and
dust in the
atmosphere

about 14%
is reflected
(highly variable
from place
to place)

10%
absorbed
and scattered
in the
atmosphere

Less than 1%
captured by
photosynthesis

roughly 34% is
absorbed and converted
into heat, winds,
ocean currents, and the
evaporation of water

Figure 2.11 **Absorption, reflection, and utilization of solar energy**

sky colour by counteracting the effect of Rayleigh scattering; it makes the sky a darker blue after rain has washed out solid particles.

Once warmed, the Earth also behaves as a black body, radiating energy in the long, infra-red waveband. All the received energy is reradiated. All the portion which is captured by green plants and subsequently passed to animals that eat the plants is converted back into heat by the process of respiration and escapes from the Earth. This must be so, because if captured energy were retained permanently the Earth would grow continually hotter, and it does not. Overall, the amount of radiation received from the Sun is equal to the amount radiated into space from the surface of the Earth, but a proportion of the outgoing energy is retained for a time in the atmosphere. This produces the 'greenhouse effect'.

Solar energy can be exploited for domestic and industrial use, as a so-called 'renewable' energy source, but none of the exploitive technologies is free from problems (RAVEN *ET AL.*, 1993, pp. 234–250).

Fast-growing crops, harvested to be burned, are being cultivated in several parts of the world as 'biomass' fuel. Willow (*Salix* species) and similar woody plants can be burned directly, after drying then chopping and compressing them, which reduces their bulk. Alcohols can be obtained from plants rich in sugar or starch and either used directly or dehydrated and mixed with gasoline to make 'gasohol'. Low petroleum prices led to a decline in the number of Brazilian cars being built to run on 'gasohol', but in 1999 car manufacturers announced an increase in production in an attempt to boost car sales. Fiat planned to raise output of these vehicles from 90 in August to 1,300 in September and Volkswagen planned an increase from 800 to 1,200. General Motors introduced a new model in September and Ford planned to relaunch its models in the spring of 2000. Methanol, an alternative liquid fuel, can also be obtained from plant material. Such fuels are renewable because they can be replaced easily by growing more of the fuel crop, and although they are based on carbon they make no contribution to the greenhouse effect, because the carbon they release when burned is absorbed photosynthetically by the plants which replace them; the carbon is recycled. Biomass crops occupy land, however, and if they are to be grown on the scale needed to supply useful amounts of fuel they could compete for space with food or fibre crops, and because they sell for a lower price than conventional crops they may be grown very intensively to maximize yields.

Solar heat is absorbed by a black surface. Manufactured solar collectors exploit this, using the absorbed heat to warm water, then transferring the heat to a hot water system. Collectors are limited geographically, because they are most efficient where insolation is greatest and do not work well in high latitudes. They have been installed on many buildings, usually attached to roofs or walls facing the Sun, but their high capital cost often makes the energy they provide more expensive than that supplied by the public utilities. In the tropics, however, direct solar heating can be used to distil water and for cooking, with real advantages.

Photovoltaic cells can be used in any latitude, because they convert light, not heat, into electrical power. In years to come this technology may provide useful amounts of energy, but at present its low efficiency (about 15 per cent) means very large arrays are required and the resulting power is much more expensive than electricity generated by other means. For many years, scientists and engineers have been discussing the technical feasibility of constructing truly vast arrays of photovoltaic cells in geostationary orbit and transmitting the power generated as microwaves to a receiving station on the surface, where they would be reconverted to electrical power. The amount of energy obtainable in this way would be great, although it would not be cheap, and at present no one can predict its environmental consequences.

Sunlight can be concentrated. A device developed in Israel by the commercial arm of the Weizmann Institute of Science and Boeing uses highly reflective mirrors (heliostats) to track the Sun and reflect sunlight to another reflector on top of a central tower. This reflector redirects the sunlight to a matrix of concentrators, which increase the intensity of the light 5000 to 10 000 times, compared with the sunlight reaching the surface outside the facility. The concentrated light is fed to a receiver, called 'Porcupine' because it contains hundreds of ceramic pins arranged in a geometric pattern. Compressed air flowing across the pins is heated and channelled to gas turbines that generate electrical power. The prototype plant was installed late in 1999.

Wind power is also exploited widely, but it, too, suffers from the fact that although solar energy, as wind, is abundant, it is variable and very diffuse. The amount of energy captured by a wind turbine is proportional to the square of the diameter of the circle described by its blades and the cube of the wind speed (ALLABY, 1992, pp. 194–202) in a 32 km h^{-1} wind, a 15-metre-diameter rotor linked to a generator operating at 50 per cent efficiency generates 24 kW of power. Most modern wind generators have a rated capacity of about 750 kW and are established in arrays ('wind farms'), each turbine occupying about 2 ha, the spacing necessary to avoid mutual interference, therefore up to 3000 turbines, occupying 6000 ha, are needed to match the output of a large conventional power station. The unreliability of the wind means conventional generating capacity must be held available for use when the wind speed is too low or so high that the blades must be feathered (turned edge-on to the wind) to stop the rotors turning. Suitable sites for such installations are limited, and in highly valued open landscapes they tend to be visually intrusive and arouse strong opposition. Were wind power to provide a substantial proportion of our energy requirements, there could be a risk that the very large installations might affect local climates by extracting a significant part of the energy of weather systems.

The vertical movement of sea waves can also be used to generate electricity. The technology is well advanced, but wave power suffers from disadvantages similar to those of wind power. The installations need to be large and both they and the cables carrying the energy they generate to shore must be able to withstand ocean storms. They must also be located in places where wave movement is large and reliable, but well clear of shipping lanes. This limits the availability of suitable sites. An alternative device, which occupies a much smaller area, extracts energy from the oscillation of waves within a cylindrical structure, and energy can also be obtained in still waters by exploiting the temperature difference between warm surface water and cold deep water.

Wind and wave power can probably be used most effectively on a small scale in places that are beyond the reach of conventional energy distribution systems, such as remote, sparsely populated islands where demand is modest and the cost of links to the mainland high. Direct solar energy, captured by solar collectors or photovoltaic arrays, may become more popular if they can be made more efficient and ways can be found to spread the high capital cost over the lifetime of the installation. Biomass conversion, exploiting energy captured by photosynthesis in green plants, is perhaps the most promising of the renewable technologies. It requires land that is surplus to rival agricultural needs; a criterion that perhaps is now being met in many parts of the European Union.

Compared with the energy our planet apparently receives from the Sun, the amount we derive from fossil and nuclear fuels seems trivial. It is tempting, therefore, to suppose that solar energy can be harnessed to provide environmentally benign power from an original source that is free. Unfortunately, the technical problems are formidable, the costs high, and the environmental consequences uncertain.

12 Albedo and heat capacity

Walk in a snow-covered landscape on a sunny day and you may feel more comfortable if you wear dark glasses. Indeed, you may be well advised to wear dark glasses, because the light may be bright enough to hurt your eyes. Once the snow has melted, and the ground is carpeted with plants, you will have less need of dark glasses. The light will not be so bright.

We can see objects because of the light reflected from them. Their colour is determined by the wavelengths of the light they reflect and their brightness by the amount. Freshly fallen snow reflects 80–90 per cent of the light falling on it, grass 18–25 per cent, and this is why you need dark glasses when crossing snow: it may be almost as bright as the Sun itself. Grass, on the other hand, is much duller.

The proportion of light reflected by a surface is called the 'reflection coefficient', or more usually 'albedo' of that surface. It can be measured and is usually expressed as a fraction or a percentage. As Table 2.1 shows, albedo varies widely from one surface to another.

Table 2.1 **Albedos of various surfaces**

Surface	Albedo (%)
Snow	
Fresh snow	80–90
Melting snow	40–60
Sea ice and old snow	30–40
Cloud	
Cumulonimbus cloud	70–90
Stratocumulus cloud	60–84
Cirrus cloud	40–50
Dry land	
Desert	25–30
Sand	30–35
Tundra	15–20
Grass	18–25
Dry savannah	25–30
Wet savannah	15–20
Deciduous forest	15–18
Coniferous forest	9–15
Tropical rain forest	7–15
Field crops	3–15
Dry ploughed field	5–25
Concrete	17–27
Asphalt	5–17
Earth:	
Overall	31
Land and water surface	14–16
Cloud	23

Table 2.2 **Effect of the incident angle of radiation on water's albedo**

Incident angle (°)	Albedo (%)
0	99 (or more)
10	35
30	6
50	2.5
90	2

About 70 per cent of the Earth's surface is covered by water. Its albedo varies according to the angle at which sunlight reflects from it. When the Sun is low in the sky much more light is reflected and the water looks brighter than when the Sun is high; in the latter case most of its light penetrates the water and is absorbed, and the water looks dark. Early and late in the day, when the water is calm, the occupants of open boats can develop sunburn quite quickly, even in cool weather. Table 2.2 compares the incident angle of radiation with the resultant albedo for water.

Reflected radiation does not warm the surface and so albedo has an important climatic effect, and one that can be modified by human intervention, although the relationship is more complex than it may seem. The clearance of tropical rain forest to grow field crops, for example, involves little change, but if that change is from forest to pasture for feeding livestock, the albedo could double. In this case, the ground would absorb less heat, so there would be less evaporation of water and less cloud would form. This would reduce the average cloud albedo, however, so increasing the amount of radiation reaching the surface and warming it again, evaporating more water and increasing cloudiness once more, but not necessarily to its original value.

This rather intricate relationship illustrates an important point. Climate is strongly subject to feedback effects. In most cases, as in our example, these tend to stabilize conditions, as negative feedback, but positive feedback also occurs. It exaggerates effects and so has a destabilizing effect which can be felt rapidly, as in the onset of glaciation and glacial melting. Eventually, positive feedback is overridden by negative feedback and a destabilized system finds a new level of stability.

Clean air reflects no light, but air is seldom clean. It contains very small particles, called 'aerosols'. In the upper troposphere and at lower levels over the open sea, concentrations range from about 100 to 600 per cm^3, but at low level over continents they are much higher and in industrial regions can reach millions. They vary in size from 10^{-3} to 10^2 μm and gravity has little effect on them, because they are so small. They tend to reflect short-wave radiation, thus increasing planetary albedo, and in saturated air water vapour will condense on aerosols smaller than about 0.5 μm, encouraging cloud formation and also increasing albedo (and removing the aerosols in precipitation). Increasing the albedo has a cooling effect, but aerosols absorb radiation at infra-red wavelengths, so those in the lower atmosphere also have a warming influence; for those reaching the stratosphere, on the other hand, the cooling influence of increased albedo is dominant. To complicate matters further, aerosols settling on to clean snow 'dirty' it, reducing its albedo.

Aerosols are released into the air by volcanoes, as salt crystals formed when droplets of sea spray evaporate, from forest fires, and as tiny soil particles raised by the wind. They are also produced by a range of human activities, especially the burning of fuels. It is difficult to separate natural sources from those linked directly to human activities, and both vary from season to season and year to year but, on average, agriculture and industry account for about one-third of the particulate matter in the air. Forest clearance and the overgrazing of marginal land in semi-arid regions, for example, leads to large injections of small particles as wind-blown soil.

From time to time people suggest altering albedos to trigger climatic change. Particles injected into the upper troposphere might increase the formation of cirriform cloud, for example, and particles injected into the stratosphere might increase planetary albedo for several years. In both cases the quantities of particles required would be huge. It might also be possible to reduce albedo in deserts, by colouring large areas black (perhaps by covering them with plastic sheeting). Such 'thermal mountains' would stimulate convection, hopefully leading to the formation of cumuliform clouds that would release rain. Most climatologists are wary of such schemes, suspecting that in the unlikely event that they worked the unanticipated consequences might be unpleasant. Happily, perhaps, their high cost makes them unattractive to governments.

Varying albedo means some surfaces absorb more solar energy than others, but there are also wide differences in the way heat is absorbed. On a really hot summer day the surface temperature of sand on a beach may be high enough to make it painful to walk across it in bare feet, whereas the water is cool, yet both sand and water are exposed to the same amount of insolation. Dig your feet into the sand, however, and you soon reach a cooler level. The differences were measured over a Saharan sand dune at 1600 hours, when the temperature reached its maximum. The air temperature was a little over 40 °C, that of the sand surface 65 °C, but 30 cm below the surface the temperature was about 38 °C and 75 cm below the surface it was 25 °C (BARRY AND CHORLEY, 1982, p. 288). Soon after sunset, of course, the sand surface would feel cool.

Different materials vary in their response to radiant energy because they have different heat capacities. Heat capacity is calculated as the ratio of the amount of energy applied, to the resulting rise in temperature. The heat capacity of water is much greater than that of rock. This means that much more energy is needed to raise the temperature of water than of rock, or any substance made from rock. It also means that water loses heat much more slowly than rock (HIDORE AND OLIVER, 1993, p. 58). Consequently, water responds to insolation by warming and cooling slowly and land by warming and cooling quickly. This explains the difference in temperature between the sand on a beach and the water beside it, but it also has profound climatic implications.

The rate at which temperature decreases below the surface depends on the conductivity of the material and its mobility, which affects the transfer of heat by convection. Sand grains conduct heat poorly, which is why a layer of cool sand lies at quite shallow depth. Although water is not a very good conductor of heat, heat moves through it readily by convection, and turbulence due to the wind mixes warmed surface water with cooler water immediately beneath it.

13 The greenhouse effect

Since the solar constant is known it is possible to calculate what the 'black-body' temperature of the Earth should be: it is 250 K (–23 °C) (HARVEY, 1976, pp. 43–44). This is what the average temperature at the surface would be were it not for the absorption of long-wave radiation by the atmosphere, which delays the loss of heat and thus warms the planet. The absorption of radiation modifies the climate, or 'forces' it into a (warmer) state than it would be otherwise. The actual average surface temperature is about 288 K (+15 °C), a difference of 38 °C. The forcing that achieves this difference is called the 'greenhouse effect' and, clearly, without it life on Earth would be very uncomfortable, if it were possible at all with surface water, including the oceans to considerable depth, frozen solid and the liquid water beneath the ice markedly more saline because it would contain the salt removed from solution as water crystallized to form ice.

Nitrogen and oxygen are almost completely transparent to electromagnetic radiation at wavelengths greater than 0.29 μm, but some of the incoming solar radiation is absorbed by other constituents of the atmosphere. Of the total, about 4 per cent is absorbed by stratospheric ozone, 20 per cent in the infra-red band by carbon dioxide and 13 per cent by water vapour (in three narrow infra-red wavebands at about 1.5 μm, 2.0 μm, and 2.5–4.5 μm), and 6 per cent by water droplets and dust. At wavelengths greater than about 4.0 μm, however, several atmospheric gases absorb radiation, each in particular wavebands related to the size of its molecules. The significance of this arises from the fact that the Earth, warmed by the Sun, behaves like a black body at a temperature of 288 K and emits electromagnetic radiation at 4–100 μm, with a peak of intensity around 10 μm. More than 90 per cent of this outgoing long-wave radiation is absorbed in the atmosphere (BARRY AND CHORLEY, 1982, pp. 33–35). The remainder, about 6 per cent, with wavelengths between those at which it can be absorbed, escapes into space. These 'gaps' in the absorption bands, at about 8.5 μm and 13.0 μm, are called the 'atmospheric window'.

Molecules which absorb radiation reradiate it in all directions (see Figure 2.12). Some returns to the surface, some is absorbed by other atmospheric molecules and some is radiated upwards, out into space. Of course, the surface is warmed by the Sun only during daytime, but its heat is radiated away by night as well as by day. Eventually, as much energy leaves the Earth as reaches it from the Sun. It must do, because otherwise the atmospheric and surface temperatures would either rise or fall steadily over time; the overall energy budget must balance, and it does, although the transfer of energy is complicated.

The 'greenhouse' metaphor is colourful but a little misleading. It is true that the glass of a greenhouse is transparent to short-wave radiation and partly opaque to infra-red radiation, so its action is similar to that of the absorbing gases in the atmosphere, but the temperature difference inside and outside a greenhouse is due mainly to the fact that air inside is prevented from being cooled by mixing with air outside. With this minor qualification, however, the

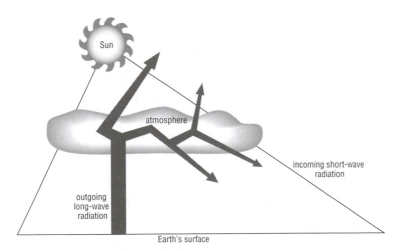

Figure 2.12 **The greenhouse effect**

atmospheric greenhouse effect is real and important, and the gases which cause it are justly known as 'greenhouse gases'.

Both the global climate and atmospheric concentrations of greenhouse gases vary from time to time. Studies of air trapped in bubbles inside ice cores from Greenland and from the Russian Vostok station in Antarctica have revealed a clear and direct relationship between these variations and air temperature, in the case of the Vostok cores back to about 160 000 years ago. The correlation is convincing, although it is possible that the fluctuating greenhouse-gas concentration is an effect of temperature change rather than the cause of it. As temperatures rose at the end of the last ice age, the increase in the atmospheric concentration of carbon dioxide lagged behind the temperature (CALDER, 1999) and so carbon dioxide cannot have been the cause of the warming. There is also evidence that the carbon dioxide concentration was far from constant prior to the start of the Industrial Revolution (WAGNER ET AL., 1999). Carbon dioxide measurements taken from air bubbles trapped in ice cores are unreliable, because carbon dioxide is soluble in solid ice.

Nor has the temperature always been linked to the concentration of carbon dioxide. The two were uncoupled between about 17 and 43 million years ago. The air then contained less than two-thirds of the present concentration of carbon dioxide (180–240 µmol mol^{-1} compared with 360 µmol mol^{-1} today), but the climate was up to $6\,°C$ warmer than it is today (COWLING, 1999).

Nevertheless, it is estimated that the atmospheric carbon dioxide concentration immediately prior to the Industrial Revolution was about 280 µmol mol-1 and that the increase since then has been due entirely to emissions from the burning of fossil fuels. This may not be the case. The solubility of gases, including carbon dioxide, is inversely proportional to the temperature. A rise in temperature, therefore, will cause dissolved carbon dioxide to bubble out of the oceans. This is called the 'warm champagne' effect. Rising temperature will also stimulate aerobic bacteria. Their respiration will release carbon dioxide. This is called the 'warm beer' effect (CALDER, 1999).

Carbon dioxide is the best-known greenhouse gas, because it is the most abundant of those over which we can exert some control, but it is not the only one. Methane, produced naturally, for example by termites, but also by farmed livestock and from wet-rice farming (present concentration about 1.7 ppm), nitrous oxide (0.31 ppm) and tropospheric ozone (0.06 ppm), products from the burning of fuels in furnaces and car engines, and the industrially manufactured compounds CFC–11 (0.000 26 ppm) and CFC–12 (0.000 44 ppm) are also important. The most important of all, however, is water vapour. This enters into the calculations only indirectly, because its concentration varies greatly from place to place and from day to day and because it is strongly affected by temperature. Its influence, therefore, tends to add to those of the other gases and generally varies as they do. Figure 2.13 shows the anticipated changes in concentration for carbon dioxide, methane, and CFC–12, which is one of the family of CFC compounds. These increases are based on the (uncertain) assumption that industrial and vehicle emissions are the only source of carbon dioxide.

All greenhouse-gas effects are usually expressed as 'global warming potentials' (GWPs) which relate them to carbon dioxide. GWPs take account of the wavelengths at which particular molecules absorb, some of which overlap, and the length of time they remain in the atmosphere before decomposing or being deposited at the surface. On this basis, over a 100-year period, with carbon dioxide given a value of 1, methane has a value of 11 (i.e. it is 11 times more effective than carbon dioxide, molecule for molecule), nitrous oxide 270, CFC–11 3400, and CFC–12 7100. The estimates of future climatic warming are based on the consequences calculated for a doubling of the carbon dioxide concentration, which includes the GWPs for all the relevant gases. Figure 2.14 shows that, depending on the sensitivity of the atmosphere to greenhouse forcing, a doubling of carbon dioxide would raise the average global temperature by 1.5–$4.5\,°C$, with a 'best estimate'

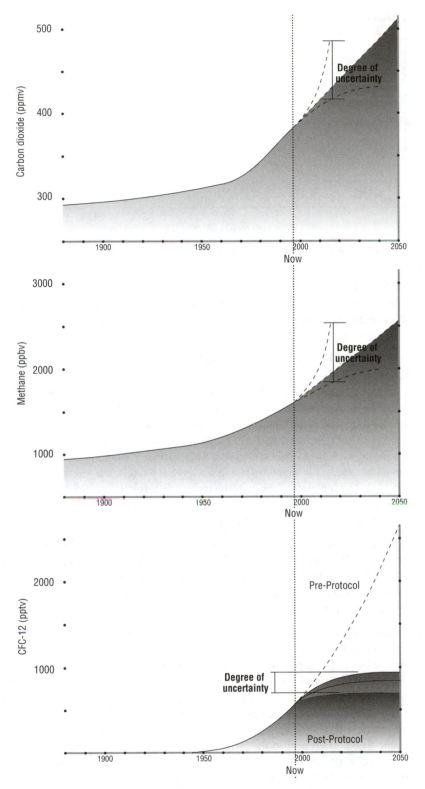

Figure 2.13 **Anticipated changes in concentration of three greenhouse gases**

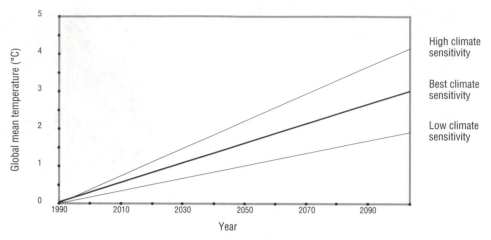

Figure 2.14 **IPCC estimates of climate change if atmospheric CO_2 doubles**

of 2.5 °C, and at the current rate of increase in greenhouse-gas concentrations these temperatures would be reached by around 2100. During the same period, warming of the oceans is
calculated to cause them to expand, producing a rise in sea level of 2–4 cm per decade (IPCC,
1992).

It is not certain that sea levels have risen, although predictions of global warming include an
assertion that they have risen world-wide by about 25 cm over the past century. In 1841, the
explorer Sir James Clark Ross visited Tasmania, where he met Thomas Lempriere, an amateur
meteorologist. The two men installed an Ordnance Survey Bench Mark, chiselling it into a rock
face at a place called the Isle of the Dead, near Port Arthur. It was positioned with great care and
precision and was meant to act as a sea-level gauge. The gauge has been rediscovered and the
Tasmanian climatologist John L. Daly visited it in August 1999. He found that it remains visible
above the water line, despite the supposed rise in sea level. It is uncertain whether the gauge was
set at a level close to the high tide mark or at the mean tide level. If it was close to the high tide
level it shows that sea level has not changed since 1841. If it was at the mean tide level it shows
the sea level has fallen.(*www.vision.net.au/~daly/*)

The mean global temperature increased by 0.37 °C between 1881 and 1940. The temperature fell
from 1940 until the 1970s, since when it has risen again, but there is no clear evidence of any
warming between 1980 and 1998. The total warming from 1881 to 1993 amounts to 0.54 °C. Two-
thirds of the warming occurred before 1940 – and before the main rise in the atmospheric carbon
dioxide concentration – and 1881 was an unusually cold year (BALLING, 1995).

Mean temperatures are calculated from three sets of data. Weather stations and ships record
surface temperatures, balloon sondes record temperatures above the surface, and TIROS-N
satellites operated by NASA on behalf of the National Oceanic and Atmospheric Administration
(NOAA) measure temperatures from space (*www.atmos.uah.edu./essl/ msu/background.html*).
Surface measurements are difficult to interpret over long periods. This is because weather stations
that are established in open country may gradually be affected by nearby urban development and
road building, which will raise the temperature by an 'urban heat island' effect, producing an
illusory warming trend. There is also the possibility that, over the years, changes in staff may
lead to unrecorded changes in the time of day when measurements are made. Ships measure the
temperature of sea water below the surface, but different ships do so at different levels. Their
measurements of air temperatures are unreliable for similar reasons. Over the course of this

century, ships have become larger, so their decks from which temperatures are measured are higher above the sea than they used to be. In any case, ships' thermometers are not calibrated against a standard.

Balloon readings are much more reliable. Weather balloons are released twice every day, usually at noon and midnight Greenwich time, from about 1000 sites. These sites are located predominantly in industrial countries, however, so records derived from them may not be typical of the world as a whole.

Satellite measurements are by far the most reliable of all. More than 30 000 measurements are made every day.

Measurements from surface stations show temperatures are rising by 0.15 °C per decade from 1979 to 1997. This is a much smaller increase than the IPCC 'best estimate' of 2.5 °C. Over the same period, NOAA weather balloons show temperatures are falling by 0.07 °C per decade and balloon data from the UK Meteorological Office show them falling by 0.02 °C per decade. The satellite measurements show them falling by 0.01 °C per decade (*science.nasa.gov/newhome/ headlines/ notebook/essd13aug98_1.htm*). The mean temperature measured by satellite since January 1998 has risen by about 0.04 °C per decade. This warming was caused by the very strong El Niño event in 1998. The warming is still very much smaller than that predicted by the IPCC.

The effects of climate forcing are being studied by teams in several countries and their results are drawn together into a scientific consensus by the IPCC. This body, involving hundreds of specialists from all over the world, was established in 1988 by the World Meteorological Organization and the UN Environment Programme to advise governments. By no means all climatologists agree with the IPCC conclusions, however.

Governments became involved following a meeting held at Villach, Austria, in 1985 under the auspices of the International Council for Scientific Unions (ICSU, now called the International Council for Science). There, research scientists, including ecologists and experts on climate and energy-demand modelling, concluded that global warming was a real threat and more research was needed and, supported by environmentalist groups, the topic quickly acquired political influence. This politicization and resultant popular dramatization of a very complex and uncertain issue has attracted criticism (BOEHMER-CHRISTIANSEN, 1994; BOLIN, 1995).

Studies of climate forcing begin with estimates of ways in which the chemical composition of the atmosphere may change in the future, to produce an 'emissions scenario'. This requires a knowledge of the sources from which greenhouse gases are released, the sinks into which they are absorbed, and ways the sinks may respond to increased loading. The oceans are the most important sink for carbon dioxide, but the behaviour of the sinks is incompletely understood and no sink has been identified for a significant fraction of the carbon dioxide known to have been emitted. Measurements of greenhouse-gas concentrations must also distinguish between genuine changes, the 'signal', and natural variations, the 'noise'. Carbon dioxide levels vary seasonally, for example, in response to the growing season for plants.

General circulation models (GCMs) are then constructed. These are based on a notional three-dimensional grid placed over the entire Earth. Atmospheric behaviour is calculated according to physical laws for every grid intersection. The input data for each calculation include the state at adjacent grid points as well as data introduced by the modeller, and so they trace the evolution of the atmosphere, simulating the climate. Using the known present state of the atmosphere, the model is used to simulate the climate over several decades and its results compared with actual climate records. If this test proves satisfactory, changes in atmospheric composition, based on the emissions scenario, are introduced to the model and their consequences evaluated.

Modelling on this scale requires massive computing power. Even when the fastest supercomputers are used, the grid must be fairly coarse to keep the number of intersection points to a manageable level. This means that some phenomena, such as cloud formation, must be greatly simplified, because they occur on a smaller scale than the $100 \times 100 \times 10$ km grid boxes. Most GCMs make similar simplified allowances for the mixing of surface and deep ocean water, although the latest 'coupled' models (CGCMs) treat the oceans as of complexity comparable to that of the atmosphere.

GCMs are being improved constantly and scientific understanding of atmospheric and oceanic processes is increasing rapidly, but much remains to be learned and estimates of the regional consequences of a general warming vary widely. The IPCC finds, for example, that warming will be reduced by 60 per cent or more over the northern North Atlantic and around Antarctica. The best illustration of the uncertainties surrounding the calculations centres on water vapour. If the temperature rises, more water will evaporate. Water vapour is a greenhouse gas, so this will accelerate the warming trend, but a more humid atmosphere will also be a cloudier atmosphere. As water vapour condenses to form cloud the latent heat of condensation, released into the surrounding air, also has a warming effect, but clouds themselves may have either a warming or a cooling effect. Generally, clouds at low level have a high albedo and, therefore, cool the surface, while high-level clouds absorb radiation and have a warming effect. Most GCMs predict an increase in middle- and high-level cloud, with a consequent warming effect, but an increase in cloud amount, with deeper cloud cover, might reduce warming. It is important to know how much cloud will form and its type, but at present this cannot be calculated for given atmospheric conditions.

Many climatologists accept there is a real possibility of global climatic warming due to an enhanced greenhouse effect. As with many large-scale changes, there would be winners and losers. Were climate belts to shift toward higher latitudes, which seems the most likely overall result, parts of the Sahara and southern Russia would receive increased rainfall. They would benefit and their agricultural output would increase. On the other hand southern Europe and the United States cereal belt might become drier. If warming produced a rate of evaporation that exceeded the increase in the rate of precipitation, soils would become more arid. It may be, however, that warming will be experienced as a reduction in the fall of temperature at night, due to increased cloudiness, with little or no change in daytime temperatures. In that event, nighttime frosts would become less frequent, soils would become somewhat moister, and agriculture would benefit.

Environmentalists favour the 'precautionary principle'. This holds that if there is a chance of adverse change we should not wait until the risk can be scientifically confirmed before taking action to minimize it. When world leaders agreed to reduce greenhouse-gas emissions at the 1992 UN Conference on Environment and Development (the 'Rio summit') they did so in accordance with this principle. There are critics of the principle, however, who point to the cost and difficulty of pursuing policies that may prove unnecessary. The principle holds that if any innovation appears to entail a risk of serious or irreversible harm to human health or the environment, then precautionary measures should be taken to avert that risk. This seems to obtain even if the link between the innovation and the harm has not been proven or if it is weak and the harm is unlikely to occur. It can be argued that this weighs the possible advantages and disadvantages of innovation, but loads the scales in favour of the disadvantages. This could have a paralyzing effect if the only way to determine whether the risk is genuine is to undertake the innovation – which the possibility of risk forbids (HOLM AND HARRIS, 1999). Not surprisingly, others disagree, maintaining that the principle does not necessarily prohibit innovation but does encourage preventive action in the face of uncertainty (RAFFENSPERGER ET AL., 1999). In this area, as in many others, the environmental science is uncertain and its translation into political action far from simple.

14 The evolution, composition, and structure of the atmosphere

When the Earth first formed, it may have had a thin atmosphere of light gases, mainly hydrogen and helium, derived from the stellar nebula. If so, this atmosphere was lost when the Sun began to radiate, supplying the gas molecules with the energy they needed to escape the planet's gravitational attraction. Volcanism then released the gases which formed a new atmosphere. No one knows the precise composition of that atmosphere, but probably it was rich in carbon dioxide and contained only traces of nitrogen and free oxygen. Our present atmosphere has evolved to its present state, partly (some would say mainly) as a result of biological processes.

These processes continue to maintain it. Nitrogen, for example, is chemically somewhat inert but, in the presence of oxygen and with a sufficient application of energy, it will oxidize to nitrate (NO_3). This reacts with water to form nitric acid (HNO_3) and is washed to the ground. Lightning supplies enough energy for the oxidation, and in the world as a whole it is estimated that there are about 1800 thunderstorms, with lightning, happening at any one time, delivering some 100 million tonnes of fixed nitrogen to the surface every year. At this rate it would not take long to strip the atmosphere of its nitrogen, were it not for the activities of denitrifying bacteria, which utilize nitrogen compounds in the soil and release gaseous nitrogen as a metabolic by-product. Were the air to be seriously depleted of nitrogen the proportion (partial pressure) of oxygen would increase and were it to increase to more than about 25 per cent exposed carbon compounds would burn even more readily than they do. This would reduce the amount of oxygen, replacing it with carbon dioxide. As it is, carbon dioxide is removed from the air by green plants as well as by dissolving in rain water.

To a considerable extent, therefore, our present atmosphere has been constructed and is maintained by living organisms. It is an essential component of the environment. It is also the route by which many organisms are disseminated and most nutrients cycled.

Air is a mixture of gases. They are not combined, so air itself is not a chemical compound, although some of its minor constituent gases are.

As the many practical uses for compressed air (in car and bicycle tyres, for example) testify, air is highly compressible. The total weight of the atmosphere (with a mass of about 5×10^{15} tonnes) compresses its lower layers, so the average density of the air decreases from about 1.2 kg m^{-3} at sea level to about 0.7 kg m^{-3} at a height of 5 km. Half of the total mass of the atmosphere lies below 5 km. The pressure exerted by the atmosphere is usually expressed in millibars (mb) in weather forecasts or in pascals (Pa) by physicists (1 mb = 100 N m^{-2} = 100 Pa). The average sea-level pressure is 1013.2 mb (or 101 320 Pa).

One consequence of its compressibility is that for all practical purposes the atmosphere is quite shallow. Although it has no clearly defined upper boundary, merging imperceptibly with the solar atmosphere at an altitude of about 80 000 km (ALLABY, 1992, p. 50), pressure and therefore density decrease logarithmically with height. At 30 km, air density is at only 0.02 per cent of its sea-level value. The part of the atmosphere in which weather occurs is confined to the lowermost 8–16 km. None of the familiar meteorological phenomena occur above this height, but they are influenced by events in the lower reaches of the overlying layer, up to a height of about 30 km.

The air is warmed convectively from below, by contact with the surface. As it warms it expands and as it expands it cools. Temperature and density determine the amount of water vapour air can hold (its humidity), and the combination of the humidity, density, and temperature imposes a layered structure on the atmosphere. Figure 2.15 illustrates this structure, relating it to height, temperature, and pressure.

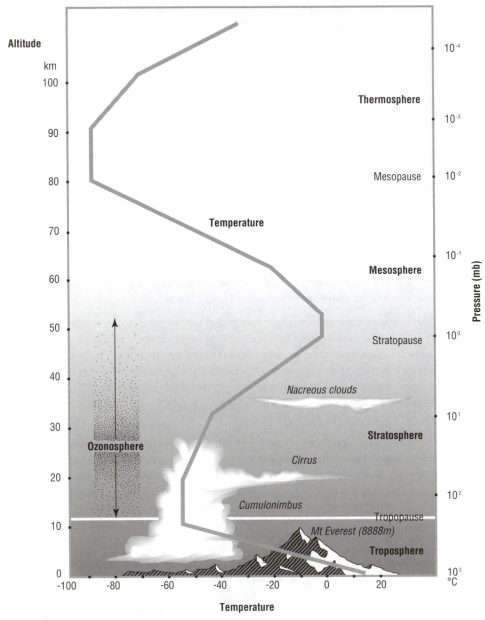

Figure 2.15 **Structure of the atmosphere**

The lowest layer, the troposphere, extends from the surface to an upper boundary, the tropopause, the height of which varies, but averages about 16 km at the equator and 8 km at the poles. Within the troposphere temperature decreases with height by an average of about 6.5 °C km^{-1} (called the 'lapse rate'). Above the tropopause temperature remains constant with height to about 20 km. Rising tropospheric air is trapped below the region of constant temperature, which forms a permanent temperature inversion, a layer in which temperature remains constant or increases with altitude, rather than falling. It is this inversion which confines meteorological phenomena to the troposphere.

Temperature then increases from a minimum of about –80 °C at the equator in summer, when the tropopause is at its highest, to 0 °C or even higher at about 50 km. This region is the stratosphere and its upper boundary is the stratopause. In the mesosphere, above the stratopause, temperature once more decreases with height, to about –90 °C at the mesopause, about 80 km, then rises again through the thermosphere. At about 350 km the temperature may exceed 900 °C, probably because of the energy imparted by absorption of ultraviolet radiation by atomic oxygen, but the air is so rarefied that objects such as satellites are not warmed by it, although it still exerts measurable drag on spacecraft moving through it.

Between about 30 and 60 km the density of oxygen molecules is high enough to intercept most of the incoming solar ultraviolet radiation at wavelengths below 0.29 µm. The energy imparted to them separates the molecules ($O_2 \rightarrow O + O$). Some of the oxygen atoms then combine with oxygen molecules to form ozone ($O + O_2 \rightarrow O_3$). Ozone is unstable and may decompose either by encountering more oxygen atoms ($O_3 + O \rightarrow 2O_2$) or by absorbing more ultraviolet radiation. Ozone is, therefore, constantly forming, decomposing, and re-forming, and the process is in equilibrium above about 40 km. There is also some transport of ozone from low to high latitudes. There is some mixing of stratospheric air, however, as a result of which a small amount of ozone is transported downward, to accumulate between about 20 and 25 km. This is the 'ozone layer'. Its density varies, being lowest over the equator and high over latitudes above 50°. Ordinarily, ozone levels are also high over polar regions in early spring. This is because ozone is neither formed nor destroyed during the polar night, when there is no radiation to drive the reactions, and ozone transported from lower latitudes is stored (BARRY AND CHORLEY, 1982, pp. 2–3). Despite being known as the 'ozone layer', if the air at that altitude were compressed to sea-level pressure the ozone would contribute only about 3 mm to it. In itself, the ozone layer does not shield the surface from ultraviolet radiation, but indicates that radiation is being absorbed at a greater height, shielding both the surface and the ozone layer.

The thickness of the ozone layer is often reported in Dobson units (DU). This unit was devised by G.M.B. Dobson, a British physicist who studied stratospheric ozone in the 1920s. It refers to the thickness of the layer that a gas would form if all the other atmospheric gases were removed and the gas in question were subjected to standard sea-level pressure. In the case of ozone, 1 Dobson unit corresponds to a thickness of 0.01 mm and the amount of ozone in the ozone layer is typically 220–460 DU, corresponding to a layer 2.2–4.6 mm thick.

The depletion of the ozone layer over Antarctica, first observed in 1986 (ALLABY, 1992, pp. 159–161) by a British scientist, occurs just as spring is commencing. During the polar night, a vortex of very still air forms over Antarctica within which the temperature may be as low as –84 °C. Clouds of ice crystals, called polar stratospheric clouds (PSC), form inside the vortex and a series of chemical reactions on the surface of the ice crystals results in chlorine monoxide (ClO) combining to form Cl_2O_2. These molecules break down when exposed to sunlight ($Cl_2O_2 \rightarrow 2Cl + O_2$) and the chlorine atoms combine with ozone in two steps which release free chlorine once more to repeat the process so that a single chlorine atom can destroy many thousands of ozone molecules ($Cl + O_3 \rightarrow ClO + O_2$; $ClO + O \rightarrow Cl + O_2$) (HIDORE AND OLIVER, 1993, pp. 74–77). CFCs are believed to be the principal source of stratospheric chlorine. Although very stable, they are decomposed by ultraviolet radiation at wavelengths below 0.23 μm, releasing free chlorine. As spring advances, the vortex disappears, ozone moves poleward from lower latitudes, and the ozone layer recovers. Seasonal depletion over the Arctic has also been reported, but it is less severe and of shorter duration, because Arctic winter stratospheric temperatures are higher than those of the Antarctic and a polar vortex rarely forms.

Ozone depletion may lead to increased exposure to ultraviolet radiation at the surface, the biological significance of which is uncertain. Ultraviolet radiation causes cataracts and non-melanoma skin cancer in fair-skinned humans (recent increases being due to the popularity of sunbathing in hot climates to which people are not acclimatized, and not to ozone depletion). It might have an adverse effect on land plants especially susceptible to it and may also affect organisms living in the uppermost few millimetres of the ocean surface; below that depth ultraviolet radiation is absorbed by sea water.

Ozone is a very minor constituent of the atmosphere. In the troposphere it occurs locally, some naturally but more commonly as a pollutant which causes respiratory irritation in humans and can damage plants, produced by photochemical reactions involving vehicle exhaust fumes. It is a constituent of photochemical smog and responsible for some of the damage attributed to acid rain.

The principal ingredients of the atmosphere are listed in Table 2.3. Water vapour comprises up to 4 per cent in the lower atmosphere, but above about 12 km it is virtually absent. The source of the water vapour from which PSCs form is unknown; it may have entered the stratosphere as water vapour and accumulated in the polar vortex or may result from the oxidation of methane ($CH_4 + 2O_2 \rightarrow CO_2 + 2H_2O$). Water vapour apart, the composition of the atmosphere remains constant to a considerable height, because of mixing caused by turbulence. Beyond the mesosphere, however, the proportions of its ingredients change. Figure 2.16 illustrates how the chemical composition changes with height.

Table 2.3 **Average composition of the troposphere and lower stratosphere**

Constituent	% by volume
Nitrogen	78.08
Oxygen	20.94
Argon	0.93
Carbon dioxide	0.035
Neon	0.0018
Helium	0.0005
Ozone	0.00006
Hydrogen	0.00005
Methane	0.00017
Krypton	trace
Xenon	trace

15 General circulation of the atmosphere

If the Earth faced it directly, the Sun would be overhead at the equator at noon every day of the year. This would have a profound effect on climates, because there would be no seasons. In fact, of course, the Earth is tilted on its axis, so we do not face the Sun directly. Our orbit traces the circumference of a plane, called the 'ecliptic', and our rotational axis is tilted to the plane of the ecliptic by 23.5° (though the angle of tilt varies between 21.8° and 24.4° over an approximately 41 000-year cycle). This means that from March to September the northern hemisphere is inclined toward the Sun and from September to March the southern hemisphere is inclined inwards, bringing summer to the two hemispheres in turn. Figure 2.17 shows how the tilted axis produces our seasons; these are labelled for the northern hemisphere and the names should be reversed for the southern hemisphere (winter becomes summer, spring becomes autumn). Only at the spring and autumn equinoxes is the noonday Sun directly overhead at the equator. At noon on 21 June it is directly above the tropic of Cancer and at noon on 21 December it is directly above the tropic of Capricorn. This is how the two tropics are defined.

Our orbit around the Sun is not circular, but slightly elliptical. We are at our closest to the Sun (perihelion) on 3 January and furthest away (aphelion) on 4 July and, therefore, we receive about 7 per cent more solar radiation in January than we do in July. This should make southern-hemisphere summers warmer than those in the northern hemisphere, and the winters cooler, but in reality the situation is reversed. This is due partly to the masking of so small an effect by the general circulation of air, and partly to the fact that at present the northern-hemisphere summer

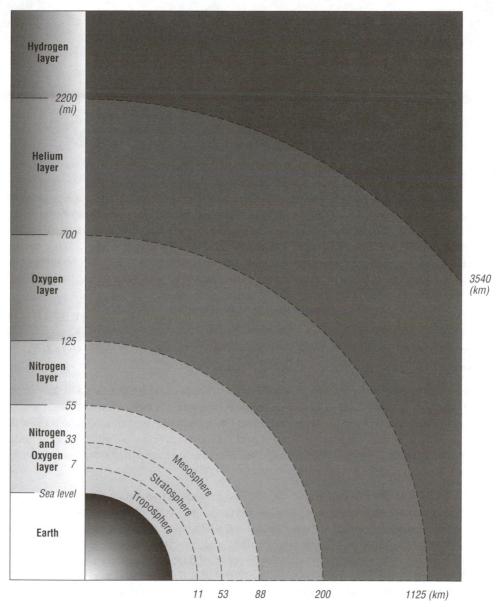

Figure 2.16 *Chemical composition of the atmosphere with height*

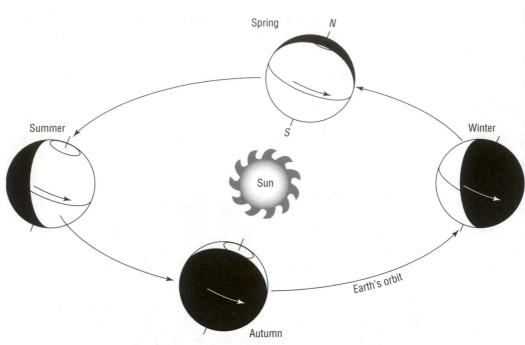

Figure 2.17 **Seasons and the Earth's orbit**

(March to September) is five days longer than the winter (September to March), a situation that changes slowly. The gravitational attraction of the Sun, Moon and, to a much smaller extent, the planets on the slight bulge around the Earth's equator cause the position of the Earth at the equinoxes to move westward by 50.27″ (″ = seconds of arc) a year, so it takes 25 800 years for them to complete a full cycle and return to their starting position. This is called the precession of the equinoxes and it alters the dates at which the Earth is at perihelion and aphelion. About 13 000 years from now we will be at aphelion (furthest from the Sun) in January.

It is within the tropics that the surface receives the most intense insolation and, therefore, is heated most strongly. The movement of air and oceans then carries heat from the tropics into higher latitudes. This transport of heat produces the main 'climates' of the world and our day-to-day weather.

To either side of the equator, the prevailing winds are from an easterly direction and they are so reliable that sailing ships made extensive use of them. The name 'Trade Wind' has nothing to do with commerce. 'Trade' used to mean 'course' and 'to blow trade' meant to blow in a constant direction. Such was their importance that eminent scientists theorized about their cause and it was from their calculations that the first understanding arose of the way the atmosphere transfers heat.

In 1686, the astronomer Edmond Halley (1656–1742) suggested that hot equatorial air rises and is replaced by cooler air from higher latitudes. He was almost correct, but he could not explain why the returning air arrived from the north-east and south-east, rather than from due north and south. This was explained in 1735 by George Hadley (1685–1768). He realized that the Earth rotates beneath the air, changing its apparent direction of flow, but a century passed before the French physicist Gaspard Gustave de Coriolis (1792–1843) discovered what really happens. In 1835 he proposed what is now known as the 'Coriolis effect' (or 'force', although no force, in a mechanical sense, is involved).

On the surface of a rotating sphere, the speed at which any point moves depends on its latitude, because that determines the distance it must travel: a point at the equator travels faster than one at a higher latitude. Air at the equator is moving eastward at the same speed as the surface beneath it. If it moves away from the equator, because it is not attached to the surface, it continues to travel eastward at the same speed (slowing steadily by friction), which is now faster than the surface beneath it, so its motion has an eastward component in relation to the surface. Similarly, air moving towards the equator is moving eastward more slowly than the surface and so it appears to drift westward, in fact because the surface is overtaking it. This is the Coriolis effect and it accounts for the fact that air does not move north or south along straight paths in relation to the surface. The strength of the Coriolis effect increases with distance from the equator and in 1865 the American meteorologist William Ferrel (1817–91) pointed out that in low latitudes the conservation of angular momentum would be more influential than the Coriolis effect (BARRY AND CHORLEY, 1982, p. 138).

What Hadley appreciated was that air, warmed at the equator, will rise, cool as it rises, and descend again. This establishes a convective cell of moving air which drifts to the east as it moves away from the equator and to the west as it returns. We identify winds by the direction from which they blow (i.e. the opposite of the direction in which they blow), so this accounts for the easterly trade winds experienced at the surface in the tropics and also proposes westerly winds at high altitude. This tropical cell, in reality a system of several cells, is known as the 'Hadley cell'.

Rising equatorial air produces a region of predominantly low surface atmospheric pressure. The air cools near the tropopause and descends in the subtropics, where surface pressure is predominantly high. Very cold, dense air also sinks over the poles, forming polar high-pressure regions. Air flows out from these regions at low level and rises again in middle latitudes, forming a second set of cells. These two drive a third, mid-latitude system of cells, comprising air flowing away from the equator at low level, rising where it meets air flowing in the opposite direction from the pole, and dividing so that some of its air feeds the polar cell and some returns equatorward, descending in the subtropics with the descending air from the Hadley cell. Figure 2.18 illustrates this flow and the winds associated with it. The intertropical convergence zone (ITCZ) is where the two tradewind systems meet. Air rises gently, producing a low-pressure belt near the surface, and this in turn often results in calm air, the region sailors called the 'doldrums', a name it acquired in the middle of the last century.

Where the cells meet, near the tropopause in the subtropics and again at about 60°, there is a sharp difference in temperature in the air to either side of a boundary. These temperature gradients produce strong westerly air flows, known as the 'jet streams'. The subtropical jet stream is fairly constant. The polar front jet stream, at an altitude of 9–15 km, is much more irregular, varying in its latitudinal location and sometimes disappearing altogether, but it also produces the strongest winds. These can reach 150–250 km h^{-1} and more than 450 km h^{-1} in winter, when the temperature gradient is strongest.

Atmospheric convection cells produce the boundaries (fronts) between bodies of air at markedly different temperatures which give rise to the jet streams, but it is the jet streams that dominate the weather conditions experienced below. This is especially true of the polar front jet stream, which develops waves and then breaks into cells. This process most commonly occurs in February and March in the northern hemisphere, each complete cycle lasting several weeks. Figure 2.19 illustrates four stages in what is known as the 'index cycle'. At first (1), the winds are zonal (i.e. they flow steadily from west to east) and there is little mixing of the air to either side. Waves start to develop (2) as the jet stream widens and its velocity increases. Air is now flowing more to the north and south and so it is influenced by the Coriolis effect and the conservation of angular

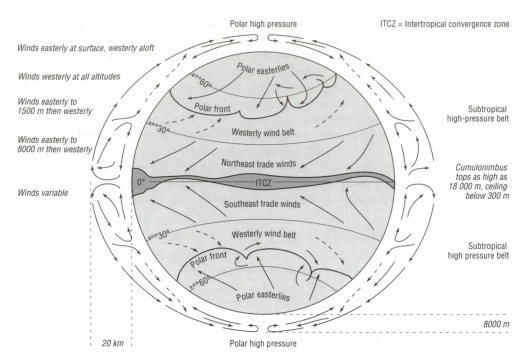

Figure 2.18 **General circulation of the atmosphere**
Note: **ITCZ = intertropical convergence zone**

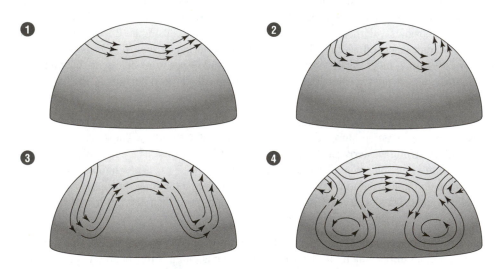

Figure 2.19 **The development of cells in jet streams and high-level westerlies**

momentum. These tend to exaggerate the undulations, which become extreme (3). Finally (4), the wind pattern fragments into cells. After this, the jet stream disappears for a short time, then re-establishes itself and the index cycle repeats. During 1, surface pressure systems and the weather associated with them move steadily in an easterly direction, but during 2 and 3 the movement becomes more irregular, with latitudinal deviations. In 4, the weather becomes almost stationary, with mild high-pressure areas in high latitudes and cold low-pressure areas in lower latitudes blocking the easterly movement.

The unsettled nature of the climate at latitudes around 50–60° is due to the proximity of the polar front and the vagaries of the jet stream associated with it: surface conditions are being determined by events at a height of some 10 km. As the front moves north and south, places in about the latitude of Britain may be exposed alternately to polar and tropical air masses and to pressure systems that move rapidly or remain stationary for weeks at a time. Should the global climate change, becoming generally warmer or cooler, there could be a more permanent shift in the location of the polar front, with quite profound implications for places in this meteorologically critical region.

Weather comprises more than temperature, of course, and temperature is dependent on more than the simple convective transfer of heat. Both air temperature and surface weather involve the evaporation, condensation, and transport of water by large masses of air that have acquired distinct characteristics over the oceans or continental interiors then moved into different areas, and on reactions between adjacent air masses with different properties.

The convection-cell model provides only a general, rather crude account of the general circulation of the atmosphere, by which heat is transported from low to high latitudes. It takes no account of the transport of heat by the oceans, for example. This is of major importance and, as the El Niño and NADW–Dryas episodes demonstrate (see the next section), apparently minor disturbances can produce dramatically different situations. The atmospheric system is nothing if not dynamic!

16 Oceans, gyres, currents

In the early 1990s, southern Africa experienced its worst drought this century. Nearly 100 million people went short of food. This was an extreme example of a change that brings abnormal weather to many parts of the world every few years, associated with a rise in the sea-surface temperature off the north-western coast of South America. That an apparently minor warming of the sea, of not much more than 3 °C, can have so profound an effect half-way round the world demonstrates the degree to which our climate is influenced by the oceans. The link between the temperature change and climate is now so well established that the surface temperature in the eastern Pacific just south of the equator can be used to predict maize yields in Zimbabwe up to a year ahead, with an even more advanced early warning of what to expect from the ability scientists now have to predict the temperature changes themselves up to a further year ahead (CANE ET AL., 1994).

The circumstances which have such drastic consequences in Africa are called 'El Niño' or, to give the phenomenon its full title, an 'El Niño–Southern Oscillation event', or ENSO. Figure 2.20 shows the parts of the world most seriously affected, but minor effects are felt even further afield. North-western Europe, for example, often has a cool, wet summer following an ENSO. In the areas most seriously affected, temperatures deviate by up to 0.5 °C from normal, but deviations of about 0.2 °C occur throughout the northern hemisphere. As well as droughts in various parts of the world, a recent ENSO event was blamed for unusually warm weather in Alaska, a remarkably warm winter in the eastern United States, a 100-mm rise in sea level and severe beach erosion in California, the death of coral reefs in the Pacific, and a variety of diseases ranging from bubonic plague to encephalitis in the United States!

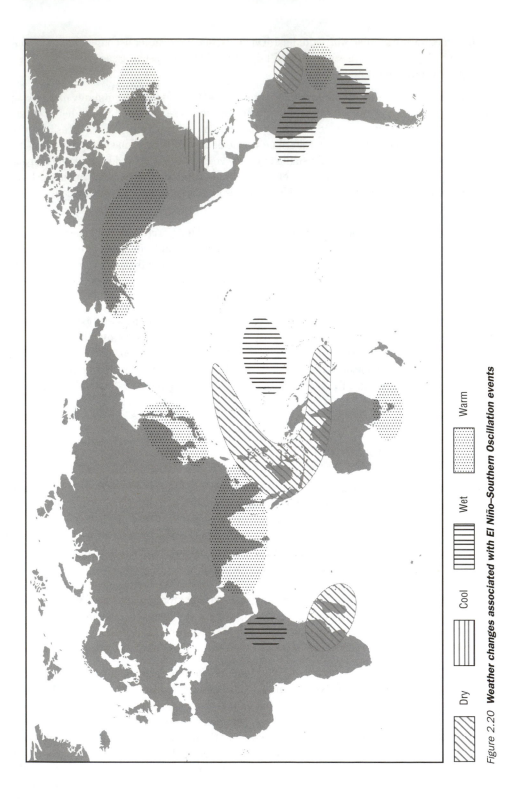

Dry Cool Wet Warm

Figure 2.20 *Weather changes associated with El Niño–Southern Oscillation events*

Ordinarily, the South Equatorial Current, driven by the trade winds, carries surface water westward, away from South America and towards Indonesia. Beneath this fairly shallow layer of surface water there is a boundary, the thermocline, below which the temperature drops sharply. The westward movement of surface water is compensated by an easterly flow along the thermocline, known as the Equatorial Undercurrent, or Cromwell Current. This system moves warm surface water westward, so the surface layer, above the thermocline, is about 200 m deep around Indonesia but very shallow off the American coast, where the thermocline almost reaches the surface. The sea-surface temperature is between 27.5 °C and 30 °C; this is close to the maximum temperature sea water can reach, because it is cooled by evaporation and the higher its temperature the more water evaporates from it.

The change begins with the distribution of atmospheric pressure over the Indian and Pacific Oceans. This causes the intertropical convergence zone (ITCZ) to move further south than it usually does in the midwinter months of December to February. This is the 'Southern Oscillation' and it causes the southern-hemisphere trade winds to weaken or even change direction. The wind driving the South Equatorial Current weakens or reverses, and so warm surface water begins to accumulate off South America, sometimes reinforced by water being driven by the wind from the west. Air moving across the warm water becomes moist and loses its moisture when it reaches the coast. The resulting rains brings abundant grazing for livestock along the arid coastal region, heralding an *año de abundancia*, and because it usually commences around Christmas the phenomenon is known as El Niño, 'the (boy) child'. In some years the opposite happens. The ITCZ stays far to the north, the trade winds strengthen, and the water off South America is colder than usual. This is known as 'La Niña' (HIDORE AND OLIVER, 1993, pp. 169–178).

Although El Niño brings abundance to the farmers of the coastal strip, most other people suffer, especially the fishermen. The Humboldt or Peru Current, flowing northward along the western coast of South America, carries nutrients collected on its long journey from Antarctica, and off the coast of Peru and Ecuador its rich, cold waters well up to the surface through the thin warm-water layer. These upwellings sustain abundant populations of marine plants and animals and support an important anchovy fishery. During an El Niño, however, the nutrient-rich water ceases to reach the surface and the fishery fails.

ENSO events happen roughly every seven years. They are not new: the first was recorded in 1541 and more recently they happened in 1891, 1925, 1953, and 1972–83. There was a particularly strong one in 1982–3, the effects of which were still being detected ten years later. Severe ENSO events also occurred in 1986–87, 1995–96, and 1998–99 (*www.elnino.noaa.gov/lanina_ new–faq.html*). The change in the flow of equatorial water caused long-period waves (Rossby waves) that crossed the North Pacific and shifted the flow of the Kuroshio Current northwards, bringing warmer surface waters to the mid-latitude Pacific that were still being detected in 1993 (JACOBS ET AL., 1994).

Ocean currents play an important part in transporting heat from the equator to high latitudes. The currents themselves are mainly driven by the prevailing winds, but as water moves away from the equator or towards it the motion is influenced by the Coriolis effect. This results in the generally circular currents, or 'gyres', of the North and South Pacific, North and South Atlantic, and Indian Oceans. Figure 2.21 shows the principal gyres and currents and their directions of flow.

With centres about 30° north and south of the equator, the gyres flow in a clockwise direction in the northern hemisphere and an anticlockwise direction in the southern hemisphere. Generally, the climatic effect is similar in both hemispheres. Water travelling from the poles towards the equator passes close to the western coasts of the continents, cooling them, and

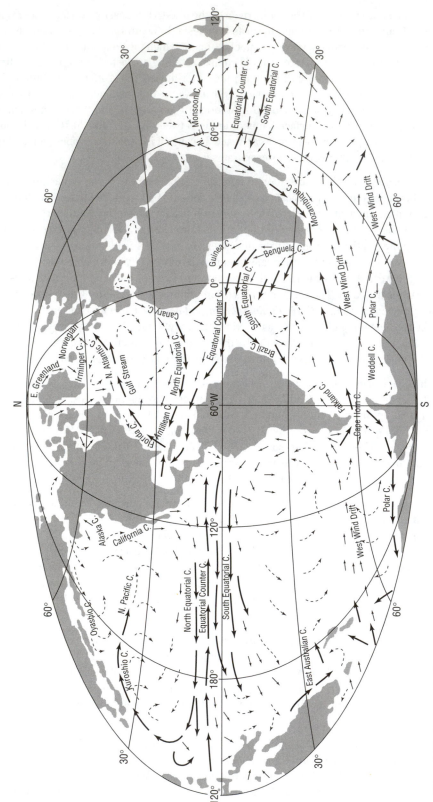

Figure 2.21 *Ocean currents*

equatorial water flowing poleward passes the eastern coasts, warming them. It is possible to see from this how a northward deflection of the warm Kuroshio Current, off Japan, could warm the northern Pacific with climatic consequences felt in California.

There is one important exception to the general rule and it provides another example of the strong link between ocean currents and climate. In the North Atlantic, the North Equatorial Current flows westward into the Caribbean, then turns north through the Gulf of Mexico, where it becomes first the Florida Current, then the Gulf Stream. The Gulf Stream flows north-eastwards across the Atlantic. It turns south again in the latitude of Spain and Portugal, but a branch continues to flow north-east, passing the shores of Britain. This is the North Atlantic Drift (or Current) and it gives Britain a much milder climate (with palm trees growing in western Scotland and almost subtropical conditions in sheltered places on the Isles of Scilly) than Newfoundland, which in fact is to the south of Britain but cooled by the Labrador Current. Should the Gulf Stream alter its behaviour so that the North Atlantic Drift ceased to break away from it and the entire current turned south together, Britain and north-western Europe would experience a much colder climate.

Atlantic conveyor

Near the edge of the Arctic sea ice, water is close to the temperature at which it is densest and as ice forms the crystallization process separates water molecules from salt molecules. This increases the salinity of adjacent water and, because it contains more salt, the density of the water also increases. This water sinks all the way to the bottom of the Atlantic, flowing south as the North Atlantic Deep Water (NADW). It crosses the equator and continues to the edge of the Antarctic Circle, where it joins the West Wind Drift, or Circumpolar Current, flowing from west to east. From there the water drives systems of currents in all the oceans, eventually returning to the North Atlantic.

The NADW removes cold water from the North Atlantic. This water remains cold until it rises in the South Pacific. During its progress through equatorial and tropical regions the water warms and it returns as warm water to the area near Greenland, where it replaces the NADW that is sinking and moving south.

This circulation is of major importance in regulating the climates of the world.

This is believed to have happened about 11 000 years ago and, apparently paradoxically, to have been caused by rapid climatic warming. At that time, the last (Devensian) glaciation was coming to an end and the ice sheets were melting. The Laurentide ice sheet, which covered much of northern North America, discharged its water into the North Atlantic, mainly through the Mississippi River system, releasing a huge amount of very cold fresh water which floated above the denser sea water. This was part of the reason for the failure of the North Atlantic Drift. The other concerned the formation of North Atlantic Deep Water (NADW). At the edge of the sea ice, sea water is very dense. When sea water freezes, its salt is removed as the ice-crystal lattice forms, so adjacent water is more saline than average. At the same time it is also at about 4 °C, the temperature at which water is densest. The dense water sinks beneath the less dense water and forms a slow-moving current close to the ocean floor, flowing towards the equator. This is the NADW and the sinking water is replaced by northward-flowing surface water. It is this system that controls the North Atlantic gyre and the North Atlantic Drift which is part of it. The warming at the end of the glaciation was associated with a retreat of the sea ice as the polar front moved as far north as Iceland, and this disrupted the formation of NADW.

About 11 000 years ago and for about 1000 years Western Europe was plunged back into ice-age conditions. When the reversal started, Scotland may have been entirely clear of ice, but before long much of the country lay beneath an ice sheet hundreds of metres thick (ROBERTS, 1989, pp. 50–53). This rapid and dramatic deterioration in climate was first detected by the presence of pollen grains of *Dryas octopetala* (mountain avens) in soils that could be dated to this period. *Dryas octopetala* is a plant characteristic of alpine and subarctic conditions and it has given its name to the climatic reversal, which is known as the Younger Dryas. This lasted from about 11 000–10 000 years BP (before present) and it is called 'Younger' because there is some evidence of an earlier cooling, prior to 12 000 years BP, known as the Older Dryas (PENNINGTON, 1974, pp. 32–34), which may have had a similar cause.

Ocean currents are clearly defined and some are fast-flowing. The Kuroshio Current, for example, flows at up to $3 \, \text{m s}^{-1}$. Their climatic effect is indirect, however, in that they affect the properties of the air in contact with them, rather than the water warming or cooling the coasts with which it makes contact, and it is the air which brings weather to the continents. The link is obviously of the greatest importance, and any long-term climatic prediction must be based on a much greater understanding of it than exists at present. Is it feasible, for example, that a rapid climatic warming in the northern hemisphere due to the greenhouse effect might disrupt the formation of NADW and trigger a severe cooling in north-west Europe? What causes ENSO events and are they likely to become more or less frequent should there be a global change in climatic conditions? At present such questions cannot be answered. Until they are, all predictions of the regional implications of climate change must be approached with great caution.

17 Weather and climate

Rain, snow, sunshine and showers, wind and storm, are among the phenomena that constitute our weather, the conditions we experience from day to day in a particular place and which weather forecasters aim to predict. The average weather conditions experienced over a large area year by year constitute the climate of that area. The two concepts, of weather and climate, are distinct and provide the subject matter of two equally distinct scientific disciplines: meteorologists study weather and climatologists study climates. Obviously the two overlap, for you cannot understand one without a fairly detailed knowledge of the other. When we discuss the greenhouse effect, for example, we base our ideas on studies made by climatologists; when we wish to know whether it would be wise to plan a picnic for the weekend we consult a meteorologist.

Weather phenomena result from interactions between bodies of air and water at different temperatures and are governed by a small number of general principles. Because air is compressible, atmospheric pressure decreases with height above sea level. If a 'bubble' of air (known technically as a 'parcel' of air) is made to rise, therefore, its volume will increase as it enters regions where pressure is lower. As it expands, the air becomes less dense. This means its molecules are further apart, and in 'making more room' for themselves molecules push one another aside, an activity which requires them to expend energy. Having less energy, the molecules move more slowly and as they slow down, so the parcel of air cools. There is no exchange of heat with the surrounding air; the cooling involves only the expanding parcel. Similarly, if air descends it is compressed, acquires energy, and warms. This cooling and warming is called 'adiabatic' and it is a version of the first law of thermodynamics (which states that in an isolated system, such as a parcel of air, the total internal energy of the molecules can be changed only by work done by them or on them). The concept of total internal energy explains why cold air high above us does not simply sink to the surface: were it to do so it would warm adiabatically to a temperature as high as, or higher than, that of air near sea level; although it is cold, its 'potential temperature' is high (HIDORE AND OLIVER, 1993, p. 112).

The rate at which air temperature decreases with height in the troposphere is called the 'lapse rate'. The average sea-level temperature is 15 °C, the average tropopause temperature –59 °C, and the average height of the troposphere 11 km, so the 'standard' lapse rate is about 6.7 °C km^{-1}. The actual lapse rate, called the 'environmental' lapse rate, varies from the standard according to local conditions, and if air is cooling (or warming) adiabatically its rate of temperature change depends on the amount of moisture it contains. For dry air, the dry adiabatic lapse rate is 10 °C km^{-1}, but if the cooling triggers the condensation of water vapour the cooling air will be warmed by the latent heat of condensation, so the saturated adiabatic lapse rate is lower than the dry adiabatic lapse rate. Its value varies with the amount of condensation, but averages about 6 °C km^{-1}.

Adiabatic cooling and warming

When air rises it expands, because there is less weight of air above it in the column reaching to the top of the atmosphere and, therefore, atmospheric pressure decreases with increasing altitude. As a gas expands, its molecules move further apart. In doing so they expend energy and, since the molecules have less energy, the temperature of the expanded air decreases.

It follows that a rising 'parcel' of air will cool without exchanging energy with the surrounding air. This is 'adiabatic' cooling (from the Greek *adiabatos*, 'impassable').

A descending 'parcel' of air warms by the same adiabatic process. As it enters a region of higher pressure it is compressed and gains energy, which heats it.

Water vapour is a gas and the amount of it in a particular body of air is expressed as the 'humidity' of the air. Humidity can be measured in several ways. Absolute humidity is the mass of water vapour in a given volume of air, and specific humidity is the mass of water vapour in a given mass of air (including the water vapour). Relative humidity, which is the measure most widely used, is the proportion of water vapour in relation to the amount required to saturate the air, and is given as a percentage. Warm air will hold more water vapour than cool air, so relative humidity varies with temperature as well as actual water-vapour content. The converse of this is that if air is cooled, a temperature will be reached at which its water vapour condenses. This is the 'dewpoint' temperature.

When water changes phase, between solid and liquid, liquid and gas, or directly (by sublimation) between solid and gas, energy is either absorbed or released, as latent heat. Latent heat warms or cools the surrounding air. It is why the air feels warmer when snow starts to fall and cooler when ice thaws, and it also governs the dynamics of storm clouds, hurricanes, and tornadoes. The amount of heat is considerable. When 1 gram of water evaporates, 2500 joules of energy is absorbed and the same amount is released when water vapour condenses; the change between liquid and solid requires the absorption or release of 334.7 J of energy; and sublimation between solid and vapour requires the absorption or release of the sum of these, 2834.7 J for every gram.

Over the continents and oceans, mixing of the air tends to equalize the pressure, temperature, lapse rate, and humidity horizontally over a large area. Such air is called an 'air mass'. Its characteristics are determined by the region in which it formed, so air masses can be labelled. The first division is between those formed over continents and those formed over oceans, the second between those formed in arctic, polar, and tropical latitudes. This yields six types: continental arctic (cA), continental polar (cP), continental tropical (cT), maritime arctic (mA), maritime polar (mP), and maritime tropical (mT). Continental air is dry, maritime air moist and, as the names suggest, tropical air is warmer than polar or arctic air; temperatures are more extreme in continental than in maritime air, owing to the moderating influence of exposure to the oceans. Air masses do not remain stationary, of course. They move, and as they do so they pass into new regions which modify their characteristics. A continental air mass originating over North America becomes a maritime air mass by the time it has crossed the Atlantic and reached Europe.

An air mass does not exist in a vacuum. It has borders, beyond which lie other air masses, with different characteristics, and as it moves the air mass ahead of it must also move. Not all air masses move at the same speed, however, so the boundaries are sites where one air mass is displacing another. The theory describing these reactions was developed during the First World War by scientists at the Bergen Geophysical Institute, Norway, led by Vilhelm Firman Koren Bjerknes (1862–1951). Because of the predominant news at the time the theory was being constructed, Bjerknes and his colleagues described the boundary between different air masses as a 'front'.

Fronts are identified as 'warm' or 'cold' depending on whether the air behind the front is warmer or cooler than that ahead of it. They meet the ground surface at an angle, the frontal slope, and warm and cold fronts slope in opposite directions (ALLABY, 1992, pp. 86–87). Warm fronts have a gradient of about 1 : 1000; cold fronts are much steeper, sometimes with a gradient of as much as 1 : 50, and they generally travel faster than warm fronts. Figure 2.22 illustrates typical conditions in the two types of frontal region. In (1), rapidly advancing cold air moves beneath warmer air, forcing it to rise; as the rising air cools, its water vapour condenses to form cumuliform (heaped) clouds, bringing heavy showers and possibly thunderstorms. In (2), a gently sloping warm front is advancing more slowly, so its air is lifted gently above the cooler, denser air and stratiform (sheet-like) clouds form, bringing drizzle or steady rain.

Cool air is denser than warm air, so its pressure is higher. Air flows from areas of high to low pressure, but because of the Coriolis effect and vorticity (the tendency of a flowing fluid to follow a spiral path) it follows an approximately circular path with a speed proportional to the pressure gradient (the pressure difference between the high and low areas). In the northern hemisphere air flows anticlockwise (cyclonically) around areas of low pressure and clockwise (anticyclonically) around areas of high pressure; in the southern hemisphere these directions are reversed. The difference in speed between cool and warm air masses results from the way air moves around them: in the northern hemisphere air flowing anticyclonically moves faster than air flowing cyclonically and in the southern hemisphere the reverse applies.

At first, a front is simply a line, with cold air on one side, warm air on the other, and air flows in opposite directions on either side. Some air crosses the front, however, and a V-shaped wave appears in the front. As this grows more pronounced, the cold air is partly behind the warm air and overtaking it. A low-pressure centre develops at the apex of the wave; this is a 'cyclone' or 'depression'. The undercutting cold air then starts to lift the warm air clear of the ground; at this stage the fronts are becoming occluded. Occlusion continues until all the warm air has been lifted above the ground; after that the depression dissolves and disappears. Figure 2.23 shows the way clouds are distributed around the depression as it develops and dissipates.

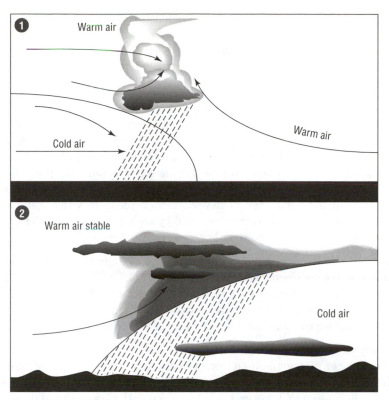

Figure 2.22 **Formation of cloud at a front. 1, Cold air is advancing rapidly, undercutting warm air and forcing it to rise. 2, Stable warm air is advancing, rising gently over cold air**

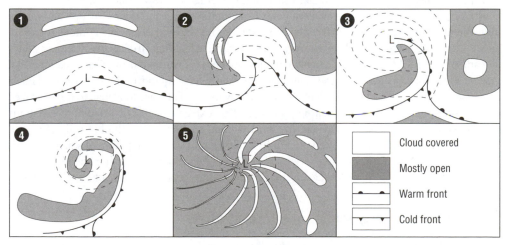

Figure 2.23 **Distribution of cloud around frontal systems. 1, The system starts to form. 2, The two fronts are still separate. 3, Warm air is rising over cold air. 4, The fronts are almost fully occluded. 5, Cloud distribution around a low-pressure centre**

Frontal systems of this general type, interspersed with periods of settled weather as the cold and warm sectors pass, typify mid-latitude climates. They can also bring severe weather, especially in the warm sector ahead of a cold front. Where warm surface air is lofted by an advancing cold front and winds at high altitude flow from a different direction, a moving line of thunderstorms, called a squall line, may occur.

Water vapour, condensing in the rising air, releases heat, so the air continues to rise and water vapour is constantly evaporating and condensing. In a fair-weather thunderstorm, caused by locally strong heating of the ground, the upward development is limited by precipitation falling from the top of the cloud, which slows the updraughts, but in a squall line the high-level wind shifts the top of the cloud to one side. This removes the limiting factor, because the precipitation falls to the side of the cloud rather than through its centre. The updraughts carry water droplets, which freeze near the cloud top, partly melt as they fall, then freeze as they rise again, growing the 'onion-skin' layers of ice that make hailstones until they are heavy enough to fall from the cloud, sometimes melting again in the warmer air between cloud and ground.

The vertical motion leads to a separation of electrical charge. Positively charged ice crystals accumulate near the top of the cloud, negatively charged particles near the bottom, and a positive charge is induced in the ground beneath the cloud. Lightning discharges these differences as soon as they are big enough to overcome the electrical resistance of the air, the energy released by the lightning causing an explosion of the air adjacent to the flash, which we hear as thunder.

There are many variations to this general scheme. It is not only a cold front that may force warm air to rise, for example. Air also rises as it crosses high ground and may lose moisture as a consequence. The western side of Britain has a wetter climate than the eastern side, because air from the Atlantic loses moisture as it crosses the hills, especially the Pennines and the mountains of Wales and Scotland. Air forced over a mountain may cool adiabatically to below the temperature of surrounding air. On the lee side of the mountain it descends again, warming adiabatically as it does so, to form a warm, dry wind, called a Föhn wind in Europe and a Chinook in North America. Cold air may also flow down a hillside, to produce 'frost hollows'. Valleys may funnel the wind, intensifying it. The cold mistral of southern Europe is caused by funnelling along the Rhône valley.

Such local and regional effects mark the difference between the weather people living in a particular place experience and the climate, which is the average of the weather conditions over an entire region. Aided by satellite observations of weather systems and their direction and rate of movement, a wide network of surface reporting stations on land and at sea, and immense computing power, it is possible nowadays to forecast weather with a fair degree of accuracy for a few days ahead. Weather systems are extremely complex, however. Two meteorological situations that appear identical can develop in radically different ways over a matter of days or weeks. This is because although they appeared identical, in fact they differed in ways too small to be measured and these differences directed their development. This extreme sensitivity to minute variations is characteristic of 'chaotic' behaviour, the most famous analogy being 'the butterfly effect', in which the flapping of its wings by a butterfly in Beijing alters the development of storm systems a month later in New York. This makes it impossible to forecast weather further ahead than a few days; indeed, such long-range forecasting may be inherently impossible (GLEICK, 1988, pp. 9–31).

18 Glacials, interglacials, and interstadials

Jean Louis Rodolphe Agassiz (1807–73) was a zoologist with a particular interest in fishes, both living and extinct. He was appointed professor of natural history at the University of Neuchâtel in his native Switzerland and there he might have remained, the most eminent ichthyologist of his generation, were it not for what began as a secondary interest.

Switzerland is noted for its glaciers and the Swiss are very familiar with them. In various parts of Europe there are boulders and gravels made of rock quite different from that of the region in which they are found, and in the 1830s some Swiss scientists were speculating that these rocks had been pushed into their present positions by glaciers. If so, it implied first that glaciers move and, second, that they once extended much further than they do now. Agassiz was sceptical, but decided to test the idea and in 1836 and 1837 he spent his summer holidays studying Swiss glaciers.

He soon discovered piles of rock to either side of glaciers and where they terminated, and noticed that some of these rocks were scoured with lines, as though small stones had been dragged across them under great pressure. Continuing with his visits to glaciers, in 1839 he found a hut, built on a glacier in 1827, a mile from its original location. His final test involved fixing a line of stakes across a glacier from side to side. Two years later, in 1841, he found they had moved and now formed a U shape, because the stakes near the centre of the glacier had moved further than those near the sides.

Already persuaded that glaciers do move, Agassiz published his ideas in 1840, as *Études sur les glaciers*, shortly before his rival, Jean de Charpentier (1786–1855), published his own version of the same theory. This proposed that in the fairly recent past all Switzerland and all those regions of Europe in which unstratified gravel occurs had been covered by sheets of ice similar to those that still cover Greenland. Agassiz extended his studies to other parts of northern Europe, and concluded that the 'Great Ice Age' had been very extensive. In 1846 he was invited to lecture in the United States, mainly because of his work with fossil fish, but he used the opportunity to deliver popular lectures about the Great Ice Age and to seek, and find, evidence for glaciation in North America. He remained in the United States, spending most of his time at Harvard, and became an American citizen.

Boulders and unstratified gravel, unrelated to the underlying rocks, have clearly been transported. Agassiz supplied an explanation for the mechanism of their transport, but an alternative explanation already existed. Many scientists believed the Earth had once lain beneath water, perhaps the biblical flood. In overturning this conjecture, Agassiz had a profound influence on our ideas of Earth history.

The Great Ice Age was soon accepted, but in modern times the original concept has been greatly modified. Figure 2.24 shows the extent to which ice sheets have covered the Earth at one time or another during the last 2 million years. As the map indicates, recent ice ages have affected both hemispheres, although glaciation has been more marked in the northern hemisphere, where land extends into higher latitudes. Much of what appears as open sea between ice sheets was in fact frozen, as the sea around the North Pole is frozen today.

Conventionally, the ice ages are said to occupy the Pleistocene Epoch, which began about 2 million years ago and ended about 10 000 years ago, with the end of the last glaciation and the commencement of the Holocene (or Recent) Epoch, in which we now live. In fact, glaciation began somewhat earlier, rather more than 3 million years ago, as glaciers advanced and then retreated again several times (GENTRY AND SUTCLIFFE, 1981), and it is possible that the date of the beginning of the Pleistocene may be revised. Indeed, it may be that our use of the word 'Holocene', or 'Recent', is premature. Most palaeoclimatologists (scientists who study the climates of the distant past) agree that we are presently living in an interglacial, called the Flandrian, and that one day (no one can say when) this will end and there will be another ice age. If this is so, perhaps the Pleistocene Epoch has not yet ended!

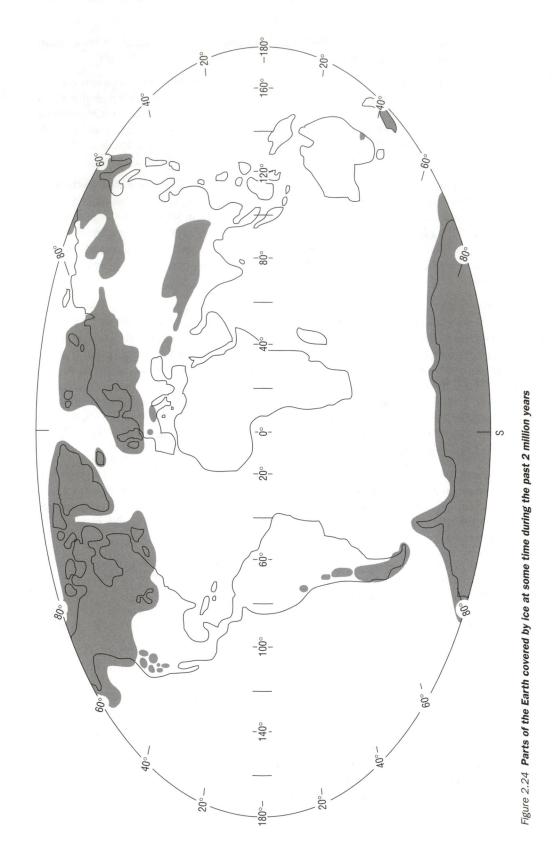

Figure 2.24 *Parts of the Earth covered by ice at some time during the past 2 million years*

Although we usually think of the Pleistocene as the 'Great Ice Age', this is misleading. There was not a single ice age (or glaciation; the terms are synonymous) but several, and in the interglacials separating them climates were sometimes markedly warmer than those of today. During the Ipswichian (or Trafalgar Square) Interglacial, for example, the average summer temperature in southern Britain was some 2–3 °C warmer than today and remains of elephants, hippopotami, and rhinoceros, dated to that time (100 000–70 000 years BP), have been found in what is now central London (hence the name 'Trafalgar Square Interglacial'). Nor is the Pleistocene the only geological period during which glaciations are known to have occurred. In various parts of the world there may have been glaciations some 2.3 billion years ago and between 950 and 615 million years ago, during Precambrian time, at the end of the Ordovician Period about 440 million years ago, and in the southern hemisphere around 286 million years ago, at the end of the Carboniferous and beginning of the Permian Periods. It does seem, however, that the Earth enjoyed an ice-free period between the Permo-Carboniferous glaciation and the onset of the Pleistocene.

During the Pleistocene, there are now believed to have been four glacial episodes and three inter-glacials in North America and five glacial and four interglacial episodes in Europe. The present epoch, the Holocene, is also known as the Flandrian Interglacial. This makes it the fifth European interglacial, and acceptance of the name implies that the Pleistocene glaciations have not yet ended; we are still living in the Pleistocene and one day the glaciers will start to grow again. If and when this happens it may do so rapidly, for glaciations can begin quickly and end even more quickly.

Even during glaciations there are periods of remission. Briefer and cooler than interglacials, these are known as 'interstades' (or 'interstadials') and are identified by the presence of pollen from plants known to require mild climates. During interglacials, too, temperatures fluctuate. Figure 2.25 shows how average temperatures have varied from 18 000 years ago, when the most recent glaciation was at its most severe, to the present day. During the climatic optimum, about 5500 years ago, mid-latitude temperatures were about 2.5 °C warmer than those of today; this was the period during which civilizations flourished in Asia Minor. A smaller optimum, peaking around AD 1000, allowed the Vikings to colonize Greenland. Between about 1450 and 1880 temperatures fell during the 'Little Ice Age'. This is when fairs were held on the frozen Thames and it may be a minimum from which our climate is still warming. As the graph indicates, the weather has been markedly warmer than it is today during several episodes in the past.

Confusingly, glaciations and interglacials have different names in different places and it can be difficult to match them (ALLABY, 1992a, pp. 145–148). The most recent glaciation is known as the Devensian in Britain, the Weichselian in northern Europe, the Würm in the Alps, and

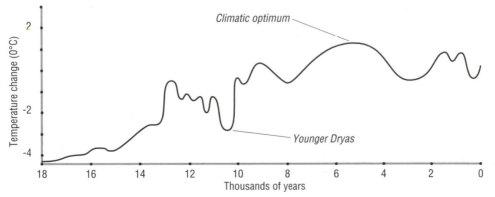

Figure 2.25 **Temperature changes since the last glacial maximum (present temperature = 0)**

is roughly equivalent to the North American Wisconsinian. It began about 70 000 years ago and ended about 10 000 years ago.

Glaciations and interglacials are associated with climatic changes more extensive than the advance or retreat of ice sheets. Precipitation patterns also change. Today, central Antarctica is possibly the most arid place on Earth and the Arctic is also dry, though less extremely so. Ice sheets accumulate because fallen snow is removed only slowly by ablation, not because precipitation is heavy. At the peak of the Devensian (Wisconsinian) glaciation, 18 000 years ago, most deserts were larger than they are today, but during the warm episode following the end of the Younger Dryas the Sahara was smaller than it is now and lakes in that part of Africa were larger (GENTRY AND SUTCLIFFE, 1981, p. 239). Generally, climatic cooling implies increasing aridity, climatic warming increasing precipitation. Indeed, the correlation with rainfall makes it possible to equate interglacials with pluvials (periods with wet climates) and glacials with interpluvials (HOLMES, 1965, pp. 715–716).

Sea levels also change with the advance and retreat of ice sheets. As more and more water reaches the frozen regions and is trapped there, sea levels fall. At times they have been 100 m below their present limit. At times during the Devensian (Wisconsinian), the North Sea was dry land, crossed by rivers, Alaska was joined to Siberia by a wide strip of land, and New Guinea was linked to Australia. The exposure of land bridges presents living organisms with opportunities to expand their ranges and it was during the Pleistocene that humans reached most of the regions in which they live now. About 60 000 years ago, people arrived in Australia (MORELL, 1995), presumably having crossed about 100 km of open sea to migrate from Asia to New Guinea–Australia, and some time later there was the first of several migrations from Asia into North America. It is only the more isolated Pacific islands, Madagascar, and New Zealand that have been colonized during the Holocene, about 3000, 2000, and 1000 years ago respectively (ROBERTS, 1989, p. 56).

People did not live on the ice sheets, of course, but the tundra landscapes bordering the ice supported abundant game and the seas provided fish and seals as well as marine invertebrates. Climate change, especially if it is rapid, might be supposed to challenge species and, indeed, many animals became extinct during the Pleistocene. These extinctions are unlikely to have been linked solely to changing climate, however. It is much more likely that they were caused by over-hunting as humans expanded their range. Vast quantities of animal bones have been found in certain places, sometimes at the foot of cliffs over which entire herds appear to have been driven, presumably by hunters who then took the meat and other materials they needed from the pile of carcasses. One such site, at Solutré, France, has the bones of more than 100 000 horses (ROBERTS, 1989, p. 59), and in North America many large mammals became extinct within 1000 years of the emergence of the earliest human culture. The extinctions were confined to the late Pleistocene and are not linked to earlier climate changes, some of which were at least as rapid; they principally affected large mammals, and they occurred on all continents. Australia lost some 81 per cent of its large mammals 26 000–15 000 years ago, South America 80 per cent 13 000–8000 years ago, North America 73 per cent 14 000–10 000 years ago, Europe 39 per cent 14 000–9000 years ago, and Africa 14 per cent 12 000–9500 years ago.

Glaciers scour away soil and their retreat leaves a barren landscape, but one to which plants and animals soon start to return. Holocene recolonization has been well documented, together with the limits species reached before rising sea levels prevented their further migration. The mole (*Talpa europea*), common shrew (*Sorex araneus*), beaver (*Castor fiber*), aurochs (*Bos primigenius*), European elk or North American moose (*Alces alces*), and roe deer (*Capreolus capreolus*) never reached Ireland, for example, because of the formation of the Irish Sea 9200 years ago (SIMMONS *ET AL.*, 1981, pp. 86–89).

Land that is now free from permanent ice was glaciated for rather more than half of the Pleistocene. It is reasonable to suppose that the global climate now fluctuates between glacial and interglacial conditions and that the present interglacial may be near its end, unless the subsequent cooling is overridden by warming induced by the greenhouse effect. The alternation between glacial and interglacial is the most extreme climate change imaginable, but the historical evidence suggests that living organisms adapt to it fairly robustly. As conditions deteriorate enough of them migrate to more favourable environments for their species to survive, from which they return when opportunity allows. In certain local areas, called refugia, communities survive even without migrating, because the older conditions prevail. There are several Pleistocene refugia in Britain, Upper Teesdale being probably the best-known example. Although undoubtedly inconvenient for humans, rapid climate change does not necessarily imply the extinction of species, merely their absence until the next change allows them to return.

19 Dating methods

Consider the world as it might appear to an intelligent mayfly. As it emerges from the stream in which it has spent most of its life, the insect sees a world where the Sun shines, trees are in full leaf, a summer world. To the mayfly, this is how the world is, the only appearance it can ever present. Long before leaves start to fall and even longer before water starts to freeze and the ground is covered with snow, the mayfly will have died.

Unlike the mayfly, we know the world changes, that there is winter as well as summer. Yet our lives, too, are brief and deny us any opportunity to observe at first hand the fact that those aspects of the world we regard as permanent are no more so than the summer sunshine and leaves. Our Earth is changing constantly. Continents move, mountains are thrust upward and then eroded into plains, ice ages come and go, species evolve only to vanish again, but these changes occur on a time-scale that seems long to us. Compared with the 4.6 billion years during which our planet has existed, a human lifespan is ephemeral indeed.

If we are to understand the environment in which we find ourselves, to observe ways in which it may be changing, and to predict future changes and our own influence upon them, we must learn to appreciate the time-scale on which such events occur. We must try to discover how the environment arrived at its present condition if we are to discern trends and compile forecasts. We must study the history of our planet and the first requirement for any historical reconstruction is a reliable means for dating events. We must know when and in what order past events happened.

The first step in reconstructing the past is fairly straightforward. Sedimentary rocks begin as sediments, most deposited beneath water, and they can be seen to form layers. Obviously, the older layers must have been precipitated before those lying above them and changes in the composition of layers must reflect changes in the depositional environment. Unfortunately, sediments seldom remain undisturbed, so although it is easy enough to recognize their layers it is more difficult to determine their relative ages; to do that it is necessary to determine which way up they were when they were precipitated.

It was Georges Cuvier (1769–1832) who first realized that fossils might be used to identify sedimentary strata. He and Alexandre Brogniart (1770–1847), an engineer, applied this idea to a study of the Paris Basin, describing their discoveries in *Descriptions géologiques des environs de Paris*, published in 1811 (BOWLER, 1992, pp. 213–220). Their scheme was based on the observed fact that fossil invertebrates found in some strata are absent from others and, therefore, that certain

Table 2.4 *Geologic time-scale*

Eon	Era	Sub-era	Period	Epoch	Began (Ma)
Priscoan					4600
Archaean					4000
Proterozoic					2500
Phanerozoic					
	Palaeozoic				
		Lower Palaeozoic			
			Cambrian		570
			Ordovician		510
			Silurian		439
		Upper Palaeozoic			
			Devonian		408.5
			Carboniferous		
			Mississippian		362.5
			Pennsylvanian		322.8
			Permian		290
	Mesozoic				
			Triassic		245
			Jurassic		208
			Cretaceous		145.6
	Cenozoic				
		Tertiary			
			Palaeogene		
				Palaeocene	65
				Eocene	56.5
				Oligocene	35.4
			Neogene		
				Miocene	23.3
				Pliocene	5.2
		Quaternary			
			Pleistogene		
				Pleistocene	1.64
				Holocene	0.01

fossil assemblages can be used to identify particular strata wherever those strata are found. In other words, animal species have come into existence, lived for a while, and then have disappeared and their places have been taken by others, newly arrived: we would say evolved. From this it is possible to construct a 'stratigraphic column', a vertical section through sedimentary rocks in which each stratum is shown in chronological order. The Cuvier and Brogniart study inspired geologists all over Europe to apply the method to their own localities and eventually to divide geologic time into distinct episodes on the basis of the animals associated with them. Though much amended, the geologic time-scale used today is derived from this work, as are most of the names.

Geologic time is divided into eons, eras, suberas, periods, and epochs. Table 2.4 shows the present arrangement, starting with the oldest, although some of the dates are revised from time to time (Ma means millions of years ago). Priscoan, Archaean, and Proterozoic together comprise the eon formerly known as the Precambrian. The term 'Precambrian' is still widely used, but not in a formal sense (all it means, after all, is 'before the Cambrian').

The order in which historical episodes should be arranged having been established, the next step is to allot dates to them. The thickness of strata is no help with this. Sedimentation is an irregular process, so a thick layer may have accumulated rapidly, a thin one more slowly, and there is no way to tell. Some sediments, however, build more regularly, and it was the record they left that allowed the retreat of Scandinavian ice sheets, starting about 10 000 years ago, to be traced. Each spring, as the ice melts, an assortment of mineral particles is washed into a lake by the melt-water. Heavier particles, such as sand grains, settle quickly. Later in the year, as water freezes again, the supply to the lake ceases and the finer particles, of silt and clay, gradually settle on top of the sandy layer. Year after year the process is repeated, each pair of layers, one pale and coarse, one dark and fine, being known as a 'varve'. These can be counted, each varve representing one year, and if varves are forming at the edge of a retreating glacier they will follow it, so that its progress can be traced and dated. The study of varves is known as varve analysis, varve chronology, or a varve count.

Varves resemble tree rings, which provide another method of measuring time. In spring, woody plants grow rapidly by producing large, thin-walled cells in the xylem, just below the bark of stems and branches. Growth slows in summer, ceasing in late summer, and consists of smaller cells with thicker walls. The large cells of spring are pale in colour, the smaller ones of summer dark, and so each year the plant produces a ring of pale wood separated by a thin, dark ring from the pale wood of the following year. A count of the rings is a count of years, but there are some risks. If conditions are very severe, a plant may produce no growth for a whole year, and if conditions are unusually favourable it may produce two or more sets of rings. For this reason, tree-ring dating (called dendrochronology) must be based on as many specimens as is practical, obtained from widely scattered locations. The fact that rings are strongly affected by growing conditions has advantages. The width of rings can be used to infer weather (dendroclimatology) and environ-mental (dendroecology) conditions at the time they formed.

Obviously, the study of tree rings can provide dates only up to the age of the living plant from which they are taken, but trees can live a surprisingly long time. There are bristlecone pines (*Pinus longaeva*), found in California, more than 4600 years old, and correlating rings from them (taken as cores, without destroying the tree) with rings from dead pines has allowed scientists to construct a chronology for arid zones going back 8600 years and, at the upper tree limit on mountains, one going back 5500 years.

These chronologies are used to calibrate radiocarbon (^{14}C) dates. Bombardment by cosmic radi-ation generates neutrons, a few of which collide with atoms of nitrogen (^{14}N), displacing a proton and converting the ^{14}N to ^{14}C. Chemically, ^{14}C behaves just like ordinary ^{12}C and living organ-isms exchange both with their surroundings. When they die, however, carbon exchange ceases. Carbon–14 is radioactive, half of any amount of it decaying to ^{12}C in 5730 ± 30 years (its half-life), so the ratio of ^{12}C : ^{14}C in dead organic matter is directly related to the time that has elapsed since it died. Radiocarbon dating rests, however, on the assumption that the rate of ^{14}C formation in the atmosphere is constant. This is now known not to be so, because the intensity of cosmic-ray bombardment is variable, but, when correlated with tree-ring series from bristlecone pine, radio-carbon analysis makes it possible to date material up to about 70 000 years old.

Dating material older than this requires other methods. These, too, are based on the decay of radioactive elements, but ones with much longer half-lives. The first to be exploited were uranium (U) and thorium (Th). Uranium occurs naturally as a mixture of two isotopes, ^{238}U and ^{235}U in the constant proportions 137.7 : 1; both decay to stable isotopes of lead (Pb). Uranium–238, with a half-life of 4510 million years, decays to ^{206}Pb, and ^{235}U, with a half-life of 713 million years, to ^{207}Pb. Thorium–232, with a half-life of 13 900 million years, decays to ^{208}Pb. Lead also occurs natu-rally as the stable isotope ^{204}Pb, so this must be deducted from lead isotopes resulting from radioactive decay before an age can be calculated.

Potassium–40, a radioactive isotope of potassium with a half-life of 1300 million years, also occurs naturally (and, because of its presence in our food, is the principal source of our own exposure to radiation). Most ^{40}K decays to ^{40}Ca, which cannot be used because calcium is so common, but about 11 per cent decays by a different route to ^{40}Ar (argon). This decay is used to date rocks more than 250 000 years old.

A radioactive isotope of rubidium, ^{87}Rb, decays in a single step to strontium (^{87}Sr) and this decay is used to date certain rocks, especially those containing mica and potassium, but there is some doubt about the half-life of ^{87}Rb. Two values are used: 4.88×10^{10} and 5.0×10^{10} years. A more recent method uses the decay of samarium (^{147}Sm) to neodymium (^{143}Nd). Samarium–147 has a half-life of 2.5×10^{11} years and this decay is used in studies of the formation of rocks in the Earth's crust and mantle (and can also be used on materials of extraterrestrial origin).

It is impossible to predict when an individual unstable atom will decay, but it is possible to calculate the probability that the atom will decay within a certain period. This is called the 'decay constant' for the isotope, from which the half-life can be calculated as the time taken for the decay of half the unstable atoms present. The process is exponential: half the atoms decay in the first half-life period, half the remainder in the second period, half of that remainder in the third, and so on (e.g. 100; 50; 25; 12.5, etc.). Most of the decays used are based on half-lives much longer than the age of the Earth, but it is not necessary to wait until a complete half-life has elapsed before calculating an age. What matters is the ratio of isotopes.

Since radioactive decay involves only the nucleus of the atom, its rate is not affected by temperature, pressure, or any other outside influence. This makes it a very reliable measure of the age of materials. Radiometric dating has allowed scientists to reconstruct the history of the Earth in some detail.

20 Climate change

Milutin Milankovich (1879–1958) spent most of his career as a mathematician and physicist working at the University of Belgrade, where he devoted thirty years to comparing the amount of solar radiation received in different latitudes over the last 650 000 years with the climates during that time. He discovered a clear relationship between solar variability and the incidence of ice ages that is now accepted by most climatologists. Presented as a graph, it is known as the Milankovich solar radiation curve (*geography.miningco.com/library/weekly/aa121498.htm*).

We picture the Earth spinning on its axis and orbiting the Sun in a very regular fashion. So it does, but within its regularity there are slow, cyclical variations. Milankovich identified three that affect climate when they coincide to maximize or minimize insolation.

The first cycle, illustrated in Figure 2.26, concerns the Earth's orbital path. Much exaggerated in the diagram, this varies from almost circular to slightly more elliptical. In other words, the path stretches, varying the distance between the Earth and Sun at perihelion and aphelion. Starting at any date, it takes about 95 000 years for the orbit to move through the full cycle and return to its initial path. Clearly, a variation in the distance between the Earth and Sun affects the intensity of radiation received at the Earth's surface and, therefore, the climates of Earth.

The second cycle occurs because the axis wobbles, describing a circle, rather like a toy gyroscope (see Figure 2.27). It is this wobble, due to gravitational attraction, mainly from the Sun and Moon, that causes the position of the equinoxes to move westward, taking 25 800 years to complete one orbit. The phenomenon is called the precession of the equinoxes. At present

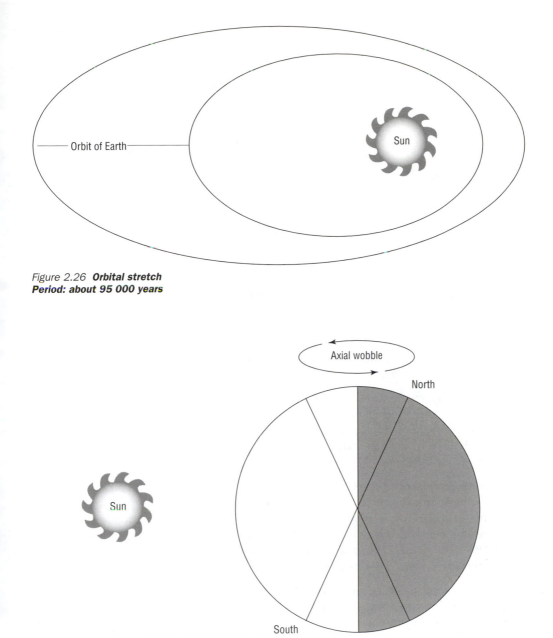

Figure 2.26 **Orbital stretch**
Period: about 95 000 years

Figure 2.27 **Wobble of the Earth's axis**

we are at perihelion in January. In AD 15 000, one half-cycle from now, we will reach perihelion in June.

The third cycle, called the 'obliquity of the ecliptic', relates to the angle of the Earth's rotational axis to the ecliptic, the plane of the Earth's orbit. Imagine the rotational axis as a straight rod projecting at both ends and forming an angle with the ecliptic. At present, that angle is 66.5° and, therefore, the axis is 23.5° from the vertical (90° to the ecliptic). Over a cycle of about 41 000 years this angle varies by about 1.5° about a mean of 23.1° (see Figure 2.28).

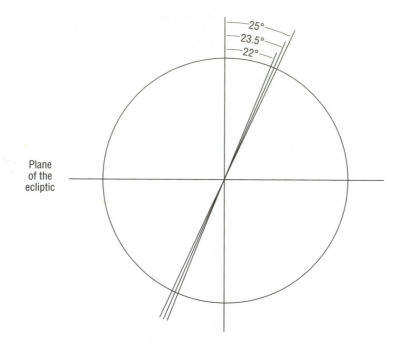

Figure 2.28 **Variations in axial tilt (obliquity of the ecliptic) in degrees**

These axial variations alter the area illuminated by the Sun. If the axis were at right angles to the ecliptic, for example, giving an obliquity of 0°, the half of the Earth facing the Sun would be lit evenly. Day and night would always be the same length and there would be no seasons. Tilt the Earth much more, on the other hand, say to an obliquity of 60°, and over almost the whole of each hemisphere the Sun would never set in summer or rise in winter.

Dramatic climate change occurs when the three cycles coincide, and the Milankovich solar radiation curve, which combines the three, is used to make long-term climatic predictions (*deschutes.gso.uri.edu/~rutherfo/milankovitch.html*). It is this that allows climatologists to assert that a cooling trend which began about 6000 years ago will continue, leading us into a new ice age (HIDORE AND OLIVER, 1993, pp. 370–371), although the solar influence may be overridden by that of greenhouse gases if these continue to accumulate in the atmosphere.

In the shorter term, the solar output itself also varies. The first person to relate this to climate change was the British astronomer Edward Walter Maunder (1851–1928). Like many astronomers, he was interested in sunspots (*es.rice.edu/ES/humsoc/Galileo/Things/sunspots.html*), dark 'blotches' on the surface of the Sun that come and go in a cycle of about 11 years. Checking through old records of sunspot activity, in 1893 he discovered that very few sunspots were reported during a period of 70 years from 1645 to 1715, and for 32 years, it seems, there were no sunspots at all. He published a paper describing his findings in 1894, but it attracted little attention, any more than did earlier papers challenging the idea of the constancy of solar output, published by Maunder and by the German astronomer Gustav Spörer (1822–95). Today, the period during which sunspots were much reduced in number is known as the 'Maunder minimum'. Its significance extends far beyond the realms of solar astronomy, because the 1645–1715 minimum Maunder identified coincides with the peak of the 'Little Ice Age', when average temperatures were about 1 °C lower than they had been previously (LAMB, 1995, pp. 69 and 321). More recently, the American solar astronomer John A. Eddy checked the Maunder and Spörer findings, added more of his own, and found a correspondence between solar activity and climate so close he described it 'almost that of a key in a lock', extending to 3000 BC (EDDY, 1977).

Again, the solar influence may be overwhelmed by that from greenhouse gases. David Thomson, a skilled statistician, has analysed data since 1659 and concluded that global temperatures are now linked more closely to atmospheric carbon dioxide concentrations than to sunspot activity or orbital effects (THOMSON, 1985), although his interpretation has been questioned by some climatologists, who think it too simple (KERR, 1995). The idea is now gaining ground that present changes in the atmosphere and climate are more likely to be due to changes in solar output and volcanic eruptions than to human intervention (CALDER, 1999).

Debate will continue for some time over what is forcing present climate change, but at least in the past it has clearly been triggered by astronomical events, and when the climate changes it can do so very quickly. At one time it was thought that ice ages begin and end gradually, it taking centuries or longer for the ice sheets to spread. This may be incorrect. According to the 'snow-blitz' theory, a slight fall in summer temperatures in high latitudes might allow some of the winter snow to survive where in previous years it had melted. The affected areas would then be white, when previously they had been dark, thus increasing albedo and lowering temperature further. In succeeding years, the snow-covered area would increase and temperatures would continue to fall, climatic forcing by the increased albedo accelerating the change by a strongly positive feedback. It might take very little time to move from our present interglacial climates to a full glaciation. Warming can also proceed rapidly, the change from glacial to interglacial perhaps taking no more than a few decades.

Stability of the polar ice sheets

If the polar ice caps were to melt, the volume of water released into the oceans would be sufficient to raise sea levels substantially. The stability of the ice caps is therefore of great importance and their condition is monitored closely.

The ice caps comprise three major ice sheets: in Greenland, West Antarctica, and East Antarctica. The Greenland ice cap is growing thicker in some areas, thinner in others, and is shrinking slightly overall. The reduction in its size is due to the rate of flow of its outflow glaciers and is not thought to be due to climatic change.

In Antarctica, the ice sheet on the eastern side of the Transantarctic Mountains is about twice the size of that on the western side. The East Antarctic ice sheet is very firmly grounded on the underlying rock. Its size remains constant and there is not considered to be any risk of it decreasing in thickness.

The West Antarctic ice sheet is less firmly grounded and the line marking the edge of the grounded sheet is retreating. It is doing so very slowly, at a constant rate, and has been retreating at this rate for about 7500 years. The retreat is due to the way glaciers within the ice sheet are moving and not to climatic change.

Evidence from the past indicates that despite minor fluctuations, the climate throughout the present interglacial, the Flandrian, has been very stable. During the last two glaciations and the Eemian Interglacial separating them, temperatures rose and fell rapidly, by 3 °C or more, bringing warmer or cooler periods lasting several centuries or a few thousand year. (GROOTES *ET AL.*, 1993). These oscillations have since been linked to changes in ocean circulation (ZAHN, 1994).

Ancient climates are reconstructed mainly from evidence obtained from ice cores, those referring to the Eemian Interglacial and the glaciations to either side having been obtained from Greenland. Ice sheets form by the compaction of snow under the weight of overlying snow, so the ice forms in seasonal layers that can be dated by counting, much like tree rings. Temperatures are inferred by oxygen-isotope analysis. There are three isotopes of oxygen, ^{16}O, ^{17}O, and ^{18}O, but only ^{16}O and ^{18}O are of importance in climatic studies. Being lighter, water containing ^{16}O evaporates more readily than $H_2^{18}O$, so fresh water is enriched in ^{16}O as compared with sea water. The degree of enrichment depends on the temperature at which the water evaporated, because the higher the temperature, the greater the rate of evaporation and the more $H_2^{18}O$ that enters the air with the $H_2^{16}O$. This allows mean surface temperature to be calculated from analyses of the ratio of $^{16}O : ^{18}O$ in dated samples of ice trapped in cores as 'fossil precipitation', the present ratio of $^{18}O : ^{16}O = 1 : 500$ providing a standard.

Astronomical climate forcing can be predicted, but volcanic eruptions are wholly unpredictable, at least at present. Some eruptions, but not all, have a climatic influence, although its scale is small and it is of short duration. If it is to affect climate, a volcanic eruption must inject material into the stratosphere, where it will remain for some time; tropospheric material is adsorbed on to surfaces or removed by precipitation in a matter of hours, days, or at most weeks. The eruption should also be in a low latitude. The convection cells governing the movement of low-latitude air allow only minor exchanges of tropospheric air between the northern and southern hemispheres. Stratospheric air is less affected and there is some interchange. Material injected into the stratosphere near the equator will be carried around the Earth and may also spill into higher latitudes in both hemispheres.

On 15 June 1991, the eruption of Mount Pinatubo on the island of Luzon, in the Philippines (latitude 15° N) caused the greatest stratospheric perturbation this century. The plume reached a height of about 30 km and released into the stratosphere some 30 million tonnes of aerosol composed of sulphuric acid and water. Within 14 days the material had spread across the equator, to about 10° S, and carried westward; within 22 days it had circled the planet. Eventually it spread as a blanket between about 30° N and 20° S. The presence of so much fine-particulate matter in the upper atmosphere increased the planetary albedo and thus reduced the amount of solar radiation reaching the surface, with the result that surface temperatures were depressed during the remainder of 1991 and for the whole of 1992; it was 1993 before they began to recover. In 1992, the mean global temperature was 0.2 °C lower than the 1958–91 average and it would have been lower still were it not for the warming influence of the 1992 ENSO event. The eruption ended the run of warm years. Because the aerosol engaged in chemical reactions, the eruption also contributed to the greatest depletion of stratospheric ozone recorded up to that time (McCORMICK *ET AL.*, 1995).

Mount Pinatubo was the biggest eruption this century, but it was not the only one. Five other eruptions were large enough to have had some climatic effect: those of Katmai (1912), Agung (1963), Fuego (1974), El Chichón (1982), and Cerro Hudson (1991), releasing 20, 16–30, 3–6, 12, and 3 million tonnes of aerosol respectively. In the last century there were two even larger eruptions, of Tambora (1815) and Krakatau (1883); these released more than 100 and about 50 million tonnes of aerosols respectively. The year 1815 was known as 'the year with no summer' and in Britain the summers of 1816 and 1817 were also wet and cold; the 1816 harvest was disastrous and there were food riots (STRATTON AND BROWN, 1978).

Our climate is changing constantly, driven by factors over which we have no hope of control. It is affected by cyclical variations in the Earth's orbit and rotation and apparently erratic fluctuations in solar output. Volcanic eruptions can depress surface temperatures and ENSO events enhance them. It may be that emissions of greenhouse gases are now overwhelming these natural forcing factors, but this does not remove them: predictions of future climate must take them into account, inherently unpredictable though some of them may be. Those attempting predictions must also bear in mind the possibility that once climate begins to change the rate of change may accelerate dramatically and that we seem to be living in unusually stable times. Predictions are concerned with the future, of course, but they must incorporate evidence gleaned from the past. Palaeoclimatologists, who study ancient climates, supply information that is vitally important to forecasters.

21 Climatic regions and floristic regions

Climates can be classified. At the simplest level, latitude, proximity to the ocean, and the convective cells transporting warm air away from the equator and cool air away from the poles provide a basic classification. Equatorial regions are warm and humid, subtropical regions, where dry air descends, are warm and dry, polar regions are cold and dry, and the mid-latitudes are mild and humid or dry with temperature extremes according to whether they are maritime or continental.

Unfortunately, it is not quite so simple as it sounds, because 'warm', 'cool', 'dry', and 'humid' are relative terms that mean little by themselves. Aridity, for example, depends not on annual precipitation, but on 'effective precipitation', which is precipitation minus evaporation, this being what determines the amount of moisture reaching the ground water. This, in turn, is related to temperature and a figure for the average annual temperature may conceal a very wide difference between summer and winter. Many attempts have been made to base a classificatory system on the general circulation of the atmosphere, the earliest dating from the 1930s.

The most successful scheme of this kind, proposed in 1950 by the German climatologist H. Flöhn, is illustrated in Figure 2.29. Flöhn took account of the global wind belts and distribution of precipitation (BARRY AND CHORLEY, 1982, pp. 358–373). In 1969, A.N. Strahler proposed an even simpler system based on the air masses which produce climates, dividing all climates into three types: low-latitude; mid-latitude; and high-latitude. These were subdivided according to variations in temperature and precipitation to produce 14 regional types, with a separate category for upland climates.

The two most widely used classifications, however, were introduced between 1900 and 1936 by the Russian-born German climatologist Wladimir Peter Köppen (1846–1940) and in 1931, with important revisions in 1948, by the American climatologist C. Warren Thornthwaite (1899–1963). The Köppen classification is widely used by geographers, that of Thornthwaite by climatologists.

Köppen took account of the distribution of vegetation, based originally on studies published in the last century by Alphonse de Candolle (1806–93), whose *Géographie botanique raisonée* (1855) considered the geographical distribution of plants in relation to their physiology. From this it emerges that a summer temperature of 10 °C marks the limit of tree growth, a winter temperature above 18 °C is necessary for some tropical plants, and if the average winter temperature is below –3 °C there will be at least some snow cover. Using these criteria and records of monthly average temperatures, Köppen defined six climatic types. In tropical rainy climates temperatures are above 18 °C throughout the year; in warm, temperate, wet climates temperature in the coldest month is –3–18 °C; in cold boreal-forest climates temperature in the coldest month is below –3 °C and in the warmest month above 10 °C; in tundra climates temperature in the warmest month is 0–10 °C; in polar climates the temperature never rises above 0 °C; and a final category

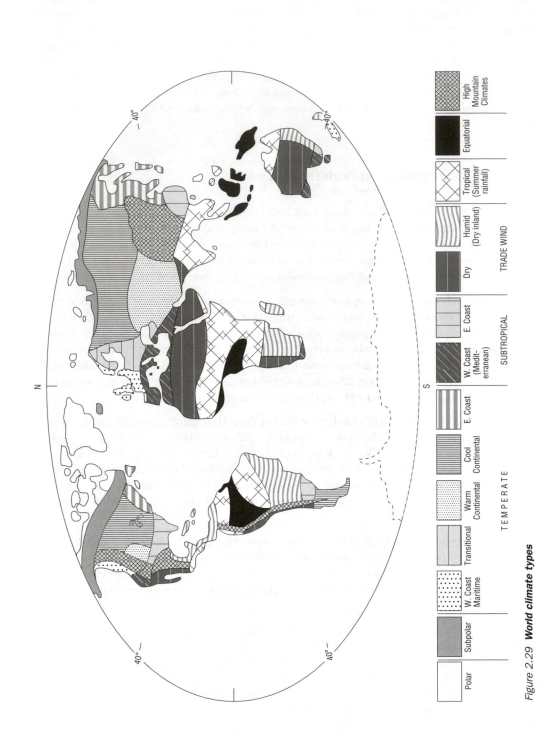

Figure 2.29 *World climate types*

of dry climates is defined by aridity. These main types were then subdivided into more detailed categories, allowing for climates with or without dry and rainy seasons, monsoon climates, and others. The relationship between temperature and plant distribution is imprecise, however, so the categories are somewhat arbitrary, with many exceptions, and his classification is rather crude, despite its popularity.

Thornthwaite adopted a different approach derived from the water required by farm crops (ALLABY, 1992a, p. 109) and based on precipitation efficiency and thermal efficiency. Both of these can be calculated. Precipitation efficiency is measured for each month as the ratio of precipitation to temperature to evaporation (as $115(r/t - 10)^{10/9}$, where r is the mean monthly rainfall in inches and t is the mean monthly temperature in °F), the sum of the 12 monthly values giving a precipitation efficiency (P-E) index. Thermal efficiency is calculated each month as the extent to which the mean temperature exceeds freezing (as $(t - 32)/4$); the thermal efficiency (T-E) index is the sum of the monthly values. The major change Thornthwaite introduced to his scheme in 1948 concerned the importance of transpiration by plants. Combined with evaporation (in practice the two cannot be measured separately in the field) this is evapotranspiration or, if water is available in unlimited amounts, 'potential evapotranspiration' (PE). It is calculated in centimetres from the mean monthly temperature in °C, corrected for changing day length.

Using his three indices, Thornthwaite defined nine 'humidity provinces' and nine 'temperature provinces', the respective index value doubling between each province and the next in the hierarchy. He then added further subdivisions to reflect the distribution of precipitation through the year, leading to 32 distinct climate types. The humidity provinces, with their denoting letters, are: perhumid (A); humid (B_4, B_3, B_2, B_1); moist subhumid (C_2); dry subhumid (C_1); semi-arid (D); and arid (E). The temperature provinces are: frost (E'); tundra (D'); microthermal (C'_1, C'_2); mesothermal (B'_1, B'_2, B'_3, B'_4); and megathermal (A'). This classification makes no assumptions about the distribution of plants, but is based wholly on recorded data.

These classifications are described as 'empirical', because they are based on data. Their disadvantage arises from the fact that divisions among sets of continuous variables are inevitably arbitrary, so the number of categories is potentially huge, and the more regional variations a scheme recognizes the more unwieldy it becomes. 'Genetic' classifications, derived from seasonal patterns of insolation and precipitation or the dominant air masses, are not widely used, but there are several of them. Indeed, there are many classificatory systems (HIDORE AND OLIVER, 1993, pp. 263–264), but those of Köppen and Thornthwaite remain the most popular.

Thornthwaite devised a scheme to classify climates independently of the vegetation each type supports, but the historical association between climate classification and plant distribution is close. Up to a point the link is obvious. Tropical rain forests flourish in the humid tropics, cacti and succulents in arid climates, conifer forests in high latitudes, and tundra vegetation borders the barren polar regions. Clearly, plants occur only where the climate suits them; bananas do not grow in Greenland (at least, not in the open). Although plant distribution is linked to climate, however, other factors also influence it. Continental drift has separated what were once adjacent landmasses supporting similar plants, producing very discontinuous distributions. The southern beeches (*Nothofagus*), for example, occur in Australasia and western South America, and pepper bushes (*Clethra*) in China and South-East Asia and from the south-eastern United States to the northern and central regions of South America, but with fossil remains in Europe. Major climate changes alter vegetation patterns, but often leave remnants of the former pattern surviving as isolated relicts. The strawberry tree (*Arbutus unedo*) belongs to a pattern of plants known as Lusitanian; these occur in south-western Europe, but also, as relicts, in southern Ireland and Brittany.

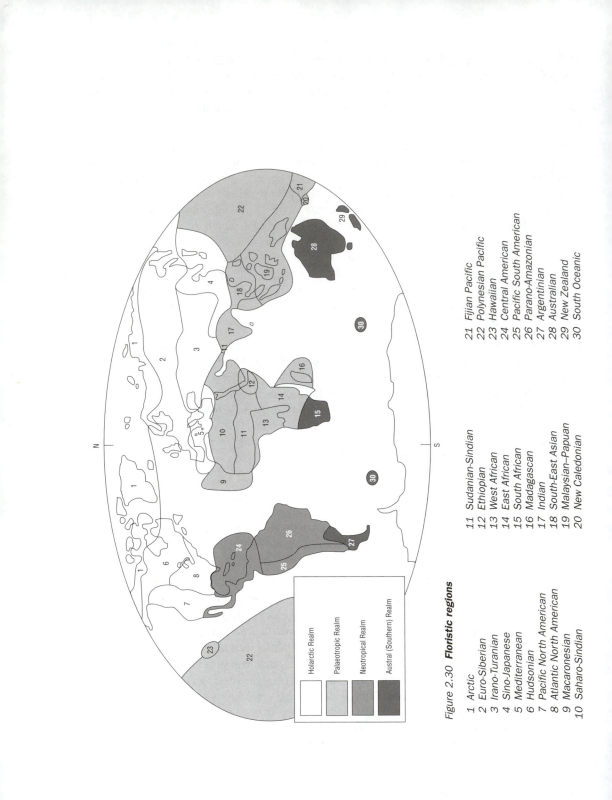

Figure 2.30 *Floristic regions*

1 Arctic
2 Euro-Siberian
3 Irano-Turanian
4 Sino-Japanese
5 Mediterranean
6 Hudsonian
7 Pacific North American
8 Atlantic North American
9 Macaronesian
10 Saharo-Sindian

11 Sudanian-Sindian
12 Ethiopian
13 West African
14 East African
15 South African
16 Madagascan
17 Indian
18 South-East Asian
19 Malaysian–Papuan
20 New Caledonian

21 Fijian Pacific
22 Polynesian Pacific
23 Hawaiian
24 Central American
25 Pacific South American
26 Parano-Amazonian
27 Argentinian
28 Australian
29 New Zealand
30 South Oceanic

Holarctic Realm
Palaeotropic Realm
Neotropical Realm
Austral (Southern) Realm

Nevertheless, regions of the world can usually be defined in terms of the plants occurring naturally within them and those regions coincide, more or less, with the climatic zones. The plants growing in a particular area comprise the 'flora' of that area and floras can be grouped into units, called 'phytochoria' (singular phytochorion), in which small unrelated floras, such as the Lusitanian in northern Europe, are designated 'elements'. Once defined, phytochoria can then be grouped further into a hierarchical system. The highest category is the floral realm or kingdom (both terms are used), which is divided into regions. Regions are subdivided further into provinces or domains (the terms are synonymous), each comprising a number of districts. Some classifications allow intermediate ranks and subdivisions of districts (MOORE, 1982, pp. 210–219). Realms are identified by the presence of particular plant families, regions by the presence of 20–30 per cent of plant genera that are not found elsewhere (i.e. endemic genera), and provinces by their endemic genera.

Most classifications recognize four floral realms: Holarctic, Palaeotropic, Neotropical, and Austral. The Holarctic Realm comprises North America, Greenland, Europe, and Asia except for India and the south-west and south-east (which became attached to the main landmass during the Tertiary). Floristically, the mountains extending from the Atlas range in North Africa across southern Asia to the Himalayas mark the southern boundary of the northern hemisphere in the Old World. With a few exceptions, coniferous trees occur north of the boundary and palms to its south.

The Palaeotropic Realm (the name means 'old' tropical) comprises Africa south of the Atlas Mountains except for southern Africa, Madagascar, Arabia, southern Asia including India, and the islands of the tropical Pacific. The Neotropical Realm ('new' tropical) comprises Central America including the southern tips of California and Florida, the Caribbean, and most of South America. Although their climates are similar, floristically these tropical realms differ from one another markedly because of the time that has elapsed since continental drift separated them. Cacti, for example, are characteristic of the New World and are one of the defining families (Cactaceae) of the Neotropical Realm; those found in the Old World have been introduced. This is why they are regarded as two distinct realms, rather than one.

The Austral (or Southern) Realm comprises the southern part of South America, southern Africa, Australia, New Zealand, and the islands of the southern Atlantic and Pacific. Here, too, the land-masses are now isolated from one another. Southern Africa and southern South America differ floristically from the rest of the continents to their north and share some plant families with Australia and New Zealand. On this basis they are grouped together as one floral realm or, in some classifications, ranked as individual Australian, Cape, and Antarctic Realms. Figure 2.30 shows these realms and the 30 floral regions of which they are composed.

Floristic realms and regions vary in size, but all are vast and difficult to comprehend. It is not until their subdivision reaches the provincial level that they become easily recognizable. Western Europe, for example, from northern Spain to Denmark and the Norwegian coast, constitutes the Atlantic Province. The Boreal Province, supporting vast tracts of coniferous forest (known in Russia as the taiga) forms a belt across Europe and Asia between the Ural River and Gulf of Finland and latitude 60° N. The North American equivalent, covering most of Alaska and Canada south of the Arctic, is called the Hudsonian Province.

Animal distribution is also described geographically and, because particular animals are often associated with particular plants, zoographical and floristic realms almost coincide. The concept of realms, with their subdivisions, should not be confused with that of biomes, which are defined ecologically. Floristic classification reflects climates, past and present, and the history as well as present geography of the planet.

End of chapter summary

All living organisms, including us, depend entirely on the materials from which the physical Earth is made and on the energy our planet receives from the Sun. It is important, therefore, to understand, at least in a very general way, how rocks and minerals form and how the landscape changes over long periods of time.

Ocean currents and all atmospheric movements are driven by energy the Earth receives from the Sun in the form of electromagnetic radiation. The movement of air and water produce the climates of the world, and the weather we experience day by day. Climate is the principal factor in determining the plants and animals that live in a particular area.

Climate is not constant, however. It has changed many times in the past. At present we are living in an interglacial and one day the ice sheets and glaciers may begin the advance that marks the dawn of a new ice age. Meanwhile, many people fear we may be inadvertently modifying the climate ourselves.

End of chapter points for discussion

Why are the oceans so important climatically?
What evidence is there that glaciations occurred?
Why are there deserts in the tropics of both hemispheres?
How does climate affect the distribution of plants?

See also

Formation and structure of the Earth (section 6)
Formation of rocks, minerals, and geologic structures (section 7)
Weathering (section 8)
Evolution of landforms (section 9)
Solar energy (section 11)
Albedo and heat capacity (section 12)
Greenhouse effect (section 13)
Structure of the atmosphere (section 14)
Atmospheric circulation (section 15)
Ocean circulation (section 16)
Weather and climate (section 17)
Glacials, interglacials, interstadials (section 18)
Dating methods (section 19)
Climate change (section 20)
Fresh water (section 22)
Soils (section 26)
Transport by water and wind (section 27)
Soil, climate, and land use (section 28)
Mining for fuels (section 30)
Mining and processing of minerals (section 31)
Biogeography (section 32)
Nutrient cycles (section 33)
Transnational pollution (section 62)

Further reading

Air: The Nature of Atmosphere and the Climate. Michael Allaby. 1992. Facts on File, New York. A popular science book providing useful information on atmosphere and climate.

Climatology: An Atmospheric Science. John J. Hidore and John E. Oliver. 1993. Macmillan, New York. Provides a broad overview of its subject, clearly written in simple language.

The Holocene: An Environmental History. Neil Roberts. 1989. Basil Blackwell, Oxford. Describes briefly the history of the last 10 000 years, with details of the methods used to interpret the past.

Principles of Physical Geology. Arthur Holmes. 2nd edn 1965. Nelson, Walton-on-Thames. Old, but still possibly the most approachable introduction to the earth sciences.

The Study of Landforms. R.J. Small. 1970. Cambridge University Press, Cambridge. Provides a broad general introduction to geomorphology.

Water. Michael Allaby. 1992. Facts on File, New York. A popular science book with useful explanations of climate and weather.

Notes

1 These are outlined in Hutchison, R. 1981. 'The origin of the Earth', in Cocks, L.R.M. (ed.) *Evolving Earth: Chance, change and challenge.* British Museum (Natural History) and Cambridge Univ. Press, London. pp. 5–16.

2 There is a brief outline of the current ideas of the mechanism underlying movements in the mantle in 'Making waves', by Sarah Simpson, *Scientific American*, August 1999, p. 10.

3 There is a brief description of the wildlife on Surtsey in Helfferich, Carla. 1993. 'Sandwort, Seabirds, and Surtsey', Article 1132, Alaska Science Forum. *www.gi.alaska.edu/ScienceForum/ASF11/1132.html.*

4 For a rather technical explanation see Thorpe, Richard and Brown, Geoff. 1985. *The Field Description of Igneous Rocks. Geological Society of London Handbook.* Open Univ. Press, Milton Keynes (published by John Wiley and Sons in the US and Canada).

5 The formation of Britain is described fairly simply in Dunning, F.W., Mercer, I.W., Owen, M.P., Roberts, R.H., and Lambert, J.L.M. 1978. *Britain Before Man.* HMSO for the Institute of Geological Sciences, London.

6 For a guide to the interpretation of sedimentary rocks, see Tucker, Maurice E. 1982. *The Field Description of Sedimentary Rocks. Geological Society of London Handbook.* Open Univ. Press, Milton Keynes (published by John Wiley and Sons in the US and Canada).

7 See Fry, Norman. 1984. *The Field Description of Metamorphic Rocks. Geological Society of London Handbook.* Open Univ. Press, Milton Keynes (published by John Wiley and Sons in the US and Canada).

8 These are described in Angel, Heather, Duffey, Eric, Miles, John, Ogilvie, M.A., Simms, Eric, and Teagle, W.G. 1981. *The Natural History of Britain and Ireland.* Michael Joseph, London. pp. 108–110.

9 Stefan's law is expressed as: $E = \sigma T^4$, where E is the amount of radiation emitted, T is the temperature, and σ is the Stefan–Boltzmann constant, which is the amount of radiant energy released by a black body. It is represented by σ and is equal to 5.67×10^{-8} W m^{-2} K^{-4} (watts per square meter per kelvin to the fourth power). The temperature is in kelvins and the energy units are watts per square meter (W m^{-2}). The law (and constant) were discovered in 1879 by the Austrian physicist Josef Stefan (1835–93), and at first they were known as Stefan's law and constant. In 1884, the Austrian physicist Ludwig Eduard Boltzmann (1844–1906), Stefan's former student, showed that the law holds only for black bodies and his name was added to that of the law and constant.

10 Wien's law can be stated as $\lambda_{max} = C/T$, where λ_{max} is the wavelength of maximum emission, C is Wien's constant, and T is the temperature in kelvins. Wien's constant is 2897×10^{-6}m (2897 μm), so the law becomes: $\lambda_{max} = (2897/T) \times 10^{-6}$m. It is valid only for radiation at short wavelengths. The law was discovered in 1896 by the German physicist Wilhelm Wien (1864–1928) and for it he was awarded the 1911 Nobel Prize for physics.

References

Adams, Peter. 1977. *Moon, Mars and Meteorites.* HMSO for the British Geological Survey.

Allaby, Ailsa and Michael (eds) 1999. *The Concise Oxford Dictionary of Earth Sciences.* 2nd edition. Oxford University Press, Oxford.

Allaby, Michael. 1992. *Elements: Air.* Facts on File, New York.
—. 1992a. *Elements: Water.* Facts on File, New York.
—. 1998. *Dangerous Weather.* Floods. Facts on File, Inc., New York.

Avery, Dennis. 1995. 'Saving the Planet with Pesticides: Increasing Food Supplies While Preserving the Earth's Biodiversity', in Bailey, Ronald (ed.) *The True State of the Planet,* The Free Press, New York, pp. 73–76.

Balling, Robert C., Jr. 1995. 'Global Warming: Messy Models, Decent Data and Pointless Policy', in Bailey, Ronald (ed.) *The True State of the Planet,* The Free Press, New York, pp 83–107.

Barry, Roger G. and Chorley, Richard J. 1982. *Atmosphere, Weather and Climate*. Methuen, London.

Boehmer-Christiansen, Sonja A. 1994. 'A scientific agenda for climate policy?', *Nature*, 372, 400–402.

Bolin, Bert. 1995. 'Politics of climate change' (letter in reply to Boehmer-Christiansen), *Nature*, 374, 208.

Bowler, Peter J. 1992. *The Fontana History of The Environmental Sciences*. Fontana Press, London.

Burbank, Douglas W. 1992. 'Causes of recent Himalayan uplift deduced from deposited patterns in the Ganges basin', *Nature*, 357, 680–682.

Calder, Nigel. 1999. 'The Carbon Dioxide Thermometer and the Cause of Global Warming', *Energy and Environment*, 10, 1, 1–18.

Cane, Mark A., Eshel, Gidon, and Buckland, R.W. 1994. 'Forecasting Zimbabwean maize yield using eastern equatorial Pacific sea surface temperature'. *Nature*, 370, 204–205.

Clark, John. 1977. *Coastal Ecosystem Management*. Wiley-Interscience, New York.

Cowling, Sharon A. 1999. 'Plants and Temperature-CO^2 Uncoupling', *Science*, 285, 1500–1501.

Dalziel, Ian W.D. 1995. 'Earth before Pangaea', in *Scientific American,* January. pp. 38–43.

Eddy, John A. 1977. 'The case of the missing sunspots,' *Scientific American*. pp 80–89.

Gentry, A.W. and Sutcliffe, A.J. 1981. 'Pleistocene geography and mammal faunas', in Cocks, L.R.M. (ed.) *The Evolving Earth: Chance, change and challenge*. British Museum (Natural History) and Cambridge Univ. Press, London, pp. 237–251.

Gleick, James. 1988. *Chaos: Making a new science*. Heinemann, London. pp. 9–31.

Graham, A.L. 1981. 'Plate tectonics', in Cocks, L.R.M. (ed.) *The Evolving Earth: Chance, change and challenge*. British Museum (Natural History) and Cambridge University Press, London. Chapter 11.

Grootes, P.M., Stuiver, M., White, J.W.C., Johnsen, S., and Jouzel, J. 1993. 'Comparison of oxygen isotope records from the GISP2 and GRIP Greenland ice cores,' *Nature*, 366, 552–554.

Hambrey, M.J. and Harland, W.B. 1981. 'The evolution of climates', in Cocks, L.R.M. (ed.) *The Evolving Earth: Chance, change and challenge*. British Museum (Natural History) and Cambridge University Press, London. Chapter 9.

Harvey, John. G. 1976. *Atmosphere and Ocean*: Our fluid environments. Artemis Press, Sussex.

Hidore, John J. and Oliver, John E. 1993. *Climatology: An atmospheric science*. Macmillan, New York.

Holm, Søren and Harris, John. 1999. 'Precautionary principle stifles in discovery', letter to *Nature*, 400, 398.

Holmes, Arthur. 1965. *Principles of Physical Geology*. Nelson. Walton-on-Thames. 2nd edition.

Hudson, Norman. 1971. *Soil Conservation*. B.T. Batsford Ltd, London.

Hunt, Charles B. 1972. *Geology of Soils: Their evolution, classification, and uses*. W.H. Freeman and Co., San Francisco.

Intergovernmental Panel on Climate Change. 1992. *1992 IPCC Supplement: Scientific assessment of climate change*. WMO and UNEP.

Jacobs, G.A., Hurlburt, H.E., Kindle, J.C., Metzger, E.J, Mitchell, J.L., Teague, W.J., and Wallcraft, A.J. 1994. 'Decade-scale trans-Pacific propagation and warming effects of an El Niño anomaly', *Nature*, 370, 360–363.

Johnson, Nicholas and David, Andrew. 1982. 'A Mesolithic site on Trevose Head and contemporary geography', in *Cornish Archaeology,* 21, 67–103.

Kempe, D.R.C. 1981. 'Deep ocean sediments', in Cocks, L.R.M. (ed.) *The Evolving Earth: Chance, change and challenge*. British Museum (Natural History) and Cambridge University Press, London. pp. 116–117.

Kendeigh, S. Charles. 1974. *Ecology With Special Reference to Animals and Man*. Prentice-Hall, Eaglewood Cliffs, New Jersey.

Kerr, Richard A. 1995. *'Sun's role in warming is discounted,'* Science, 268, 28–29.

Lamb, H.H. 1995. *Climate, History and the Modern World*. 2nd ed. Routledge, London.

McCormick, M. Patrick, Thomason, Larry W., and Trepte, Charles R. 1995. 'Atmospheric effects on the Mt Pinatubo eruption,' *Nature,* 373, 399–404.

Moore, David M. (ed.) 1982. *Green Planet: The story of plant life on Earth.* Cambridge University Press, Cambridge.

Morell, Virginia. 1995. 'The earliest art becomes older – and more common', *Science,* 267, 1908–1909.

Pennington, Winifred. 1974. *The History of British Vegetation,* 2nd ed. Hodder and Stoughton, London.

Raffensperger, Carolyn, Tickner, Joel, Schettler, Ted, and Jordan, Andrew. 1999. ' . . . and can mean saying "yes" to innovation', letter to *Nature,* 401, 207.

Raven, Peter H., Berg, Linda R., and Johnson, George B. 1993. *Environment.* Saunders College Publishing, Orlando , Florida.

Roberts, Neil. 1989. *The Holocene: An environmental history.* Basil Blackwell, Oxford.

Simmons, I.G., Dimbleby, G.W., and Grigson, Caroline. 'The Mesolithic', in Simmons, Ian and Tooley, Michael (eds) 1981. *The Environment in British Prehistory.* Duckworth, London.

Small, R.J. 1970. *The Study of Landforms.* Cambridge Univ. Press.

Stratton, J.M. and Brown, Jack Houghton. 1978. *Agricultural Records AS 22–1977,* 2nd edn. John Baker, London.

Thomson, David J. 1985. 'The seasons, global temperature, and precession,' *Science,* 268, 59–68.

Tolba, Mostafa K. and El-Khody, Osama A. 1992. *The World Environment 1972–1992.* UNEP and Chapman and Hall, London.

Trimble, Stanley W. 1999. 'Decreased Rates of Alluvial Sediment Storage in the Coon Creek Basin, Wisconsin, 1975–93', *Science,* 285, 1244–1246.

Wagner, Frederike, Bohncke, Sjoerd J.P., Dilcher, David L., Kürschner, Wolfram M., van Geel, Bas, and Visscher, Henk. 1999. 'Century-Scale Shifts in Early Holocene Atmospheric CO_2 Concentration', *Science,* 284, 1971–1973.

Windley, Brian F. 1984. *The Evolving Continents.* John Wiley and Sons, Chichester.

Zahn, Rainer. 1994. 'Linking ice-core records to ocean circulation,' *Nature,* 371, 289.

Physical Resources

When you have read this chapter you will have been introduced to:

- the hydrologic cycle
- the life cycle of lakes
- salt water, brackish water, and desalination
- irrigation, waterlogging, and salinization
- soil formation, soil ageing, and soil taxonomy
- soil transport
- soil, climate, and land use
- soil erosion and its control
- mining and processing fuels
- mining and processing minerals

22 Fresh water and the hydrologic cycle

In the sense used here, a 'resource' is a substance a living organism needs for its survival. There are also non-material resources, such as social contact and status, which may be essential to a feeling of well-being or even to survival itself, but these are not considered here.

Non-humans as well as humans make use of the resources available to them; animals need such things as food, water, shelter, and nesting sites, all of which are resources, as are the sunlight and mineral nutrients required by plants. Human biological requirements are similar to those of other animals. Like them, we need food, water, and shelter, although we differ from other species in the means we have developed for obtaining them. It is because human and non-human requirements often coincide that sometimes we find ourselves in direct competition for resources with non-humans. It is not only we who find crop plants edible and nutritious, for example, and before we can build houses to shelter ourselves we must clear the land of its previous, non-human occupants.

Water is, perhaps, the most fundamental of the resources we require. Without water, as the cliché has it, life could not exist on land. Our bodies are largely water (by weight), and if you add together the ingredients listed on many food packets you will find they seldom amount to more than about half the total weight: the remainder is water.

It is not any kind of water we need, of course, but fresh water. Sea water is of only limited use to us, and out of reach for people living deep inside continents, and drinking it is harmful, although it can be rendered potable by the removal of its dissolved salts. For the most part, therefore, we humans must obtain all the water we need from rivers, lakes, and underground aquifers. In the world as a whole, it is estimated that by the year 2000 we will be using about 4350 km^3 (4.35×10^{15} litres) of water a year. Of this, almost 60 per cent will be needed for crop irrigation, 30 per cent for industrial processes and cooling, and 10.5 per cent for domestic cooking, washing, and drinking (RAVEN ET AL., 1993, p. 273).

Of all the water in the world, 97 per cent is in the oceans, so our freshwater needs must be met from the remaining 3 per cent. It is not even that simple, however, because of all the fresh water, more than half is frozen in the polar icecaps and glaciers and about 0.5 per cent is so far below ground as to be beyond our reach. Atmospheric water vapour, falling rain and snow, and flowing rivers contain no more than about 0.005 per cent of the planet's water (KUPCHELLA AND HYLAND, 1986, pp. 222–223). Stated like this, the amount available to us sounds alarmingly small, but it is so only as a proportion of the total. The quantity available to us, including that in lakes and inland seas, is in the region of 15×10^{18} litres.

Water can exist as either gas or liquid at temperatures commonly encountered near the surface and consequently it is constantly evaporating and condensing again. Each year, some 336×10^{15} litres evaporates from the oceans and 64×10^{15} litres from the land surface (including water transpired by plants). About 300×10^{15} litres falls as precipitation over the oceans and 100×10^{15} litres over land, and 36×10^{15} litres flows from the land back to the sea (HARVEY, 1976, p. 22). This movement of water between oceans, air, and land constitutes the hydrologic cycle, and by dividing the quantity of water at each stage of the cycle by the amount entering or leaving, it is possible to discover approximately the average time a water molecule remains in each: its residence time. This reveals that a molecule spends about 4000 years in the ocean, 400 years on or close to the land surface, and 10 days as vapour in the atmosphere.

Most of the water falling on land evaporates again almost immediately or is taken up by plant roots and returned to the atmosphere by transpiration. Some flows directly over the surface, down slopes and into lower ground where it may enter lakes, rivers, or marshes. What remains drains downward through the soil until it encounters a layer of impermeable clay or rock, then flows laterally, very slowly, through the soil. Were it not to flow, but simply accumulate, the ground would soon be waterlogged and water would lie at the surface. Above the impermeable material a layer of soil is saturated with water. This is ground water and its upper limit, above which the soil is not saturated, is the water table. Permeable material through which ground water flows is called an 'aquifer' and it may lie deep beneath the surface. Aquifers are permeable because the particles of which they are composed, such as gravel or sand, are not packed together so tightly as to leave no spaces between them. They are said to be 'unconsolidated' and allow water to flow through them. Other aquifers are made from material, such as chalk or sandstone, which are consolidated (solid) but nevertheless have fissures, or pore spaces within their granular structure, through which water can flow.

It is obviously most convenient to obtain our supplies of fresh water from the nearest river or lake, but this may be too distant or insufficient. In that case it may be possible to obtain water from an aquifer, by sinking a borehole into it and pumping out the water. Figure 3.1 illustrates this and also shows what happens: abstraction lowers the water table around the borehole,

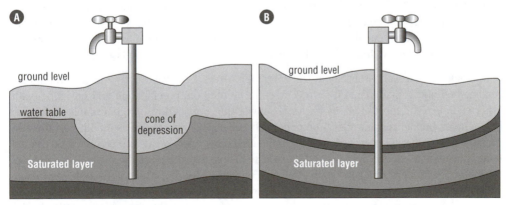

Figure 3.1 **Water abstraction**

forming a 'cone of depression'. If the rate of abstraction exceeds that at which the aquifer is recharged, the water table will fall over a wide area, eventually to a level at which the yield from the borehole decreases and the aquifer is exhausted. In the United States, there are parts of the Great Plains, California, and southern Arizona where the severe depletion of aquifers for irrigation now threatens future water supplies and also reduces water quality. Quality is affected because, in coastal regions, as the water table falls salt water enters to recharge it, and anywhere that toxic mineral salts dissolve in ground water, reducing the volume of water may increase their concentration, so the water requires more extensive, and therefore costly, processing to render it drinkable (RAVEN ET AL., 1993, pp. 279–281). Pollution of this kind is natural, although caused by human over-exploitation of a resource, but ground water can be polluted by industrial or domestic wastes.

Lowering the water table can also cause ground subsidence due to the reduction in volume of the material comprising the saturated layer as this dries. Between 1865 and 1931, groundwater abstraction in London caused the ground to subside at 0.91–1.21 mm yr^{-1}, producing a total subsidence of 0.06–0.08 m. In Tokyo, the ground subsided 4 m between 1892 and 1972, at a rate of 500 mm yr^{-1}, and Mexico City is sinking at 250–300 mm yr^{-1} for the same reason (GOUDIE, 1986, p. 207).

Not all aquifers require pumping. An unconfined aquifer is one into which water drains freely from above, but where two approximately parallel impermeable layers are separated by a layer of porous material, the resulting aquifer is said to be confined. Natural undulations in a confined aquifer produce low-lying areas in which water is under pressure from the water at a higher level to either side (see Figure 3.1B). This water will flow without pumping from a borehole drilled into the aquifer through the upper impermeable layer and it will continue to flow provided the aquifer is constantly recharged by water draining into the hollow. The result is an 'overflowing' or 'artesian' well.

Where the water table reaches the surface, water will flow spontaneously, as a spring, and on sloping ground it will form a stream and eventually, through the merging of many small streams, a mighty river. Rivers also supply water, but since long before our ancestors invented wheeled vehicles and built roads for them they have also been used to convey people and goods. It is no coincidence that most of the world's major inland cities are located beside large rivers. Almost any river might serve as an example, but the Rhine is an especially good one, because it flows across a densely populated continent for a distance of 1320 km. Figure 3.2 shows the river together with some of its more important tributaries and the principal cities that border it.

Over the centuries the cities along the Rhine prospered and grew, and as Europe industrialized several of them became important manufacturing centres. Most industries use water and produce liquid wastes, and humans produce sewage, a mixture of urine, faeces, and water that has been used for washing and cooking. At one time all this was poured into the river, which removed it, and wastes discharged into the Rhine were joined by those discharged into its tributaries, including the Emscher, which drains the Ruhr and enters the Rhine north of Düsseldorf.

Rivers have a remarkable capacity for cleaning themselves, because their waters are continually replenished and contaminants removed by extreme dilution, precipitation and burial beneath other sediment, or, most of all, by bacterial activity that breaks down large, organic molecules into simpler, biologically harmless compounds. In the case of rivers such as the Rhine, however, transporting foul water merely delivers it to the next city downstream, where it must be treated before it can be used, and the further downstream people live, the more their drinking water will cost them. In modern times the problem has been addressed, but it was not simple. As Figure 3.3 shows, water drains into the Rhine from an area of 220 150 km^2 in six countries. Why should the Swiss pay more to treat effluent prior to discharge for the benefit of the distant Netherlands? Why should the French regulate discharges from their chemical industries in Alsace when the

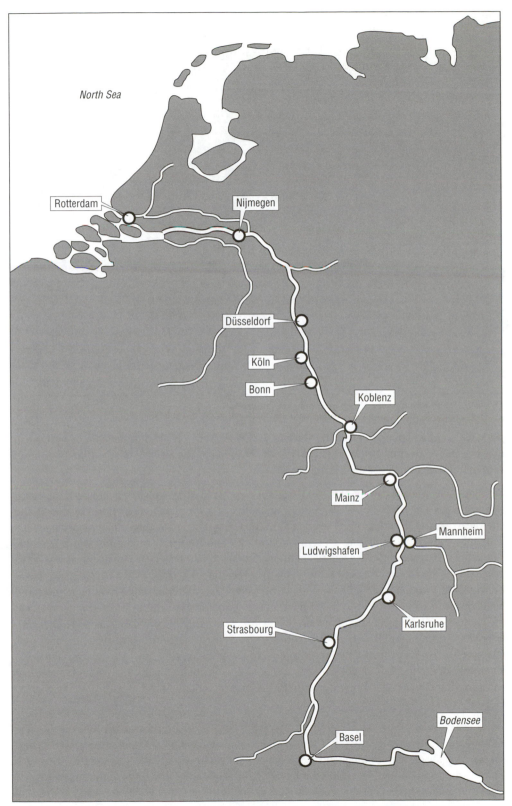

North Sea

Rotterdam

Nijmegen

Düsseldorf

Köln

Bonn

Koblenz

Mainz

Mannheim

Ludwigshafen

Karlsruhe

Strasbourg

Bodensee

Basel

Figure 3.2 **Principal cities bordering the Rhine (not to scale). Total length of the Rhine 1320 km**

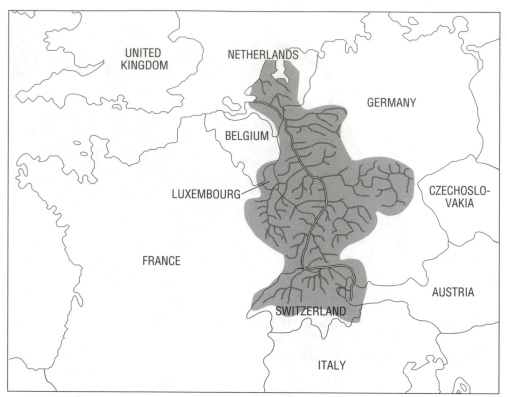

Figure 3.3 **The Rhine basin, draining land in six countries**

principal source of pollution was the German Ruhr? Fortunately, such transnational issues can now be resolved within the European Union, where mechanisms exist to ensure that the costs of antipollution measures are shared equitably.

Regulations are necessary, but accidents cannot be prevented by legislation and they can cause serious harm. On 1 November 1986, there was a fire at a warehouse near Basel owned by the chemical company Sandoz. Water used to fight the fire washed an estimated 30 tonnes of chemicals into the Rhine, including mercury and organophosphorus compounds and a red dye, rhodamine, that allowed the progress of the pollutants to be observed. The accident was exacerbated by a smaller spillage of herbicide on the preceding day from a Ciba-Geigy plant, also at Basel. By 12 November pollution was severe between Basel and Mainz, the river being declared 'biologically dead' for 300 km downstream from Basel, and by the time the affected water reached the Netherlands its mercury content, of 0.22 µg litre^{-1}, was three times the usual level. Drinking water had to be brought by road to supply several cities. Despite the severity of the incident, however, the river had almost recovered one year later (MASON, 1991, pp. 2–3). Switzerland is not a member of the EU, but its government accepted responsibility for the 1986 pollution incident and promised to consider bringing its antipollution regulations into line with those of the EU (ALLABY, 1987).

Water is a so-called 'renewable' resource. After it has been used it returns to the hydrologic cycle and in time it will be used again. It is also abundant globally and the oceans are so vast that their capacity for absorbing, diluting, and detoxifying pollutants is immense. Despite this, the provision of wholesome fresh water and the hygienic disposal of liquid wastes in the impoverished semi-arid regions of the world is woefully inadequate. It is there that fetching water for ordinary domestic use involves arduous hours of walking and carrying, mainly by women and children, and where debilitating water-borne diseases are common.

The resource is renewable, but distributed unevenly, and its efficient management requires an elaborate infrastructure of reservoirs, treatment plant, pipelines, and sewerage, coordinated within an overall strategy by an authority with the power to prevent abuses. For people in those regions, improvements in living standards depend crucially on the establishment of such strategies for water management, and once living standards begin to rise it is inevitable that the demand for water will increase substantially. As rising demand encounters limits in the supply available, conflicts may ensue, as they have already between Israel and Jordan over abstraction from the river Jordan. This is one of the most formidable challenges facing us. It is encouraging to note, however, that throughout history, competition between nations for scarce water resources has almost invariably been settled peacefully.

23 Eutrophication and the life cycle of lakes

In the late 1960s there was widespread popular concern over the pollution of rivers, lakes, and ground water by nitrate from sewage, farm effluents, but most of all by leaching from farmed land. It was feared that high nitrate levels in water might lead to health problems (principally methaemoglobinaemia, or 'blue-baby' syndrome) in infants less than 6 months old. Methaemoglobinaemia is very rare, but between 1945 and 1960 about 2000 cases were reported in the world as a whole, causing the deaths of 41 infants in the United States and 80 in Europe. The fear was not unreasonable. Today, when nitrate levels in water exceed a permitted maximum parents are advised to use bottled water for mixing infant foods and drinks. There were also fears that nitrates might form nitrous acid (HNO_2) in the body and react with amides (derived from ammonia by the substitution of an organic acid group for one (primary amide), two (secondary), or all three (tertiary) of its hydrogen atoms) or amines (also formed from ammonia, when one or more of its hydrogen atoms are replaced by a hydrocarbon group). Amines and amides are common and the product of the reaction would be N-nitrosamines and N-nitrosamides, which are known to cause cancer in experimental animals. In fact, there is no evidence that nitrate causes cancer in humans (ROYAL COMMISSION ON ENVIRONMENTAL POLLUTION, 1979, pp. 87–92). Indeed, dietary nitrates have no adverse effect whatever on human health. Although nitrites remain implicated in infant methaemoglobinaemia, it is now known that they are formed in feeding bottles by bacterial action on nitrates contained in the food in the bottle. Nitrates in the water are not involved (L'HIRONDEL, 1999). In parallel with this there was also concern that the nitrate loading of waters would cause their widespread over-enrichment (eutrophication).

Nitrogen is an essential plant nutrient and plants take it up readily in the form of nitrate (NO_3) ions, because all nitrates are highly soluble in water. Grass is present throughout the year, so its roots are always absorbing nitrate. Arable fields, on the other hand, are bare for part of the year, often at times of heavy rainfall. With no plant roots to intercept the nitrate, it is washed (leached) from the soil. Nitrate pollution was perceived as a problem in the 1960s because of agricultural changes that had taken place in Britain in the preceding years.

In 1938, the area of land growing arable crops in Great Britain was smaller than it had been at any time since the middle of the last century. The depression of the 1930s had so reduced the profitability of farming that large areas were almost abandoned, and as the Second World War began, with the likelihood of a sea blockade to restrict the import of food, the British people faced real hunger. Drastic steps were taken to increase agricultural output and after the war these continued as farming modernized. A major consequence of these changes was a substantial reduction in the area growing grass and a corresponding increase in the area growing cereals. In 1938, less than 1.2 million ha was sown to barley and wheat; in 1966 those crops occupied 3.3 million ha. During the same period, the area devoted to permanent and temporary grassland fell from 8.4 million ha to 6.8 million ha. The 2.1 million ha increase in the cereal area was achieved by reducing the grassland area. (MAFF, 1968, p. 34)

Thus the change from grassland to cereal cropping led inevitably to an increase in the movement of nitrate from the soil and into surface and ground water. The widespread introduction of soluble, nitrogen-based fertilizers exacerbated the problem, especially when heavy applications were followed by very wet weather, but the fertilizer contribution should not be exaggerated. In 1964, for example, nitrogen runoff was measured following 114 mm of rain in two falls in Missouri (SMITH, 1967). Bare soil, which had received no fertilizer, lost 0.9 kg N ha^{-1}; unfertilized maize and oats lost 0.3 kg N ha^{-1}; and continuously grown maize, fertilized with 195 kg N ha^{-1}, lost 0.1 kg N ha^{-1}.

This is not the only source of nitrogen reaching both land and water. Substantial and increasing amounts also arrive from the air. Elemental nitrogen is oxidized by lightning, in the course of burning plant materials, and in high-compression internal combustion engines, and biologically by the action of nitrogen-fixing soil bacteria. Urine from farm livestock releases ammonia, also a soluble compound. It has been found that in the mid-1970s much of Europe received 2–6 kg N ha^{-1} yr^{-1} and that some areas now receive 60 or more kg N ha^{-1} yr^{-1}. This level of fertilization may be altering the composition of certain ecosystems, especially those established on nitrogen-poor soils (MOORE, 1995).

Plants have similar physiological requirements whether they grow on dry land or in water. If plant nutrients enter water, therefore, they will stimulate the growth of aquatic plants. Nitrate alone is not enough, of course. The full range of nutrients must be supplied and plant growth is limited by the availability of the nutrient in shortest supply (in water this is usually phosphorus); this is the 'law of the minimum' first stated in 1840 by the German chemist Justus von Liebig (1803–73). Other nutrients are less mobile than nitrate, so nitrate leaching has less effect on plant life than might be supposed.

Agricultural change apart, the movement of nutrients from the land and into water is an entirely natural process, an inevitable consequence of the drainage of rain water. As water moves downward through the soil to join the ground water, soluble soil compounds dissolve into and are carried by it. Were this not so, freshwater aquatic plant life would be severely restricted. Water draining into surface waters, such as rivers and lakes, also carries fine particulate matter that is deposited as sediment when the power of the stream falls below a certain threshold. Fast-flowing streams rapidly remove material that enters them and accumulations occur only in slow-moving rivers and still water. It is there, and only there, that sedimentation and eutrophication may cause difficulties.

Eutrophication leads to the proliferation of aquatic plants, especially algae, and cyanobacteria, organisms that derive nutrients directly from the water, rather than through roots attached to a substrate. A eutrophic lake or pond can usually be recognized by its surface covering of green algae. The life cycles of such organisms are short and as they die their remains sink and are decomposed by aerobic bacteria, whose populations increase in proportion to the food supply available to them. The bacteria obtain the oxygen they need from that dissolved in the water, and under eutrophic conditions the amount they remove exceeds the amount being introduced, so the water is depleted of dissolved oxygen. A common measure of water pollution is its 'biochemical oxygen demand' (BOD), calculated from the reduction in the amount of dissolved oxygen in a water sample incubated in darkness for 5 days at a constant 20 °C; it is also a measure of bacterial activity.

If the water body is used for water abstraction, angling, or navigation, eutrophication is likely to reduce its value. The cost of treating water to bring it to potable standard will increase, navigation may be impeded by plants, and preferred species of fish may disappear. At high densities, some algae and cyanobacteria produce potent toxins. The alga *Prymnesium parvum* is highly toxic to fish, and toxins produced by such cyanobacteria as *Microcystis*, *Aphanizomenon*, and *Anabaena* attack the liver and may be neurotoxic. In 1989 there were outbreaks of toxic cyanobacteria in some British lakes and a number of dogs died after swimming in them and ingesting their water. Not surprisingly, eutrophication also brings about marked changes in the populations of aquatic organisms. The water supports fewer plant and animal species, but more individuals, the water becomes more turbid because of the large amount of organic matter suspended in it, the water becomes increasingly anoxic, and the rate of sedimentation increases.

A eutrophic lake is an old lake, and eutrophication is an ageing process. When it first forms, a lake typically supports little plant life, but fish such as trout, which feed on insects caught at the surface, may thrive. Its water is clear and well oxygenated, but very deficient in nutrients. There is little or no sediment at the bottom and plants grow beside it, but well clear of the water. A lake in this condition is said to be 'oligotrophic' (the Greek *oligos* means 'small' and *trophe* 'nourishment').

Rivers flowing into the lake bring nutrient and particulate matter, and in time the lake becomes 'mesotrophic' (Greek *mesos*, 'middle'). Its water is still clear enough for light to penetrate deeply, so algae flourish, but without proliferating uncontrollably because they are grazed by a diverse population of invertebrate and vertebrate animals, including fish. Sediment is accumulating on the bottom. This provides anchorage and nutrient for rooted plants, which now extend from the banks and into the lake margins, colonization by plants that have to reach the air being limited only by the depth of water. The accumulation of sediment also raises the bottom, so the lake has become shallower. In a eutrophic lake (Greek *eu-*, 'well') the sediment is deep and the lake shallow. Plants rooted in the sediment extend far from the banks. The three drawings in Figure 3.4 illustrate this life cycle.

Life cycles, which paradoxically are linear so far as individuals are concerned, end in death, and the life cycle of a lake is no exception. It is the fate of all lakes and ponds eventually to become dry land or, if they occupy low-lying ground where the water table is at or very close to the surface, a bog, marsh, or fen. Accumulating sediment makes the water shallower, but its colonization by plants also removes water, by transpiration. Once plants are established across the whole area of a lake, its demise is fairly rapid. Aquatic plants give way step by step to land plants that can tolerate waterlogging around their roots, and then these are replaced by true dryland or wetland plants. As the sediment dries and becomes soil, it is the acidity of the soil that determines whether the lake evolves into lime-loving grassland and, over much of north-western Europe, from there to scrub followed by woodland and forest, or to acid-loving heath. Figure 3.5 illustrates this development.

Such eutrophication is natural, but the life span of a lake should be measured in thousands of years. Artificial eutrophication, caused by discharging sewage and other wastes into lakes, shortens it greatly. Untreated human sewage may have a BOD of 300 mg litre^{-1}, paper-pulp effluent 25 000 mg litre^{-1}, and silage effluent 50 000 mg litre^{-1}. Deoxygenation is by far the commonest type of freshwater pollution. Bacteria decomposing the faeces from one human use 115 g of oxygen a day; this is enough oxygen to saturate 10 000 litres of water (MELLANBY, 1992, p. 88). Halting natural eutrophication may be undesirable, even if it is practicable, but artificial eutrophication should be prevented or, if it is too late for prevention, cured.

It can best be remedied, of course, by finding alternative means of waste disposal or at least by reducing the nutrient content of the discharges, especially of phosphates, which are the limiting nutrient in most waters. This can be done by reducing the phosphate content of detergents, which

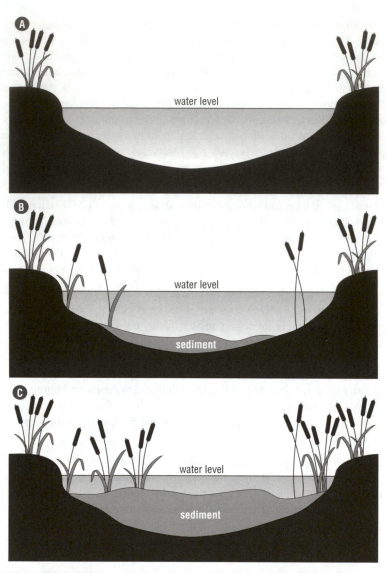

Figure 3.4 **The life cycle of a lake. A, Oligotrophic. Little bottom sediment; water nutrient-poor; plants grow on banks only. B, Mesotrophic. Mud accumulating on the bottom; plants rooted in mud extending into the lake; moderate nutrient supply. C, Eutrophic. Deep bottom sediment; plants rooted in mud far into the lake; water very rich in nutrients; depth of lake decreasing owing to accumulation of sediment and evapotranspiration**

are the principal source, or by stripping the phosphate from sewage before it is discharged. This is possible, with 90–95 per cent efficiency (MASON, 1991, p. 131). but there have been cases of a reduction in phosphate input being followed by the release of phosphate from sediment by mechanisms which are not well understood. In extreme cases it may be feasible to remove the sediment itself by dredging. Where land drainage is the main source of sediment and nutrient, reducing soil erosion may be effective. If oligotrophic water is available, using it to recharge a eutrophic lake may bring benefits. Beyond such measures as these, remediation usually involves manipulating the plant and animal populations. Obviously, no two water bodies are precisely similar and remedial measures must be appropriate to the particular conditions encountered.

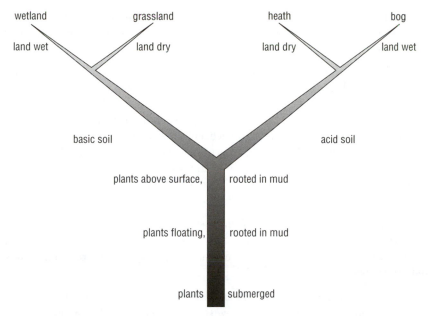

wetland   grassland   heath   bog

land wet   land dry   land dry   land wet

basic soil   acid soil

plants above surface,   rooted in mud

plants floating,   rooted in mud

plants   submerged

Figure 3.5 **Evolution of a lake into dry land, marsh, or bog**

It is easy to over-dramatize the problems of eutrophication. They are confined to still or slow-moving waters, which limits their extent. Nevertheless, remediation is often necessary, because the affected water body represents a valuable resource, and it is always complicated and expensive. Prevention being better than cure, control of discharges into surface waters, introduced primarily to improve the quality of river water that is not liable to eutrophication, will nevertheless reduce eutrophication in lakes fed by the improved rivers. The principal cause of river pollution is identical to that which produces artificial eutrophication.

24 Salt water, brackish water, and desalination

Water is a scarce resource in many parts of the world. Even in regions where rainfall is usually adequate, periodic droughts can bring shortages, and restrictions on water use are fairly common in Britain, despite its generally moist, maritime climate. These restrictions have never been so severe as to direct serious attention to alternative sources of supply, however, except on some offshore islands, such as the Isles of Scilly, in the Western Approaches off Land's End, where a desalination plant has been proposed.

Since almost all the water on Earth is in the oceans, sea water is the most obvious place to seek supplies and, after all, nowhere on the Isles of Scilly is more than a mile or so from the sea. The disadvantage of sea water, of course, is its salt content. Industrial plants located in coastal areas can use sea water directly for cooling, which is why many British nuclear power plants are located at the coast, but sea water is useless for agricultural or domestic purposes. The cells of living organisms are contained within membranes that are partially permeable, allowing water molecules to pass, but blocking the passage of larger molecules, in a process known as 'osmosis'. If a partially permeable membrane separates two solutions of different concentrations, an osmotic pressure will act across the membrane, forcing water molecules to pass from the weaker to the stronger solution until the concentrations equalize. When cells are exposed to sea water, its salt concentration is higher than the concentration inside the cell, and water moves out of the cell. Salt water thus has a dehydrating effect and its salts must be removed before land-dwelling plants or animals can use it.

This is expensive, and there is another source of fresh water: the polar icecaps. The idea may sound absurd, but probably it would be technologically and economically feasible to tow large icebergs into low latitudes, moor them close to the shore, and 'mine' them for fresh water. An iceberg would begin to melt as it entered warm water, but the rate of melting would be low enough to ensure the survival of the great bulk of the ice and the loss would be acceptable. Clearly, the resource is vast and possibly self-renewing. There is a major disadvantage, however. Because the communities needing the water are located far inland, but the iceberg is at the coast, water must still be transported over a long distance. Combined with the cost of towing, this would probably make the operation prohibitively expensive.

'Iceberg mining' has not yet been attempted and neither has a rival scheme, suggested by Walter Rickel, Governor of Alaska, to construct a submarine pipeline to carry water 3220 km to California from the headwaters of Alaskan rivers. The scheme was considered, but rejected because of its estimated $100 billion cost (REINHOLD, 1992).

Desalination (*www.ce.vt.edu/enviro2/wtprimer/desalt/desalt/html*), on the other hand, is used widely in the Near and Middle East. It is also used in the United States. For some years the Office of Saline Water, of the Department of the Interior, has maintained a demonstration desalination plant at Freeport, Texas, and there is a large plant in Arizona. More recently, water shortages in California led to the construction of a plant at Catalina yielding 580 280 litres of fresh water a day, and plants are also to be built at Santa Barbara and Morro Bay (REINHOLD, 1992).

The purpose of desalination is the removal of salts from sea water, but not all sea water is equally saline. Together, temperature and salinity determine the relative densities of different water bodies, which form water masses analogous to air masses. Plotted on a graph, seawater masses can be identified by their position along a temperature–salinity (T–S) curve. Salinity is conventionally reported in parts per thousand (per mille). In the centre of the North Atlantic, for example, the T–S curve ranges from 8 °C and 35.1 per mille to 19 °C and 36.7 per mille; around Antarctica the seawater temperature is 2–7 °C and salinity 34.1–34.6 per mille (HARVEY, 1976, pp. 61–63). Elsewhere, salinity may be markedly higher or lower. The Mediterranean loses more water by evaporation than it receives from inflowing rivers and precipitation; it also loses water at depth and gains inflowing water near the surface through the Straits of Gibraltar. This regime results in a salinity higher than that of the Atlantic, ranging from about 37.0 per mille near Gibraltar to about 39 per mille at the eastern end. The Black Sea, in contrast, has an average salinity of about 19.0 per mille, the Caspian 12.86 per mille, and the Red Sea 41.0 per mille (DAJOZ, 1975, pp. 126–128). Variable though the salinity of sea water is, it remains true that all sea water is too salty to drink: fresh water has a salinity of less than 0.3 per mille.

Water that is neither fresh nor sea water is known as 'brackish' and its salinity is even more variable. Oligohaline water is only slightly more saline than fresh water, with a salinity of 0.5–5.0 per mille; mesohaline water has 5.0–16.0 per mille; polyhaline water has 16.0–40.0 per mille; and saline water has more than 40.0 per mille. Water in the Great Salt Lake has a salinity of 170 per mille and that in the Dead Sea 230 per mille. Yet all these are 'brackish' waters.

Salinity is measured by titrating a sample of water with silver nitrate until all the chloride ions have been precipitated, and adding potassium chromate, which reacts with silver nitrate when all the chloride has been precipitated, forming potassium chromate, which is red. The reaction is:

$$Cl^- + AgNO_3 \longleftrightarrow AgCl \downarrow + NO_3^-$$

In other words, what is being measured is *chlorinity*.

Regardless of its salinity, or chlorinity, the composition of sea water is fairly constant (Table 3.1). Some of the 'other' (listed at the foot of Table 3.1) is of commercial importance, actually or potentially. It contains about 3 parts per million of uranium, for example, and about 0.003 per cent of all water, including sea water, is deuterium oxide, or 'heavy water', used as a moderator in the Candu (Canadian deuterium–uranium) fission reactors and, in years to come, as a fuel in fusion reactors. Interpreted ionically, the percentage composition of sea water is shown in Table 3.2. Fresh water has a much more variable composition, but one dominated by carbonates (79.9 per cent) and sulphates (13.2 per cent), with chlorides contributing only 6.9 per cent.

Removing the dissolved salts from sea water leaves a highly concentrated brine. Those salts for which industrial markets can be found can be extracted and sold. Common salt, metallic magnesium, magnesium compounds, and bromine are obtained in this way. Indeed, nearly 30 per cent of the world supply of salt is obtained by evaporating sea water. In this process, calcium sulphate and calcium carbonate precipitate first; when they have been removed the brine is moved to another pond, where salt crystallizes. The remaining brine, called 'bitterns', is removed, fresh, concentrated brine is added, and this is repeated until the layer of crystalline salt is thick enough to be harvested. Bromine can then be extracted from the bitterns. Where no market for by-products can be found, however, disposal of the brine is difficult, and for every 30 000 litres of fresh water produced by desalination, 1 tonne of salts remains.

Water may be separated from its dissolved salts by distillation, freezing, electrolysis, or reverse osmosis. Distillation is the most widely used method. In low latitudes, the Sun may supply enough energy to evaporate sea water. The evaporate is then condensed and after several cycles of evaporation and condensation the water is sufficiently pure to be fed into the public supply. More usually, however, energy must be provided. Several distillation methods are used. Figure 3.6 illustrates multistage flash evaporation, which is one of the most efficient. Incoming sea water is heated under pressure, to prevent it from boiling, then released into a chamber where pressure is lower. It boils instantly ('flash boiling') and the vapour rises, to condense on the pipe carrying cold, incoming sea water. The latent heat of condensation warms the incoming water, reducing the amount of heating required. The condensate is collected and removed and the remaining brine fed to the next chamber where the process is repeated.

Ice contains little salt and so freezing sea water purifies it. In this technique, the sea water is chilled almost to its freezing temperature, then either sprayed into a partly evacuated chamber or mixed with a volatile hydrocarbon, such as butane, and poured into a chamber. The low pressure, or high volatility of the hydrocarbon, causes immediate evaporation of the hydrocarbon or some of the water and the chilling caused by the latent heat of evaporation causes some of the remaining water to freeze. The slurry of ice and brine is then pumped into another chamber, fresh water is added to separate ice from brine, and the fresh water is removed.

Table 3.1 **Composition of sea water**

Constituent	%
Sodium chloride	77.8
Magnesium chloride	9.7
Magnesium sulphate	5.7
Calcium sulphate	3.7
Potassium chloride	1.7
Calcium carbonate	0.3
Other	1.1

Table 3.2 **Ions in sea water**

ions	%
Chloride	55.04
Sodium	30.61
Sulphate	7.68
Magnesium	3.69
Calcium	1.16
Potassium	1.10
Bicarbonate	0.41
Bromide	0.19
Other	0.12

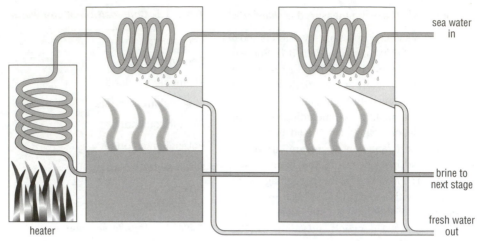

Figure 3.6 **Multistage flash evaporation**

Osmosis

If two solutions of different strengths are separated by a membrane that allows molecules of the solvent to pass, but not those of the solute (the dissolved substance), solvent molecules will cross the membrane from the weaker to the stronger solution until the two are of equal strength. The membrane separating them is called 'differentially permeable' if it allows water molecules to pass but slows the passage of larger molecules or prevents some of them, or 'semi-permeable' if it is completely permeable to molecules of solvent and completely impermeable to those of the solute. Cell membranes are differentially permeable. Membranes that allow the passage of some but not all molecules are now often described as 'partially permeable'.

The passage of water through a membrane requires energy. Pure water is considered to possess zero energy and a solution to have a negative energy value. Osmosis occurs when there is an energy difference between two solutions and the energy involved, known as the 'osmotic pressure' or 'water potential', can be measured.

In reverse osmosis sufficient pressure is applied to a solution to overcome the water potential and force water molecules to cross a semi-permeable membrane from the higher to lower concentration. The pressure required is about 25×10^5 Pa (25 times ordinary sea-level atmospheric pressure).

Electrolytic desalination involves pumping sea water into a chamber containing electrodes. Some ions are attracted to the positive electrode (anode), others to the negative electrode (cathode) and partly purified water is extracted from the middle.

As its name suggests, reverse osmosis is based on a natural process. A partially permeable membrane separates fresh from sea water and the pressure of the sea water is increased. The high pressure required makes reverse osmosis difficult to apply on a large scale, but advances made in recent years have reduced the energy needed to below that required for distillation and the technique is becoming commercially attractive.

In years to come, rising demand for fresh water will lead to greater reliance on desalination. At present, the high energy requirement makes all industrial-scale desalination technologies too expensive for many of the less developed countries, where the increased demand will be felt most acutely, but this situation could change. More efficient techniques for exploiting solar energy might reduce costs in low latitudes, and in high latitudes waste heat from coastal industrial plants, especially nuclear power stations, might be used to the same end.

As the production of fresh water by desalination grows, however, so will the amount of highly concentrated brine for which no economic use can be found. It would be as well to develop satisfactory means for its disposal before proceeding rapidly along this path.

25 Irrigation, waterlogging, and salinization

Deprived of water, before long any plant other than a cactus or other succulent will begin to look very sick indeed. Its leaves will become flaccid and if it lacks a woody stem the entire plant will grow limp and collapse. It will wilt. The condition may be temporary, the plant recovering when its access to water is restored, but if it continues for too long the wilt will be permanent and the plant will die.

Plants need water to give rigidity to their cells, but water stress also produces other, more subtle effects. The stressed plant will spend more time with its stomata closed. These are the pores, each opened and closed by the expansion and contraction of a pair of guard cells, through which gases are exchanged and from which water evaporates. Keeping stomata closed reduces water loss, but a reduction in the rate of gas exchange implies a reduction in the rate of photosynthesis. The plant will grow more slowly and will be smaller than it would otherwise be, and growth is inhibited before the plant is so short of water that it wilts visibly. When an adequate amount of water becomes available to a formerly stressed plant it will increase its production of foliage, but in the case of a crop plant its final weight will never be greater than that of an unstressed plant and usually it will be smaller.

Water shortage is an obvious problem facing farmers in semi-arid climates, or in climate types with pronounced wet and dry seasons, such as that of the Mediterranean. Less obviously, it can also reduce agricultural production where rainfall is distributed fairly evenly through the year. The monthly extent of water surplus or deficit can be calculated by comparing the amount of rainfall with the amount of water lost by evaporation and transpiration from grass supplied with abundant water. Such calculations show that in central England a water deficit may occur during the summer and autumn, from June to October, when evaporation exceeds precipitation (WINTER, 1974, p. 7). If water is provided in addition to that received as rainfall, field experiments at the National Vegetable Research Station in England have shown that crop yields increase dramatically: those of maincrop potatoes rose from 37 t ha^{-1} to 50 t ha^{-1}, an increase of 13 t ha^{-1}, and those of cabbage from 41 t ha^{-1} to 59 t ha^{-1}, an increase of 18 t ha^{-1}. For every 25 mm of irrigation per hectare, yields of main-crop potato increased by 3 t and those of cabbage by 18 t (WINTER, 1974, p. 117).

Irrigation is clearly beneficial, even in much of Britain, but this is hardly news. Farmers were irrigating their crops seven thousand years ago in Mesopotamia and irrigation techniques were developed independently in China, Mexico, and Peru. In some countries unirrigated agriculture would be impossible; all farm land is irrigated in Egypt, for example. In the world as a whole, about 15 per cent of all farmland is irrigated, ranging from 6 per cent in Africa and South America to 31 per cent in Asia. Between 1970 and 1990 this area increased by more than a third, from 168 million ha to 228 million ha, most of the increase being in developing countries, and the output from irrigated land is more than double that from unirrigated land; one-third of the world's food is grown on irrigated land (TOLBA AND EL-KHOLY, 1992, p.290).

Water for irrigation is often provided by damming rivers to fill reservoirs, the flow of water from the dams also generating electrical power, but large dams can produce adverse environmental effects. Their reservoirs flood large areas, destroying existing plant and animal communities and often displacing many people, and silt carried from upstream tends to accumulate, gradually filling the reservoir. Where rivers formerly flooded land downstream at a certain time of the year, the silt deposit containing plant nutrients is lost to farmers, who must buy fertilizer to replace it. In seismically active regions, large dams may also be linked to increases in the number of earthquakes. An earthquake exceeding magnitude 5 on the Richter scale occurred while the first large dam in the world, the Hoover Dam on the Colorado River, was being filled in 1936 and there was another of comparable magnitude in 1939. There have also been earthquakes greater than magnitude 5 associated with the Koyna Dam, India (1967), Kremasta Dam, Greece (1966), Hsinfengkiang Dam, China (1962), and Marathon Dam, Greece (1938), each of them accompanied by foreshocks and aftershocks (GOUDIE, 1986, pp. 243–244).

Land can be irrigated simply by flooding it and allowing the water to sink into the ground. A somewhat more sophisticated method is to dig parallel furrows down the slope of a field and fill them with water from a ditch or pipe across the upper edge of the field. A more familiar technique involves the use of sprinklers. These are versatile, in that they can be moved to where they are most needed and the amount of water they deliver can be controlled closely. In some places, irrigation is supplied by subsurface pipes.

Environmentalists used to be fond of saying 'everything has to go somewhere'. This is as true of water as of anything else and water supply is only one side of the water management equation: water must also be removed. In some places, wet ground can be rendered cultivable only by making it drier; in others, irrigation must be accompanied by improved drainage.

Land drainage is a farming practice probably as ancient as irrigation. On sloping ground, a ditch along the upper boundary of a field, at right angles to the direction of slope, will collect water draining from higher land before it flows into the field. A network of communicating ditches can then carry the surplus water to the nearest stream.

On level ground, or where the construction of ditches is insufficient, drains may be laid below ground. The simplest technique is to install 'mole' drains, so called because the implement that makes them tunnels through the soil like a mole. The 'mole' itself is a metal cylinder fixed to the lower end of a bar, buried to the desired depth, and then towed through the soil. Figure 3.7 illustrates the device and shows that it makes a hole parallel to the surface. In most soils the hole will remain open for some years before the operation needs repeating. More permanent drains are made from short lengths of perforated piping laid end to end by a machine that digs the trench into which it lays them, then buries them as it passes. In both cases the drains feed into a stream or system of ditches. The land area drained is proportional to the depth of the drain, so it is a simple matter to plan a drainage system that will serve a whole field without leaving wet patches.

It is easy to see why farmers find it desirable to remove surplus water from wet ground. The need for a drainage system to accompany irrigation is less self-evident, but lack of drainage on irrigated land is a major cause of soil degradation.

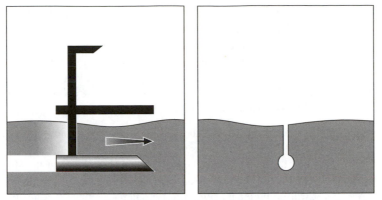

Figure 3.7 **Mole drainage. Left, mole plough; right, cross-sectional view of mole drain**

If more water is abstracted from an aquifer than flows in to replenish it, the amount of water available will gradually diminish. In coastal regions, such over-exploitation of the resource brings an additional hazard. Beneath the sea bed, the sediment is permanently saturated with salt water. The salt water moves inland, beneath the freshwater aquifer, with a boundary of brackish water separating the two water bodies. As the freshwater aquifer is depleted, this boundary moves further inland and closer to the surface, allowing salt water to penetrate the soil. As Figure 3.8 shows, a point can be reached at which water abstracted for irrigation starts to become brackish and the more that is abstracted the saltier it is. Since most crop plants are very intolerant of salt, the effect can be to sterilize the affected land. It is a problem in many coastal areas, but especially serious in low-lying islands, such as coral atolls (TOLBA AND EL-KHOLY, 1992, p. 117).

This form of contamination is known as 'salinization' (or in the USA as 'salination'). Salt water intrusion can occur only in coastal areas, but salinization quite unconnected to the proximity of sea water affects regions far inland. According to the UN Environmental Programme (UNEP), 7 million ha is affected in China, 20 million ha in India, 3.2 million ha in Pakistan and the Near East, and 5.2 million ha in the United States. Parts of southern Europe also suffer from this problem (TOLBA AND EL-KHOLY, 1992, p. 290). It arises because of the way water moves through soil.

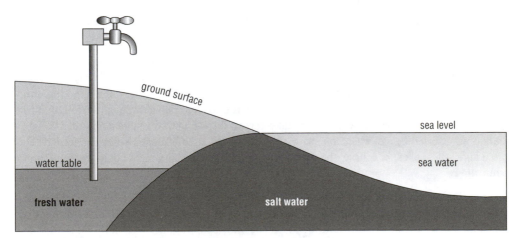

Figure 3.8 **Salt water intrusion into a freshwater aquifer**

Some of the rain falling on the ground sinks vertically through the soil, as 'gravitational water', until it reaches the ground water, the region where the soil is saturated, the upper boundary of the saturated region being the water table. Above the water table, particles comprising the unsaturated soil are coated by a very thin film of 'adhesion water' held by attraction between water molecules and the electrically charged surfaces of soil particles. Even the driest dust is usually coated with adhesion water. This film is covered by an outer film of 'cohesion water', held by the attraction of hydrogen bonds between water molecules themselves.

Water molecules at the bottom of a pot of water, or adjacent to the impermeable material underlying the ground water, are subject to a pressure equal to the weight of water above them. The higher they are in the pot, the less pressure bears down on them, until, at the surface, the pressure is zero. Any water in tiny but connected spaces above the surface will be under even less (i.e. negative) pressure: it will be under tension, a force pulling it upward rather than downward. Molecules will be easily attracted by the adhesive charge on soil particles and further molecules will join them because of the cohesive attraction of the molecules already in place. This is capillary attraction. It has little effect on adhesion water, which moves very little, but cohesion water is less tightly bound and can move. Under soil moisture tension, it moves to coat dry soil particles (becoming adhesion water) and to equalize the thickness of the layer of cohesion water throughout the soil. Very slowly, the water will rise through the unsaturated layer, and a very small suction by a plant root hair will be sufficient to dislodge cohesion water and move it into the plant (FOTH AND TURK, 1972, pp. 64–74).

Water rises through both plants and soil, evaporates, and is replaced by more water rising through the capillary pore spaces in the soil. Water vapour is almost pure H_2O, and any substances dissolved in the liquid are precipitated as it evaporates. Soil water is far from pure. Salts dissolve into it as it moves through the soil and some soils contain quite large amounts of soluble salts. Irrigation water itself is seldom pure; farmers do not irrigate their land with water fit for human consumption. The water they use commonly contains between 750 g m^{-3} and 1.5 kg m^{-3} of dissolved salts (FOTH AND TURK, 1972, p. 407). These may be left as evaporates near the soil surface, deposited from water that evaporated before soaking into the soil or from water that descended gravitationally and then rose again by capillary attraction. Gradually, the salinity of the upper soil increases until plants begin to suffer, the most salt-intolerant first, but eventually most crop species.

Salinization most commonly occurs in arid or semi-arid climates, where the rate of evaporation is high, but it is under these conditions that irrigation is most urgently needed and where it may bring its greatest benefits. The risk may be avoided by installing adequate drainage to remove surplus water before it can evaporate and by controlling the dissolved-salt content of irrigation water, especially on saline soils.

Should it occur, the remedy is slow, difficult, and expensive. The first area ever to have been irrigated, in the Tigris and Euphrates valleys, suffered from salinization and to this day much of it remains barren because its reclamation would be too costly. Fresh water, containing little or no dissolved salts, must be used to flush the salts from the soil and into a drainage system that will remove them, and it may be necessary to take care in disposing of the salt-laden water. If salinization was caused by salt-water intrusion, the freshwater aquifer must also be recharged. The old adage still applies, of course: the water used to clean saline land must come from and go to somewhere.

Over-zealous irrigation on poorly drained land can lead to a quite different problem. If more water is added to the soil than can evaporate or be transpired by plants, the water table will rise. It may do so for some time before the consequences become apparent, but eventually soil around the roots of crop plants will be saturated and, being saturated, airless. No water may be visible lying on the surface, but nevertheless the land is waterlogged and crop yields will fall dramatically. In this case the remedy is simpler. Adequate drainage must be installed and irrigation suspended until the water table has been lowered.

As the demand for food intensifies it is likely that the total area of irrigated land will increase. Some say it may double between about 1990 and the early part of the next century (PIEL, 1992, p. 216). In Asia, where such an increase is likely to be concentrated, this will allow two or even three crops to be grown each year on land that presently produces only one. The advantages will prove enduring, however, only if irrigation schemes are planned with care to avoid the hazards attendant on them.

26 Soil formation, ageing, and taxonomy

From the moment it is exposed at the surface, rock is subjected to persistent physical attack. Water fills small fissures and when it freezes it expands, exerting a pressure of up to 146 kg cm^{-2}, which is sufficient to split the toughest rock (DONAHUE ET AL., 1958, p. 28). In summer, the rock warms during the day and cools again at night, expanding as it warms and contracting as it cools, but it is heated unevenly. The surface is heated more strongly than rock beneath the surface; some parts of the surface are exposed directly to sunlight, others are in shade. As a consequence, some parts of the rock expand and contract more than others. This, too, causes the rock to break. Often flakes are loosened or detached from the surface, a process called 'exfoliation'. Detached particles then grind against one another as they are moved by gravity, wind, or water. This breaks them into still smaller pieces.

The smaller any physical object, the greater its surface area in relation to its volume: a sphere with a diameter of 4 units has a surface area of 50 units^2 and volume of 33.5 units^3, giving an area:volume ratio of 1 : 0.7; if the diameter is 2, the surface area is 12.5 units^2, volume 4.2 units^3 and ratio 1 : 0.3. As the rock particles grow smaller, therefore, the total surface area exposed to attack increases. Still vulnerable to abrasion, they are now subject to chemical attack.

This takes several forms. Some of the chemical compounds of which they are composed may be soluble in water; wetting dissolves and drainage removes them. Other compounds may react chemically with water. The process is called 'hydrolysis' and can convert insoluble compounds to more soluble ones. Orthoclase feldspar ($KAlSi_3O_8$), for example, a common constituent of igneous rocks, hydrolyses to a partly soluble clay ($HAlSi_3O_8$) and very soluble potassium hydroxide (KOH) by the reaction:

$$KAlSi_3O_8 + H_2O \rightarrow HAlSi_3O_8 + KOH$$

Hydration is the process in which compounds combine with water, but do not react chemically with it. The addition of water to a compound's molecules makes them bigger and softer and so increases their vulnerability to breakage. Oxidation also increases the size and softness of many mineral molecules and may also alter their electrical charge in ways that make them react more readily with water or weak acids. Reduction, which occurs where oxygen is in short supply, also alters the electrical charge on molecules and may reduce their size.

Compounds may also react with carbonic acid (H_2CO_3), formed when carbon dioxide dissolves in water. This reaction, called 'carbonation', forms soluble bicarbonates. Barely soluble calcium carbonate ($CaCO_3$), for example, becomes highly soluble calcium bicarbonate ($Ca(HCO_3)_2$).

Physical and chemical processes thus combine to alter radically the structure and chemical composition of surface rock. How long it takes for solid rock to be converted into a layer of small mineral particles depends on the character of the original rock and the extent of its exposure; in arid climates it proceeds more slowly than in moist ones, for example. Yet the process is remorseless. At widely varying speeds it dismantles mountains.

It does not proceed far before living organisms accelerate it: respiration and the decomposition of plant remains are the main source of the carbon dioxide engaged in subsurface carbonation. The chemical changes release compounds useful to organisms in soluble forms they can absorb, and their metabolic wastes and dead cells add to the stock of reactive compounds as well as providing sustenance to still more organisms. Bacteria are usually the first to arrive, forming colonies in sheltered cracks, invisible to the naked eye. Lichens often follow, composite organisms comprising a fungus and alga or cyanobacterium. The fungus obtains water and mineral nutrients from the rock, the alga or cyanobacterium supplies carbohydrates that it photosynthesizes and oxygen as a by-product of photosynthesis. Each partner supplies the other and the fungus protects them both from drying out and provides firm attachment to the rock, which it grips tightly with filaments that grow into the tiniest crevices. This remarkable partnership allows lichens to flourish where no plant could survive.

Organic material, derived from wastes and the decay of dead cells, accumulates beneath the lichen, mixing with the mineral particles and accelerating chemical reactions. This mixture is better at absorbing and holding water, and in time there is enough of it to provide anchorage and nutrients for plants. Mosses may arrive and small herbs may root themselves in the deeper cracks.

As the layer of mixed organic and mineral material thickens, some of it begins to be washed to deeper levels, a few centimetres below the surface. The material is starting to form two distinct layers: an upper layer from which soluble compounds and particles are being washed (the technical term is 'leached') and a lower layer in which they are accumulating. This is the first stage in the formation of soil.

From this point, vegetation becomes part of the developing soil and contributes greatly to its formation. Plant roots penetrate the material and when they decay leave channels that assist aeration and drainage. Dead plant material contributes fresh organic matter to the surface, which decays to release compounds that drain into the soil. In detail, however, this process can vary widely over a small area, in large part because of the efficiency with which the soil drains and the depth of the water table below the surface. If the soil is derived from similar mineral particles all the way down a slope, a hydrologic sequence may occur (CRUICKSHANK, 1972, pp. 47–48), illustrated in

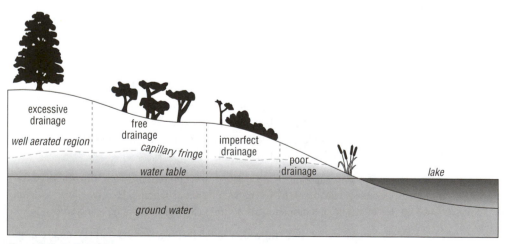

Figure 3.9 **Soil drainage**

Figure 3.9. Where drainage is excessive, the soil will be generally dry, favouring trees that can root to considerable depth. As the distance narrows between the surface and water table, the region hospitable to plant roots becomes shallower, and the plants smaller. Most of the organisms engaged in the decomposition of plant material require oxygen for respiration, so the decreasing depth of the aerated zone is accompanied by a slowing of the rate of decomposition until, where the soil is waterlogged, partly decomposed material may form acid peat.

The stage of this process that involves purely physical and chemical mechanisms constitutes weathering; as living organisms become the predominant agents it is called 'pedogenesis'. Plants growing at the surface penetrate the soil with their roots and supply a topmost layer of dead organic material, called 'litter'. This provides sustenance for a diverse population of animals, complete with their predators and parasites, fungi, and bacteria. These break down the material, which enters the soil proper, much of it carried below by earthworms, where it feeds another population. Compounds released by decomposition dissolve in water draining through the soil and are carried to a lower level, where they accumulate. At the base of this layer, the 'subsoil', rocks and mineral particles, detached from the underlying rock, are being weathered, and below this layer lies the bedrock itself.

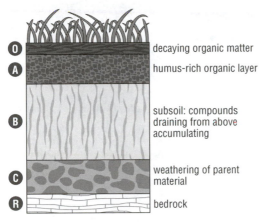

decaying organic matter

humus-rich organic layer

subsoil: compounds draining from above accumulating

weathering of parent material

bedrock

Figure 3.10 **Profile of a typical fertile soil**

If a vertical section, called a 'profile', is cut through the soil from surface to bedrock, it may reveal this structure as layers, called 'horizons', clearly differentiated by their colour and texture. Figure 3.10 shows the principal horizons, but in any particular soil there may be more or fewer and in some soils horizons are not easily distinguished at all. Conventionally, the horizons are identified by letters: O for the surface layer of organic matter; A for the surface horizons; B for the accumulation layer; C for the weathering layer; and R for the bedrock. The horizons are classified further by the addition of numbers: A2 is a mineral horizon somewhat darkened by the presence of organic matter; A3 is a transition zone between the A and B horizons. Letters are then added in subscript to denote particular characteristics: ca means the soil contains calcium and magnesium carbonates; g (for 'gleying') means the soil is poorly aerated and frequently waterlogged; m means the soil is strongly cemented together, like a soft rock.

Soils vary according to the rock from which they are derived. This affects the size of their mineral grains, ranging from coarse sand (600–2000 µm) to silt (2–60 µm) and clay (less than 2 µm), and their chemical characteristics. Soils derived from granites, for example, develop slowly, are usually sandy, and contain relatively few plant nutrients; those developed from limestones are usually fine-grained and relatively rich in plant nutrients.

Once formed, soils begin to age. The rate at which they do so depends principally on the climate and vegetation. Desert soils age slowly, and so do those in polar regions, but in the humid tropics soils age much more quickly as luxuriant plant growth extracts nutrients and returns them for decomposition into soluble forms, which are leached rapidly by the abundant water. It is possible, therefore, to describe as 'young', 'mature', or 'ancient' soils that may have been in existence for the same length of time.

We obtain our food from soil, we erect buildings of varying weight upon it, and we use clay taken from it as construction material that may or may not be fired to make bricks. Clearly it is of great importance to us and if we are to use it the more we know about it the better. It is so variable that we cannot be satisfied in calling it simply 'the' soil. It must be classified.

There have been many attempts at soil classification, the first in classical times, but it was not until the latter part of the last century that a school of Russian scientists at St Petersburg, led by Vasily Vasilievich Dokuchaev (1840–1903), proposed a theory of pedogenesis on which a formal classification could be based. It is because of this Russian origin that many soil types have Russian names, such as 'podzol' and 'chernozem'. The Russian work laid the foundation for what is now known as 'soil taxonomy', but work has continued ever since.

The system most widely used at present was devised by the US Department of Agriculture. This divides all soils into 11 orders (*www.explorer.it/aip/keytax/content.html*). The orders are divided further into sub-orders, great groups, families, and soil series. The orders, with brief descriptions, are listed in the box.

The 11 soil orders of the US soil taxonomy

Alfisols Soils of climates with 510–1270 mm annual rainfall; most develop under forests; clay accumulates in the B horizon.

Andisols Volcanic soils, deep and light in texture; contain iron and aluminium compounds. (This order is sometimes omitted.)

Aridisols Desert soils with accumulations of lime or gypsum; often with salt layers; little organic matter.

Entisols Little or no horizon development; often found in recent flood plains, under recent volcanic ash, as wind-blown sand.

Histosols Organic soils; found in bogs and swamps.

Inceptisols Young soils; horizons starting to develop; often wet conditions.

Mollisols Very dark soils; upper layers rich in organic matter; form mainly under grassland.

Oxisols Deeply weathered soils; acid; low fertility; contain clays of iron and aluminium oxides.

Spodosols Sandy soils found in forests, mainly coniferous; organic matter, iron and aluminium oxides accumulated in B horizon; strongly acid.

Ultisols Deeply weathered tropical and subtropical soils; strongly acid; clay accumulated in B horizon.

Vertisols Clay soils that swell when wet; develop in climates with pronounced wet and dry seasons; deep cracks appear when dry.

These names and descriptions are reasonably straightforward, but the system becomes much more abstruse below the level of orders. Suborders have such names as 'Psamments', 'Boralfs', and 'Usterts'; the great groups include 'Haplargids', 'Haplorthods', and 'Pellusterts'; and among the subgroups are 'Aquic Paleudults', 'Typic Medisaprists', and 'Typic Torrox'.

The classification may be powerful, but there are attractions in calling Mollisols 'prairie soils' (or chernozems), Histosols 'peat' or 'muck', and, given the widespread environmentalist concern over the degradation of some tropical soils, calling Oxisols 'lateritic soils', which are the names by which they used to be known. There is also a Canadian classification system that divides soils into two orders, Brunisolic comprising 4 Great Groups and Chernozemic with 3 Great Groups and a total of 42 Subgroups (*burgundy.uwaterloo.ca/bio1446/chap2bsm.htm*).

27 Transport by water and wind

Once it has developed, soil does not necessarily remain in the place where it formed. It can be transported, sometimes over very long distances, by wind and water, and over short distances by gravity.

Occasionally, rain in northern Europe leaves everything exposed to it coated by a thin layer of reddish dust. It is Saharan dust, lifted from the desert, carried some 2500 km by air movements, then washed to the ground. Some of that dust remains to form a very minor constituent of European soil. Fine soil particles lofted by wind in the North American Great Plains during the dust-bowl years of the 1930s fell as dust in New York and discoloured the Atlantic for hundreds of kilometres from the American coast.

The Dust Bowl

An area of about 390 000 km^2 in south-western Kansas, south-eastern Colorado, north-eastern New Mexico, and parts of Oklahoma and Texas that was originally prairie. The climate is semi-arid and prolonged droughts are common. After 1918, US grain prices rose steeply, encouraging farmers to plough the prairie, and for several years they produced satisfactory yields. A drought began in 1933 and lasted until 1939, being especially severe in 1934 and 1935. It was in those two years that topsoil, exposed by ploughing and reduced by aridity to a fine dust, blew away, the lighter particles forming clouds 8 km high. Farmers, most of whom were already poor, were ruined and thousands of them migrated to seek other work.

In 1935 the US Department of Agriculture founded the Soil Conservation Service to promote sound soil conservation practices throughout the country, and with the return of the rains some farming resumed, although much of the area was returned to grass. Drought has continued to afflict the region at intervals of about 20–22 years.

Where aeolian (wind-blown) deposits accumulate they make a soil known as 'loess'. Obviously, the smaller the particles the further the air will carry them, and so the material is graded, becoming finer with increasing distance from its source. Loess soils are extensive, covering much of the central United States, where in places they are several metres thick (HUNT, 1972, p. 138). There are also extensive deposits in Argentina, various parts of Europe, and in China; in the northern and eastern highlands they are believed to be 300 m thick (DONAHUE ET AL., 1958, pp. 24–25), with the underlying rock projecting through it locally as hill and mountain ranges.

Although deserts are a source of wind-blown dust, modern dust storms make only a minor contribution to such deposits; most are ancient, dating from past ice ages. When glaciers thawed, they released meltwater that flowed as rivers and flooded low-lying areas. The waters carried suspended particles, which settled as mud. Then, as the temperature fell, the meltwater flow ceased, the flooded areas dried, and wind blew the dust far away from the valley bottoms. The process was repeated at intervals, some of them long enough for new soil to begin to form above the loess before being buried beneath a later deposit. Loess soils are usually yellowish in colour and when young they are rich in mineral plant nutrients and calcium. This makes them inherently fertile, although their later history may exhaust that fertility. They are fine-grained, and rivers cutting through loess commonly have very steep, almost vertical, banks. Much of south-east England was once blanketed by such loessic soils.

As you will know if you have walked over a dry, sandy beach in a strong wind, sand grains also blow. Being much larger than the silt-sized particles that comprise loess, they are not carried far, but repeated lifting and dropping by a prevailing wind can transport them a considerable distance. Where they accumulate, dunes form, sometimes with a characteristic shape from which the wind direction can be determined. Crescent-shaped (called 'barchan') dunes, for example, are aligned across the direction of the prevailing wind, the convex side facing the wind behind a long 'tail' of sand being blown little by little up the tail to the top, where the dune collapses on the sheltered side to produce the face; strong, steady winds erode valley-like troughs and linear dunes; dunes forming straight ridges may be parallel (seif dunes) or at right angles to the wind direction (aklé dunes); and changing winds produce star dunes, of radiating ridges.

Blown sand is unsuitable for cultivation, but it can be stabilized if hardy plants become established on it. In temperate climates marram grass (*Ammophila arenaria*) is often used for initial colonization in coastal areas. Its underground stems (rhizomes) form networks that help hold the sand in place and allow other plants to obtain purchase. As a diverse plant community develops the marram grass is unable to compete and disappears. Then soil will form above the sand, burying it. Unstabilized sand dunes migrate slowly and can bury fertile land downwind.

Water is a much more powerful agent of transport than wind. Gravel and small stones cannot be carried far by even the strongest wind, but a river can move them long distances. A soil that is mineralogically unrelated to the bedrock beneath it and that contains sand mixed with stones of various sizes has been placed in its present position by water. It may be a marine deposit, formed on the bed of a sea that has long since disappeared. If the particles are sorted by size into layers, the deposit is more likely to be lacustrine, marking the location of a former lake. Many lacustrine deposits contain a high proportion of clay particles; if the material is more than 50 per cent clay it will be almost impermeable to water and thus prone to waterlogging or flooding.

Glaciers also transport material, but they seldom carry it very far. Their action is mainly to mix the soil already formed beneath them and to transport large pieces of rock that become frozen into the ice. Then, when the glacier retreats by melting at its lower end, these stones join the mixed soil to make 'till' (which was formerly known as 'boulder clay'). Till deposits cover substantial areas in Europe and North America. Although glaciers rarely carried the till more than 10 km, large stones entrained in the ice were sometimes carried much further and deposited as 'erratics'. Their orientation and that of stones in the till itself can be used to determine the direction in which the ice was moving. Because glaciers filled or made broad, flat-bottomed valleys, glacial till often occurs in gently rolling 'till plains'. Material pushed to the sides of a glacier and ahead of its upward-curved front, or 'snout', was left as a moraine, now visible as ridges or hills that are often too rocky to be easily cultivable.

Permafrost occurs in the vicinity of glaciers and ice sheets. This is ground where the temperature below the surface remains below freezing throughout the year. In summer the surface layers may thaw, and if the soil is on a slope greater than about 2°, the resulting mud will flow, carrying with it stones released by frost-shattering from adjacent rock. These stones, most of very irregular shape, will tend to become aligned with their long axes indicating the direction of slope and parallel to the surface. At the base of the slope the flowing material will collect as 'head', often forming a thick deposit. The sliding of material down a slope over ground that is frozen a little way below the surface is called 'gelifluction' or 'congelifluction'. It is the cold-climate variety of solifluction, the downslope creep of material lubricated by water.

Glacial meltwaters often flowed with great force. They carried huge volumes of water, sometimes through confined spaces so they were under pressure, and they carried particles of all sizes, including quite large stones. As the flow slowed, the heaviest particles were deposited first and such 'outwash' material is usually coarse-grained and sorted by grain size into beds. The finer-grained material travelled further and settled as mud on land that dried as the water flow ceased or on the bottom of glacial lakes, many of which have since disappeared. Glacial lacustrine deposits consist mainly of clay of little agricultural value, but in some places used for brick-making.

Soils derived from river-borne sediment are called 'alluvial' and occur on land that is, or was, repeatedly flooded. Many rivers burst their banks occasionally, but for the flooding to affect soil formation they must do so often, and this is most likely if their waters are periodically augmented by drainage following very heavy rain or seasonally by the melting of deep snow. Charged with a greater volume of water to transport, they flow faster and this gives them the energy to transport more material. When the river overflows its banks, the water escapes to the sides and the pressure on it is greatly reduced. It loses energy and deposits its load, the heaviest particles first. These may collect close to the point where the overflow occurred, after many floodings forming a distinctive raised bank, called a 'levee'. This may become covered with soil, and small particles will be trapped within it, but basically it is made from gravel and larger stones, and water drains through it freely. Silt and clay are precipitated beyond the levee. They pack tightly together and drain poorly, but are rich in plant nutrients and the flood plain they produce as they fill natural depressions and make a level surface is usually very fertile.

Downstream, where the river, fed by many tributaries, flows as a wide stream across land with a very low gradient, it may form meanders. A system of meanders may also create a flood plain, but by a quite different mechanism that does not require the river to burst its banks. On the outside of each meander bend, the stream flows against the bank. This increases its turbulence and also its speed, since it has slightly further to travel, and material is drawn into the water from the bank, eroding it. Some of this material may flow across the stream, near the bed, where it enters water flowing against the inside bank. Here the flow is slower and calmer. The river loses energy and deposits some of its load, extending the bank into the channel. Figure 3.11 illustrates the process.

At the same time, the entire meander tends to migrate in a downstream direction and since this movement affects all the meanders, the entire system migrates downstream. The land behind a migrating meander is covered by river-bed material, so an alluvial plain forms that is the same width as the widest meander. Meanders migrate slowly, so the fertile flood plain can be used, although the ground may be wet for most of the time.

Till and alluvium, as well as loess, are commonly buried beneath soils developed since they were deposited, so they may not be visible at the surface. They will be detectable, however, because they form the parent material for the soil covering them. This soil may not resemble its parent, but will betray its origin by being unrelated to the underlying bedrock.

A section through the soil will reveal its character, of course, and a river may cut a suitable section. As Figure 3.12 shows, a stream channel quickly penetrates the surface material to expose the

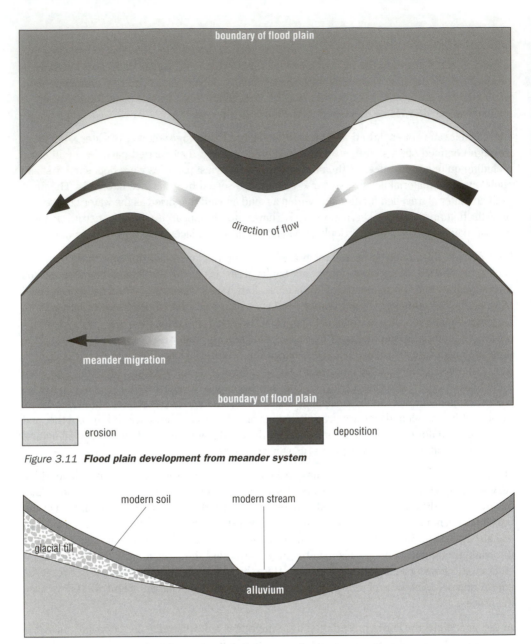

boundary of flood plain

direction of flow

meander migration

boundary of flood plain

erosion

deposition

Figure 3.11 **Flood plain development from meander system**

modern soil

modern stream

glacial till

alluvium

Figure 3.12 **Modern soil developed over flood plain alluvium and glacial till**

underlying deposit, in this case suggesting that a much larger river once flowed, probably fed by glacial meltwater which also left a layer of till.

Soil is formed by processes at or very close to the surface. Once formed, it is affected by other processes, some of which tend to transport it to new locations. The consequences of these processes may be serious, catastrophic even, for humans living where they occur. In the central plains of North America farmers from the east cleared away the natural prairie grassland and ploughed the land to grow wheat in a climate drier and more prone to drought than that to which they were accustomed. Following a series of dry years, in 1934 and 1935 the crops failed and soil reduced to little more than powder was carried away by the wind, creating the Dust Bowl (ALLABY, 1998, pp. 76–82). This brought tragedy to countless American farming families in the 1930s, most of whom were already impoverished as a result of the economic depression. After that experience it was recognized that the land was unsuitable for arable farming and much of it has been returned to grassland. To take another example, unexpected floods destroy crops and livelihoods, and may cause the deaths of livestock and people. Yet the disasters may also bring benefits to those farming the fertile loess or alluvial soils.

Such events occur naturally. Clearly, the soils of the Dust Bowl were unsuitable for the type of farming practised on them. By removing the natural vegetation cover and cultivating the soil, farmers reduced the soil to a condition in which it would blow away, but the drought was a natural occurrence. Human activities can interfere more directly. When a river is dammed, for example, silt will accumulate behind the dam. This progressively reduces the volume of water held in the artificial lake, but it also interrupts the natural sedimentation process further downstream. On the flood plain or delta, farmers may depend on the seasonal floods for the silt they bring, rich in the plant nutrients that have drained into it the entire length of the river's course. Deprived of their 'natural fertilizer', farmers may be forced to buy factory-made fertilizer, which in many cases they can ill afford, and the farming techniques they have developed may be inappropriate for alluvial soils that are not regularly replenished. Soil structure and fertility may deteriorate. Similarly, the clearing of vegetation from upper slopes may increase the transport of sediment to lower levels, polluting rivers.

28 Soil, climate, and land use

Climate is by far the most important factor controlling the development of soils (pedogenesis) from their parent materials, but it is not the only one. Pedogenesis is also affected by the type of vegetation (which is also climatically governed), the activities of animals including humans, the parent material, and topography. Most of the chemical reactions by which the mineral constituents of rock are modified do not commence until the temperature rises above $10\,^{\circ}C$, and their reaction rates double for every further $10\,^{\circ}C$ rise above that. It follows that, provided water is available, soils develop and age faster in warm than in cold climates and it is in warm climates that biological organisms are more active.

Some of the early schemes for classifying soils were based on climate, grouping soils into cold, cool–temperate, subtropical, and tropical zones, within which many of the names were descriptive. In the cold zones, for example, are tundra and mountain meadow soils, and in the cool–temperate zones prairie soils (CRUICKSHANK, 1972, pp. 155–166). This led to the concept of 'zonal soils', typical of the zones in which they occur. The zones might also contain soils formed under the influence of some local factor and therefore atypical. These were called 'intrazonal soils'. Soils that had not developed at all might be found in any climatic regime, and were called 'azonal'.

Such classifications were based on examinations of only the A and C soil horizons, the B horizon being considered merely transitional between those above and below it. In time, soil scientists came to realize that zonal schemes were classifying not so much the soils as the environments in which they form. Modern soil taxonomy is based on the soils themselves, defined in terms of more than 20 surface and subsurface 'diagnostic horizons', called 'epipedons' and 'endopedons' respectively. An anthropic epipedon, for example, is a surface horizon formed where people have lived for a long time or have grown irrigated crops; cultivation over many years might lead to the formation of an agric endopedon just beneath the ploughing depth, where clay and organic matter have collected.

Although soils are now classified according to their composition, their surface horizons are formed biologically, by the mixing of organic and mineral material. Since natural vegetation is usually typical of the climate in which it grows, it is impossible to dissociate pedogenesis and climate.

As Figure 3.13 shows, the relationship between plants and soil is intimate. Beneath a conifer forest, there is usually a deep layer (A0 horizon) of organic material, mainly needles. This decomposes only slowly, partly because conifer needles have a thick, waxy, outer layer that is not easily broken. This is an adaptation to climate: conifers grow in climates with a pronounced dry season or a long winter when water is frozen and so unavailable. The dry or cold season also reduces the rate of decomposition. The A1 horizon, dark and rich in humus (decayed organic material), is thin and the somewhat thicker A2 horizon is very pale, because its humus has leached into the B horizon. This soil is classifed in the order Spodosols.

Broad-leaved forests produce a much thinner A0 horizon, because the more delicate leaves, shed in the autumn, decompose fairly quickly during the mild, wet winter. The resulting humus forms a deep, dark, A1 horizon, a thinner, leached, A2 horizon, and a deep B horizon, where plant nutrients accumulate well within the reach of tree roots. This is an Alfisol.

Mollisols, found beneath temperate grassland, also have a thin A0 horizon, because grass produces a dense but shallow mat of roots. Organic matter decomposes rapidly. The humus-rich A1 horizon is deep and the leached A2 horizon correspondingly thin, with a deep B horizon, where nutrients accumulate.

It is the Aridisols that provide dramatic contrast. Developed under desert conditions, they support almost no vegetation and consequently have no surface litter at all. Because there is no A0 horizon, there can be no A1 horizon either, because no humus is being produced. Occasional rains produce weathering, and soluble compounds are leached into a deep B horizon, beneath which there is a further horizon where calcium carbonate accumulates.

The combination of climate and vegetation has further effects on soils. Spodosols are prone to 'podzolization'. The slow decay of organic matter releases acids that drain downward, removing carbonates as they do so, and making the whole of the A horizon acid. In extreme cases it may be so acid as to cause the leaching of clays and their accumulation as a hard, impermeable layer (a 'hardpan') in the B horizon. In permafrost regions, the winter freezing of the surface layer causes it to expand, compressing the soil beneath against the underlying permafrost. This inter-

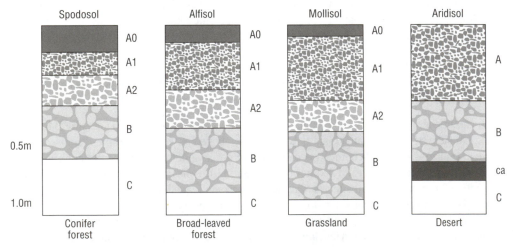

Figure 3.13 **Profiles of four soils, with the vegetation associated with them**

mediate layer, called 'gley', is wet, sticky, and blue because of the reduced iron compounds it contains, and compression forces it upward through cracks and mixes it with the overlying material in the process of 'gleying' (or in the United States 'gleyzation').

Oxisols, the predominant soils of the humid tropical lowlands, are the deepest of all soils. In many places the total soil depth, from surface to bedrock, may be 10 m and all the lower horizons are very thick. Scale apart, they are similar to Spodosols, except that their surface horizons may be badly eroded, so the B horizon is close to the surface, and they contain very little humus, because of the rapidity with which organic matter decomposes under the humid tropical climate, its nutrients being reabsorbed by plants. Almost all plant nutrients are contained within the living vegetation itself and soluble compounds have been leached from the soil, leaving it acid and inherently infertile. Clays, mainly comprising kaolinite (an aluminium–silica mineral, $Al_2Si_2O_5(OH)_4$) and ferric and aluminium oxides and hydroxides, may accumulate near the top of the B horizon, cemented together to form nodules or more extensive layers of 'laterite'. Laterite is extremely hard and impermeable and, in soils prone to it, laterization, the process by which it forms, is sometimes accelerated by clearing vegetation and leaving the ground exposed to heavy rain and thus to increased leaching.

Given the close association between climate and pedogenesis, it is not surprising to discover that the global distribution of soil orders broadly conforms to climatic zones. As Figure 3.14 illustrates, Oxisols are found in the humid tropics, Alfisols in temperate regions, Mollisols in the prairies, pampas, and steppes, and Spodosols in a belt around northern America and Eurasia.

Simply because an apparently deep, dark soil occurs in a climate favourable for agriculture, however, it does not necessarily follow that the land will sustain farming. Farmed soils are 'domesticated' by years of careful management and are markedly different from the 'virgin' soils that preceded them. Early farmers settled on the most promising land and when, after a few seasons, their crop yields began to decline, they moved elsewhere and started again. There is only a limited store of plant nutrients in any soil and it is depleted by the removal of crops, which reduces the amount available for recycling. Fertilizers and lime (to restore leached calcium) replenish the store, but if they are unobtainable (or unknown) farmers may have no alternative but to adopt some form of shifting cultivation. This still remains a common type of farming in many parts of the tropics.

Soil fertility is not determined by the amount of plant nutrients contained within reach of plant roots, because presence and proximity do not guarantee access. The roots must be able to absorb the nutrients they require and their ability to do so depends on the chemical characteristics of the soil.

Humus and silicate clays consist of masses of microscopic soil particles, each about 2 µm across. They are called 'colloids', because they change between a gel-like consistency and liquid according to the chemical environment around them. Soil colloids have negatively charged surface sites on to which cations (positive ions) can be adsorbed. Humus has many more of them than clay, which in turn has far more than sand. The commonest soil cations are calcium (Ca^{2+}), magnesium (Mg^{2+}), sodium (Na^+), potassium (K^+), and aluminium (Al^{3+}), but they are changing constantly as one replaces another, often in the order $Al \rightarrow Ca \rightarrow Mg \rightarrow K \rightarrow Na$. Fertilizers add cations, such as ammonium (NH_4^+), and lime adds calcium.

Anions (negatively charged ions) are also exchanged, but to a much lesser extent. Several important plant nutrients commonly occur as anions, including sulphate (SO_4^{2-}), nitrate (NO_3^-), phosphate ($H_2PO_4^-$ or HPO_4^{2-}), and molybdate (MoO_4^{2-}). These are not held at cation-exchange sites, but dissolve in soil water and are absorbed by plants directly from solution.

While the colloid remains saturated with exchangeable cations it retains its structure, but as these are replaced by hydrogen (H^+) the structure weakens. The soil becomes more acid (a measure

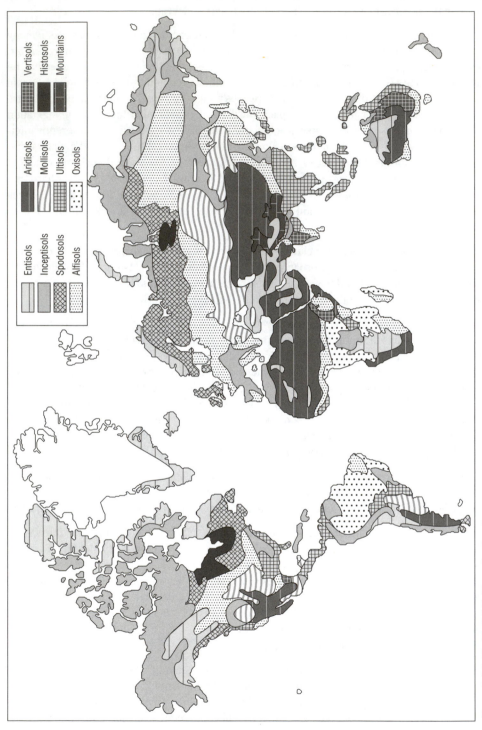

Figure 3.14 *World distribution of soil orders*

Legend: Entisols, Inceptisols, Spodosols, Alfisols, Aridisols, Mollisols, Ultisols, Oxisols, Vertisols, Histosols, Mountains

of the hydrogen ion concentration) and when it is nearly saturated with hydrogen the colloid breaks down into its constituent compounds, which move downward through the soil.

The amount of exchangeable cations in a unit weight of dry soil is known as the 'cation exchange capacity' (CEC) of that soil and it provides an important measure of soil fertility. It is measured in milliequivalents (me) per 100 g (a milliequivalent being that quantity chemically equal to 1 mg of hydrogen). Sandy soils typically have a CEC of 1–5 me $100 g^{-1}$, loams 5–15 me per 100 g, and clays more than 30 me per 100 g. The CEC for a typical Mollisol ranges from just over 24 me per 100 g at the surface to 25.7 me per 100 g in the lower part of the B horizon. The proportion of the cation-exchange sites occupied by calcium, magnesium, potassium, and sodium ions is known as the 'percentage base saturation'; the lower its value the more acid the soil. Most crop plants prefer a base saturation of 80 per cent or more, producing a neutral soil (pH 6.0 or higher), but some grow best in more acid soils, with a lower percentage base saturation (FOTH AND TURK, 1972, pp. 171–175).

Those cations that are plant nutrients are absorbed by roots while dissolved in soil water. Some are replaced from other exchange sites in the soil colloid, but these are generally insufficient to replace all of them, and cation-rich fertilizers must replenish the store, the frequency and amount of fertilizer application depending on the CEC of the soil. In other words, soil fertility can be expressed as CEC and sandy soils (low CEC) generally need more fertilizer more often than clay soils (high CEC).

Cation exchange also makes soils with a high CEC excellent purifying filters for water percolating through them. Positively charged pollutants, such as lead (Pb^{2+}) and cadmium (Cd^{2+}), are quickly adsorbed to exchange sites and thus immobilized, so water contaminated with them is purified by the time it reaches the ground water.

It is technologically possible to grow any plant anywhere in the world by supplying it with the environment it requires. Construct and heat a glasshouse, supply it with an appropriate soil, illuminate it artificially to achieve a suitable light intensity and length of day, and tropical crops could be raised in Greenland. It is possible, but hardly sensible when tropical crops can be produced so much more easily and cheaply in the tropics. Soils appropriate to particular plants are most likely to develop where the climate also suits those plants or their close relatives, but there are dangers. Soil fertility is usually reduced by cropping and must be replenished, and some soils, especially in the tropics, are much less fertile than the luxuriant vegetation they naturally sustain makes them appear.

29 Soil erosion and its control

Central Belgium has large areas of loess soils and it is estimated (KUPCHELLA AND HYLAND, 1986, p. 447) that each year these are losing between 10 and 25 tonnes of soil from each hectare by erosion. The United States is losing about 18 t ha^{-1} yr^{-1} and the Yellow River basin in China 100 t ha^{-1} yr^{-1}. In 1939, wind-blown soil from Texas fell in Iowa, 800 km away. It landed on top of snow, so the soil could be collected, weighed, and analysed. The amount deposited was 450 kg ha^{-1} and the deposit contained more than three times more organic matter and nitrogen than the soil from which it had been removed (HUDSON, 1971, p. 257). Clearly, soil erosion is a serious problem affecting cultivated land in almost every part of the world. As was mentioned earlier (Section 2) there is reason to hope that the introduction of modern farming methods may reduce the rate of soil erosion dramatically. Using herbicides to clear the ground of weeds as an alternative to ploughing is the most promising method. This makes it possible to reduce tillage to a minimum and to sow seeds directly into land covered with, and protected by, dead weeds. At the same time, increasing the productivity of the best land reduces the need to cultivate the poorer, more erosion-prone land.

In fact, soil erosion is an entirely natural process. Unconsolidated surface material is transported by wind and water from the moment it is exposed, whether the land is cultivated or not, and this transport has been continuing throughout the history of our planet. Many sedimentary rocks are made from eroded soil, after all, but the Earth is still blanketed by soil, which is constantly being formed. Erosion need not become a cause for concern until its rate exceeds the rate of soil formation; after that threshold is crossed, soil is actually being lost.

Under natural conditions, soil probably forms at an average rate of about 8 mm per century. Ploughing the land aerates the soil and increases the rate of leaching. This accelerates soil formation to perhaps 80 mm per century (HUDSON, 1971, p. 36). The planned reclamation of spoil heaps from mining can produce several centimetres of organically enriched surface soil within 5 years, but soil formation is considered to commence when substances leached from the A horizons start accumulating in the B horizons and the soil begins to acquire its layered structure. This takes much longer.

On cultivated land, therefore, soil is ordinarily forming at a rate of about 2 t ha^{-1} yr^{-1}, which is the weight represented by the 0.8 mm forming each year. A loss greater than this indicates a net loss by erosion, but just how undesirable this is considered to be depends on the soil itself. If the soil is no more than a thin layer overlying the bedrock, for example, its erosion is much less acceptable than a similar rate of erosion from a very deep, fertile soil.

Whether a particular rate of erosion is acceptable is a matter of judgement, but the susceptibility of a field to erosion under different farming regimes can be estimated and the erosion rate predicted. The technique requires a number of factors to be calculated. The erosivity of the rainfall (R) is calculated by measuring the amount and type of rainfall, converting this into an index number, and reading R from a scale. The erodibility of the soil (K) is a number representing the liability of a particular soil to erosion. The length factor of the field (L) is the ratio of the length of the field to the length of a standard field (of 22.6 m). The slope factor (S) is the ratio of the soil lost to the amount lost from a field with a 9 per cent gradient. The crop management factor (C) is the ratio of soil loss to that from a field under cultivated bare fallow. The conservation practice factor (P) is a ratio of soil loss to that from a field where no care is taken to prevent erosion. The amount of soil lost from that field each year (A) is then calculated from the universal soil-loss equation:

$$A = R \times K \times L \times S \times C \times P$$

Wind erosion can be spectacular and sometimes frightening, but water is probably the more important agent. Quite ordinary rain will initiate erosion, known as 'splash erosion', if it falls on bare soil. Raindrops fall at about 9 m s^{-1}, giving them a kinetic energy 13.6 times their own weight (DONAHUE ET AL., 1958, pp. 323–325). This is sufficient to detach soil particles, in the case of fine sands and silt splashing them to a height of up to 60 cm and a distance of up to 1.5 m. Larger particles do not move so far and clay particles are held together by their strong cohesive attraction, but even these will travel some distance. Splashing does not remove soil directly, but moves particles around and the water carries them into every tiny opening. At the same time, the pounding of the rain packs the surface particles together. When the rain ceases the surface dries as a tough, impermeable crust sealing the soil beneath. The next time it rains, water is unable to drain vertically and must flow across the surface, carrying soil particles with it in a thin mud, down the slope and away.

The effects of splash erosion can sometimes be seen on banks beside roads or ditches, where tree roots or a large rock have sheltered the soil. These are partly exposed, because the soil has been washed away from around them. This is known as 'pedestal erosion'.

Water flowing across the surface carries particles in suspension, which never touch the ground, particles that slide or roll over the surface, and particles that are repeatedly lifted and dropped. Some of the water finds its way into natural depressions and flows along them. This widens and deepens them, forming 'rills'. Rills are small enough to be removed by ordinary cultivation, so they are temporary, but unless they are removed they may grow much larger, into 'gullies'. These are more difficult to remove, because ordinary farm implements cannot cross them.

Routine cultivation destroys rills and ensures that gully erosion is rare on cultivated land, but on land that is never ploughed it can be serious. Tracks used by vehicles can turn into streams in wet weather, and from streams into badly eroded gullies. Paths much used by walkers suffer in the same way. A vicious spiral develops, in which the worn-down path becomes wet, walkers avoid the mud by walking to the side, and the gully widens as well as deepens. Some years ago in England the southern end of the Pennine Way national trail had to be rerouted, so severe was the erosion on Kinder Scout.

Two steps are involved in the removal of soil: the detachment of particles, by rain splash or wind; and their transport by water or wind. The remedy lies in minimizing both.

When rain falls on to vegetation, or wind blows across ground covered by vegetation, particle detachment is greatly reduced. The raindrops and wind are dispersed and their energy dissipated as leaves absorb the shock and rebound, like springs. For arable farmers this may be more difficult than it sounds; land is usually bare between the time of sowing and the emergence of the crop. Nevertheless, some sort of cover is valuable on vulnerable soil. In some places this is achieved by planting crops in alternate strips, such as grain and grass, or by leaving stubble lying on the surface after harvesting. In areas with a temperate climate, such as Britain, cereals are sown in autumn, as soon as possible after the completion of the previous harvest. This allows the seed to germinate and provide a vegetation cover through the winter, and a crop that starts growing rapidly in spring. It minimizes erosion, but is possible only where winter temperatures do not fall so low as to kill the young plants. In regions with a more extreme climate, cereals must be sown in spring and the soil must remain bare through the winter, although in this case erosion is reduced by the freezing of the surface or by a covering of snow.

Contour ploughing, in which the plough follows the contours of the land, produces parallel furrows oriented at right angles to the slope. Soil eroded from the plough ridges is trapped in lower furrows. In a field ploughed up and down the hill, with furrows parallel to the slope, soil can be swept downslope in furrows that quickly become rills.

Where slopes are steeper and the land is farmed intensively, terraces may be desirable. There are many types and Figure 3.15 shows cross-sections of two of the most common. Broad-base terraces are made by cutting trenches not quite at right angles to the slope and at intervals down it, using the excavated soil to build ridges on the downslope sides. Soil being washed down the hillside is held before it has travelled very far and the gently sloping trenches carry the water into drains or ditches. This avoids the situation in which soil erodes from the top of a field and accumulates at the bottom, making the soil quality uneven.

Bench terraces require more drastic engineering, because the hillside must be converted into a series of level strips, like a staircase, excavated soil being used to construct a bank along the downslope edge of each terrace. The technique is very effective, and is widely used in the tropics, but because the terraces must follow the contours they can have irregular shapes that make machine cultivation awkward.

Surplus water can be removed by 'grass waterways'. These are wide strips running down the slope and sown to grass. In effect, they are controlled gullies and can follow the routes of actual

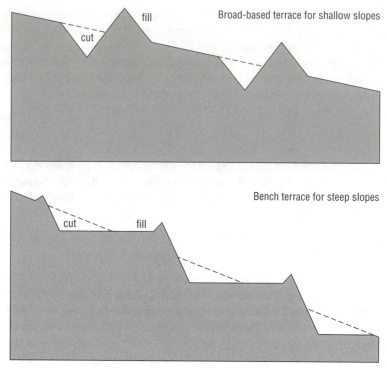

Figure 3.15 **Two types of terracing for reducing runoff**

gullies, but much more productively. During dry weather the gully is widened, soil added, and grass seed sown beneath a mulch held in place by netting. When the grass is established the waterway will continue to carry water, but will trap soil, and the grass itself can be grazed or cut for hay or silage.

Netting can also be laid over a bare surface after seed has been sown to achieve temporary erosion control. This holds the soil in place until plants have emerged to bind it more permanently. The technique is sometimes used on roadside verges and road-centre reservations, and the netting can be made from material that slowly decomposes and is incorporated into the soil.

Techniques for preventing wind erosion are designed to minimize particle detachment by reducing the speed and thus the energy of the wind. Maintaining a vegetative cover achieves this by creating a relatively calm ground-level microclimate. Strips of short stubble about 3 m wide will catch and hold almost all the soil lifted by the wind (FOTH AND TURK, 1972, p. 371).

Imagine the direction of the prevailing wind as a slope, as though it were blowing downhill, and a version of contour ploughing is also very effective. Where rows of crops are planted at right angles to the prevailing wind, each row shelters those downwind. Crops that are not grown in rows, such as cereals, can be protected by strips of row crops planted at intervals across the wind. As well as reducing erosion, this also helps prevent the soil from drying.

On a larger scale, trees and shrubs are often used as shelter belts or windbreaks. As Figure 3.16 shows, the wind is deflected over the top of the windbreak and its speed is reduced for a considerable distance downwind. The advantage of this is obvious, the disadvantage less so. Reducing wind speed also reduces the drying and chilling of the soil surface and can produce microclimatic conditions that vary markedly with distance from the windbreak. The resulting uneven growth and ripening of the crop can cause serious harvesting difficulties.

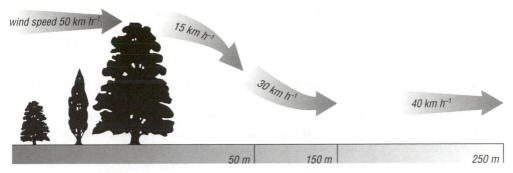

wind speed 50 km h⁻¹

15 km h⁻¹

30 km h⁻¹

40 km h⁻¹

50 m   150 m   250 m

Figure 3.16 **Effect of a windbreak in reducing wind speed**

Erosion by water may remove more than just the soil. If the land has been tilled, sown, and fertilized, the seed and fertilizer may also be lost, possibly to the advantage of the next field downhill, but more likely to the detriment of the water body into which it drains. Wind erosion may do even worse: it may carry away young plants that have only just emerged and are not yet anchored by strong root systems. The eroded soil also causes damage. It pollutes rivers, for example, and forms sediments in reservoirs, reducing their capacity. Wind-blown soil particles can severely batter crop plants and the soil can accumulate rapidly, sometimes to considerable depth, around fences and buildings, and may cover roads.

It is somewhat scandalous that soil erosion should still be a problem of such magnitude. Ancient as farming itself, its causes and remedies are very well known. It continues where they are not applied, through ignorance, rural poverty, or indolence, or where farmers who could afford to implement remedial measures perceive the cost to them as greater than their occasional loss.

30 Mining and processing of fuels

Originally, the word 'fossil' described anything dug up from below ground; zoologists still describe burrowing animals as 'fossorial'. Later, the word came to be applied to the preserved remains or traces of organisms that lived long ago (technically, more than ten thousand years ago). What we call 'fossil fuels' warrant their name on both grounds, but it may be more informative to describe them as 'carbonaceous', or 'carbon-based' fuels, because their combustion represents the rapid oxidation to carbon dioxide of the carbon they contain, which is an exothermic (heat-releasing) reaction.

Ordinarily, metabolic wastes and dead organisms are decomposed more or less rapidly. Most of the organisms responsible for decomposition require oxygen for respiration, however, and in anoxic environments their activities are curtailed. Under these circumstances it is possible for organic matter to become trapped, compressed beneath the weight of material that continues to accumulate above it, and subjected to rather different processes. Suitably airless environments are found, for example, in seafloor muds and below the surface of bogs and some swamps.

Plant material buried below the surface of a bog may be compressed into peat. If, later, the bog partly dries, the peat remains and can be dug for use as fuel. In some countries, such as Ireland, it is used in electricity power stations. Peat is the first stage in the formation of coal, into which it is converted by being subjected to much greater pressure and then heated: a 1 m seam of coal probably began as a 12 m layer of peat. The conditions necessary for the formation of coal occur only in the swamps found beside tropical rivers and seashores. Some of the coal being mined now formed around 400 million years ago, during the Silurian Period, but most dates from the Carboniferous, about 300 million years ago. Tectonic movements have since transported it to most parts of the world from Pangaea, the former supercontinent in which all the present continents were united and where it formed (ALLABY, 1993, pp. 143–151).

Coal and peat contain 'volatiles', substances that are given off as gases when the material is heated in the absence of air, and the quality of the fuel is determined by the proportion of volatiles it contains: the lower the proportion the more energy the fuel will release when burned. Peat contains more than 50 per cent volatiles, lignite (a soft, brown coal) about 45 per cent, and anthracite about 10 per cent. Anthracite is the highest quality, and very hard. Bituminous coal, the most abundant type and the one most widely used domestically, has 18–35 per cent volatiles.

Petroleum forms by a somewhat similar process. Organic material is buried by sediment, usually in a river delta, and is then trapped between two layers of impermeable rock. Many oil deposits are found beneath anticlines: rock strata that have been folded upwards into domed shapes. A similar structure occurs where a large mass of salt, deep below the surface, rises slowly through the less dense material surrounding it and the dense rock sinks to replace it. The process is called 'diapirism' and the salt dome it produces a 'diapir'. Oil is often found in 'salt-dome traps'. The material is then strongly compressed and heated. The resulting fluid fills all the pore spaces within the porous rock around it (ALLABY, 1992, pp. 162–163).

Some of the carbon and hydrogen comprising the organic matter form methane (CH_4), associated with both coal and oil. In coal mines, methane can cause fires, but when associated with oil it can be extracted and used as the fuel known as 'natural gas' (to distinguish it from 'town gas', mainly carbon monoxide (CO), obtained by heating coal and formerly an important industrial and domestic fuel).

Coal occurs in seams of varying thickness and at varying depths. There are four ways in which access can be gained to the seams and the coal extracted, illustrated in Figure 3.17. Seams at great depth are approached by sinking a shaft vertically from the surface, with associated shafts for ventilation. Most traditional British mines were shaft mines of this type, working deep seams.

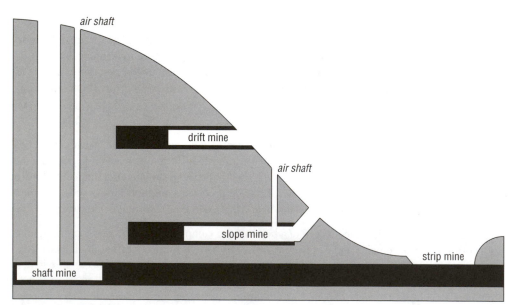

Figure 3.17 **Types of coal mines**

A slope mine approaches the seam through an angled shaft and a drift mine approaches horizontally. Where seams lie too close to the surface for a shaft to be cut to them, the overlying material is removed to expose the coal. This is an open-cast mine if it exposes a substantial part of the seam all at the same time and a strip mine if the seam is exposed and worked in sections.

All coal mining produces wastes, consisting of soil and rock that must be removed to gain access to the coal and rock mixed with the coal that must be separated from it. It is possible to store this waste, in 'spoil heaps', until the mine is exhausted then return it to below ground, but this is by no means the general practice and mining more commonly produces large, black spoil heaps. These are composed of finely crushed material with almost no soil and little in the way of plant nutrients. The heaps often contain large amounts of iron pyrites (FeS), producing very acid conditions (pH 2.0–4.0), and acid liquor, also containing metals, may leach from the heap into nearby watercourses, where it causes severe pollution. It is possible to reclaim mining spoil heaps. If they are treated with lime to reduce the acidity and soil and fertilizer supplied, a cover of grass can be established, leading in time to a more diverse plant community (MELLANBY, 1992, pp. 44–45).

Open-cast and strip mining can be even more destructive. In the past, large areas of attractive countryside were stripped of their soil (the 'overburden'), which was dumped in large heaps, and when the seam was exhausted the site was just left, utterly devastated. In some countries this is still the practice, but in many the planning consent stipulates that when operations cease the overburden must be returned to the surface and the site restored to a state better than its original condition. The effect is not always so destructive as it may appear in the older industrial countries, such as Britain. Coal seams suitable for open-cast or strip working often occur close to deeper seams that have been mined in the past to feed nearby industries, leaving land already in a state of industrial dereliction that restoration can improve once mining has ceased. In strip mining, restoration begins long before mining ceases, the reclamation of each strip starting as soon as the extractive machinery has moved on to the next strip. Indeed, British planning regulations are now stringent, and open-cast mining has little adverse long-term effect on areas of conservation or wildlife importance.

As well as the gaseous pollutants released when it is burned, coal combustion also produces ash. This can cause disposal problems, not least because it contains heavy metals.

Oil and natural gas are held in their traps under pressure. A hole drilled through the cap rock releases the pressure and they rise to the surface. The depth of drilling determines the part of the reservoir that is tapped, since the gas lies above the oil. Figure 3.18 shows how oil and gas are held within structural traps. Because all operations are conducted from the surface and no overburden has to be removed, oil and gas mining generate no spoil heaps. Such environmental damage as they cause arises from spillages of oil around the well or in transit to the refineries where it is processed.

Coal and oil may well be forming at the present time, but at a rate much lower than that at which they are being consumed. They are, therefore, non-renewable for all practical purposes. This being so, it is commonly assumed that one day they will be economically exhausted, oil first because it is much less abundant than coal. Certainly oil is being used rapidly, and in 1994 the United States imported more petroleum and its products than it produced from its own resources, for the first time becoming a net importer (ABELSON, 1995).

Impending shortages, combined with the environmental problems arising from the combustion of fossil fuels, have stimulated a search for alternatives, but all is not necessarily as it seems. Some people suggest that valuable resources should be conserved for the benefit of future generations, but consider what has happened to the coal industry. There is probably more than 45 billion tonnes of coal lying beneath Britain and a century ago it was being mined intensively and much

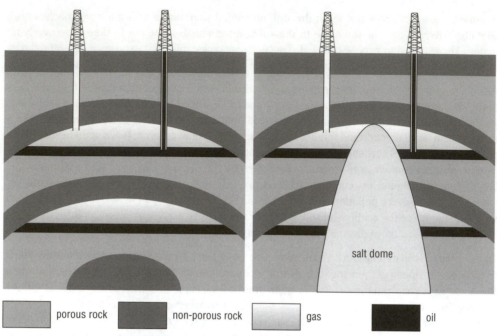

| porous rock | non-porous rock | gas | oil |

Figure 3.18 **Structural oil and gas traps**

of it exported. Coal is plentiful, but clearly not inexhaustible, and it might have occurred to people in the early years of this century that if the country continued to export this strategically important resource at that rate, the day would come when Britain had no choice but to import the fuel needed to power its industry and heat its homes. They might have felt it more sensible to restrict production and conserve coal for future needs. Today, matters appear rather different to those who might have benefited from such a conservation policy. In 1982, British mines produced about 125 million tonnes (Mt) of coal and the country consumed 111 Mt. In 1995, 52.6 Mt was produced and 76.2 Mt consumed (the difference was imported). Despite Britain's vast reserves, production and consumption declined, mainly because of a switch from coal to natural gas for power generation. Had the British decided years ago to restrict mining in order to conserve resources, the decision would have been wrong-headed. The loss of export earnings would have been economically damaging and the reduction in mine output would have caused unemployment. The decision would have caused real hardship, all to conserve a material for which a later generation had little use.

Relate this to the case of the United States, where coal mining and use are still increasing. There, between 1982 and 1992, coal production increased from 756 Mt to 905 Mt and consumption from 639 Mt to 808 Mt. Rising demand stimulated the search for and identification of new deposits that could be mined under the prevailing economic, political, and environmental conditions. Such identified deposits constitute 'reserves' and they increased, from 223 Mt in 1982 to 240 Mt in 1992.[1] Remembering the British experience, how sensible would it be for the United States to restrict coal mining in order to conserve coal for future generations? If the case for conserving coal is dubious, how many other materials and fuels might be of similarly doubtful future value?

Tonne for tonne, natural gas yields more energy than coal and releases fewer gaseous and no particulate pollutants. Fears of its imminent exhaustion may be unfounded. Conventional reserves may be much larger than was once supposed, and there may be a novel source: natural gas hydrate. This is a solid substance comprising a cage of water ice containing methane, technically a water clathrate of methane. It has been found in continental permafrost regions and in the oceans at all latitudes. Methane has been recovered from several sites. It is estimated that the total amount of natural gas hydrate may be more than double that of all other fossil fuels combined (KVENVOLDEN, 1994).

Natural gas is burned in stationary installations in power stations, factories, and homes, its principal rivals for power generation being nuclear power, coal, and hydroelectric power. Solar heat and light make a minor contribution, as does wind power, the difficulty with such 'renewables' arising from their extremely dispersed nature. It would require more than 3000 of the present generation of 450 kW wind generators, for example, occupying not less than 6000 ha, to produce as much power as one modern, 1.5 GW, conventional power station, and when conditions were calm and during storms the wind generators would not function at all. In Europe, fast-growing trees, such as willows, are being grown experimentally for fuel. After harvest, the material is dried and chopped for use in power generation.

Vehicles require a liquid fuel, although they can use gas. Biomass fuels, derived from crops grown for the purpose, have the advantage of making no contribution to atmospheric carbon dioxide, because the carbon dioxide their combustion releases precisely equals the amount absorbed by the crop plants during their growth. Ethyl alcohol has been used in this way, most notably in Brazil and the United States, although it is more costly to produce than petrol. Oilseed crops are now being developed from which 'biodiesel' fuel can be obtained, one of the most promising being rape, with seeds that are 40 per cent oil. Again, production costs are high, but may be reduced through a combination of economy of scale as output increases, and genetic engineering to increase the oil content.

Electrically powered vehicles have also been the subject of much research, but they, too, present difficulties, mainly arising from the speed and range limitations imposed by the weight and power of their batteries. Nor may they be so clean as many people suppose. It has been calculated that an electric car may release into the air 60 times more lead than a car running on leaded petrol (LAVE ET AL., 1995).

Fuel cells are also under development. These are devices comprising two electrodes separated by an electrolyte, which is a substance that permits the passage of ions (charged atoms or molecules) but not of electrons. A fuel containing hydrogen flows to the anode (positive electrode), where electrons are stripped from the hydrogen atoms. This leaves positively-charged hydrogen ions that diffuse through the electrolyte, while the electrons travel through an external circuit as an electric current. As the hydrogen ions arrive at the cathode (negative electrode) they are rejoined by the electrons and combine with oxygen to produce water, which is the only exhaust product. Unfortunately, fuel cells are extremely costly. The British physicist William R. Grove discovered their underlying principle in 1839 and NASA uses them to power spacecraft, but it will be several more years before they drive production cars (APPLEBY, 1999).

Energy conservation is often proposed as a partial alternative to increased exploitation of reserves or the search for new sources. If appliances and vehicles used energy more efficiently, it is argued, our demand for fuel would be correspondingly reduced. Unfortunately, the equation may not be so simple. If energy is used more efficiently, effectively it will be made cheaper and this could encourage an increase in the use of appliances to restore the balance. People would be able to obtain more use for the same price and energy consumption would not decrease. When US cars became more economical in their fuel use, during the 1970s and 1980s, consumption remained fairly constant; people drove their cars more for the same cost (INHABER AND SAUNDERS, 1994).

Nuclear power provides about 27 per cent of the electricity used in Britain and in some countries it provides much more. In Belgium, for example, it supplies 55 per cent of all electricity, in France 30 per cent, and in Lithuania 85 per cent.

Steam-driven turbines generate the electrical power and nuclear reactors produce the heat to turn water into steam. The core of the reactor comprises a structure with vertical holes or channels, some containing rods of fuel, others containing rods of cadmium or boron, all the rods being embedded in a substance called a moderator.

The fuel consists of an isotope of uranium, uranium-235 (often written as ^{235}U), that occurs as one part in 140 in natural uranium. When a slow-moving neutron collides and merges with the nucleus of an atom of ^{235}U that nucleus splits into two with the release of two or three neutrons. This is fission. If these neutrons also strike ^{235}U nuclei, the number of neutrons and nuclear fissions proliferate exponentially. This is a chain reaction and as the particles come to rest much of their energy is converted into heat.

At least one neutron from each fission must merge with a ^{235}U nucleus in order to sustain a chain reaction. Only slow-moving neutrons are able to cause fission; more energetic neutrons are not absorbed. Fission releases neutrons at many speeds and so fast-moving ones must be slowed. This is the purpose of the moderator. Different reactor designs use different moderator materials. Graphite, deuterium oxide (heavy water), and ordinary (light) water are widely used, light water being the most popular of all.

Cadmium and boron absorb neutrons, removing them from the reaction process. This means rods made from these elements can be used to regulate the speed of the chain reaction. The rods can be raised or lowered, accelerating or slowing the power output.

Surrounding the core, a coolant carries away the heat. In the most popular reactor design, the pressurized water reactor (PWR), the coolant is water under pressure. There are also designs that use boiling water. Magnox reactors, of which eight were built in Britain, the first in 1956 at Calder Hall (now Sellafield), Cumbria, use carbon dioxide as a coolant. The name 'Magnox' refers to the magnesium oxide alloy that is used to clad the uranium fuel rods. The advanced gas-cooled reactor also uses carbon dioxide as a coolant. Molten sodium can be used as a coolant.

Heat from the coolant passes to the water that is turned into steam to drive the turbines. In the event of a failure in the cooling system, not only would the turbines cease to function, the core would overheat, perhaps dangerously. Backup cooling systems are fitted that come into operation should this occur. These are just one of the many safety systems.

Nuclear fuel cycle

Uranium fuel for nuclear reactors is obtained from ores, in the same way as other metals. The ore is mined and milled to extract the uranium from it.

Natural uranium contains 140 parts of ^{238}U to every part of ^{235}U. It is only ^{235}U that can sustain the chain reaction necessary to generate heat. Consequently, for most reactors (but not all) it is necessary to increase the proportion of ^{235}U to about 3 per cent. This is called 'enrichment'.

Enriched uranium is then made into fuel pellets. These are placed in metal canisters. They are then fuel rods and ready to be used.

When the ^{235}U fuel is depleted, the rods must be removed from the reactor and replaced. They are very hot and highly radioactive. They are stored under water for a time, while their temperature and level of radioactivity fall.

They are then sent for reprocessing. This operation separates fuel that can be used from the waste, including radioactive products of uranium fission. Fuel is returned for use.

The remaining waste is converted into cylindrical blocks of a glass-like solid, packaged, and sent for disposal. This involves secure storage for a number of years before final disposal. No decision has yet been made on the method to be used for final disposal, but it is most likely to be below ground in a facility constructed in a geologically stable environment.

This sequence of operations, from mining to final disposal, constitute the nuclear fuel cycle.

Chernobyl

In the course of the history of civil nuclear power there has been only one accident in which a significant amount of radioactive material was released into the environment outside the plant. That accident occurred at about 1.23 a.m. Moscow time on April 26, 1986, in the No. 4 unit of the Chernobyl nuclear power station. The station is located in the eastern part of the Belorussian–Ukrainian Woodland about 4 km from the town of Pripyat, where most of the workers from the station lived, and about 18 km from the town of Chernobyl. It was reactor of the RMBK 1000 type, using slightly enriched uranium dioxide fuel encased in cans of zirconium alloy, water as a coolant, and graphite as a moderator. The coolant water flowed through the channels into which the fuel rods were inserted. Chernobyl-4 entered service at the end of 1983.

Chernobyl-4 was to have been shut down for routine maintenance on April 25. As this happened the management planned to carry out experiments to test the safety equipment. One experiment aimed to determine how long a single turbogenerator would continue to supply sufficient power for the plant to continue operating safely once its supply of steam had been cut off and it was spinning by inertia. This necessitated switching off the reactor emergency cooling system, a procedure strictly forbidden by the authorities, to prevent it from automatically restoring the steam supply. The experiment was begun by the day shift, which included the station managers and specialists, but conducted mainly by the night shift of less qualified workers.

By about 1.20 a.m. the reactor was running at about 6 per cent of its normal power and to maintain a sufficient flow of neutrons to sustain the reaction almost all of the 211 control rods were withdrawn completely. This also broke a rule on reactor safety. The steam supply was then withdrawn from the turbine. The head of the shift realized the situation was dangerous and ordered the control rods to be reinserted. They moved very slowly, however. The number of neutrons increased suddenly (a condition called 'prompt criticality') and within one second the power surged to several hundred times its normal level. This caused the first explosion in which some of the fuel cans burst.

Coolant water then came into contact with hot fuel, producing steam that reacted with the graphite moderator. This caused the second, very much bigger explosion that blew away the top of the reactor, allowing radioactive material to escape.

More than 30 fires began. The station fire brigade arrived within five minutes and the brigade from Chernobyl a few minutes later. Helicopters were used to drop boron through the hole in the roof to capture neutrons. When the fires had been extinguished, the reactor was sealed inside a concrete casing, called a 'sarcophagus'. Some years later, fears were expressed about the condition of the sarcophagus.

A total of 31 workers were killed by burns, falling masonry, and radiation, and about 150 people on the site suffered from radiation sickness. The radioactive plume was detected in Sweden.

Within 24 hours 47 000 local residents had been evacuated and by May 7 everyone living within a radius of 30 km had been removed. In all, 116 000 people from 186 settlements were evacuated. In subsequent years still more people were moved out of the affected area. Although radiation was detected further afield, outside the 30 km zone its effect on human health was predicted to be too small to be statistically detectable.

In the following years the health of exposed people was monitored closely under the leadership of the World Health Organization. Soon after the event, the most serious problem was caused by the psychological and social trauma of the accident and its aftermath of compulsory evacuation. Later there was an increase in thyroid cancer among children in Belarus.

Our production and use of fuels raises many environmental and economic questions. Posing those questions is not difficult, but finding answers to them is.

31 Mining and processing of minerals

A 'mineral' is a naturally occurring inorganic substance with a crystalline structure and characteristic chemical composition. Rocks are composed of minerals.

Whole rock is obtained by quarrying. Blocks of suitable rock are used for construction, in the case of slate after being split into thin sheets for roofing and cladding. Sand and gravel is also used for building, mainly of roads. Clay, won by a type of open-cast mining, is used for brick-making. High-pressure hoses (called 'monitors') are used to wash kaolin, or china clay, from the granite matrix in which it occurs, and it is removed as a slurry for purifying and drying. It was used originally to make fine ceramics (porcelain) but its principal use now is as a filler and whitener in paper and other materials.

Rock and building stone are quarried on a huge scale. Each year, over the world as a whole, rivers deliver to the sea about 24 billion tonnes of naturally weathered rock. Humans remove about 3 billion tonnes a year (ALLABY, 1993a, p. 150). This means we are now quarrying amounts comparable to those removed by natural processes.

Most modern quarries and open-cast mines are very large and, because they exist to detach and remove rock, cannot avoid devastating their sites. Nowadays planning consents require such sites to be restored when operations cease, but many older, abandoned quarries remain. The disfigurement they cause is not permanent, although it is only fair to point out that most older quarries were much smaller than modern ones and produced building stone, sand, or gravel in modest amounts for local use. Quarries scar the land, but they do not poison it and in time plants colonize the bare ground. An unrestored quarry site is rarely of any agricultural use and so usually it remains undisturbed and eventually may mature into a place of considerable interest to naturalists and conservationists.

Mineral mining, as opposed to the quarrying of rock, is much more disruptive, because it involves separating the desired minerals from the valueless minerals with which they are associated. The minerals themselves may be gemstones. Sapphires, oriental emeralds, and rubies are all aluminium oxides (Al_2O_3), differing from one another because of colours imparted by impurities; beryl is a compound of beryllium, aluminium, silicon, and oxygen ($Be_3Al_2Si_6O_{18}$); and diamond is a form of pure carbon. All are minerals and their high monetary value indicates their rarity: if they were common they would be cheap. When a rare substance is separated from the commoner substance containing it, a residue remains and this can cause environmental difficulties.

Metals are separated from their ores, an ore being a body of rock containing that metal in a compound, called an ore mineral, in a concentration high enough to be extracted economically. The concentration of the metal within its ore mineral may be quite high. Chalcocite (Cu_2S), for example, is 80 per cent copper and the best-quality uranite, or pitchblende (UO_2), is 85 per cent uranium, but the concentration of the metal within the ore (the rock containing the ore mineral) is very different. Iron ores are widespread and abundant, and iron is seldom extracted from ores containing less than 25 per cent of the metal, but scarcer metals commanding a high market price can be extracted economically from ores containing as little as 1 per cent or even less, for example in the case of copper. This means that in the case of iron, up to 75 per cent of the rock, and with some metals as much as 99 per cent of it, is useless waste for which some means of disposal must be found.

Mine waste cannot be returned easily to the hole whence it came. Unless the mine is exhausted, the waste would bury extractable ore, but in any case the waste no longer fits the hole. As rock, the minerals were tightly compressed; once they have been broken, crushed, and processed further to remove the desired ore mineral they consist of small particles with spaces between them. This greatly increases the bulk of the material, and there may be a truly vast amount of it. At Bingham, Utah, for example, copper is extracted from a hole 3.2 km in diameter and 900 m deep; this is large enough to accommodate two Empire State Buildings, one on top of the other, with room to spare. A lead mine in Missouri has two underground machinery repair shops 10 km apart, one with a floor area of 1.5 ha and the other 2.2 ha. Dry rock removed from these holes and expanded by processing is usually tipped to build hills of 'tailings'. Wet process residues are stored in ponds.

Tailings must be treated with respect, because the minerals they contain are in the form of tiny fragments, with a vastly increased surface area, and they are exposed to water. Subsequent chemical reactions can release very acid liquids, sometimes containing other toxic metal compounds. Dry tailings can be blown as dust, also causing contamination. Today, most governments set stringent regulations for the containment of mine tailings, so they cause little pollution, but it was not always so. In 1945, Britain had more than 1000 ha of land left totally derelict after overlying topsoil was mixed with tailings and dumped following iron-ore mining in what was otherwise good-quality agricultural land (MELLANBY, 1992, p. 47), and in south-eastern Tennessee an area of 145 km^2 remains barren to this day because of copper mining in the last century (RAVEN ET AL., 1993, p. 331).

Mineral processing begins with separation of the ore mineral from the crushed rock. If the ore mineral is denser than the unwanted rock mixed with it, called 'gangue', water will separate them, the mineral being precipitated first from a suspension. Other minerals are separated by froth flotation. A compound with a strong affinity for the mineral is mixed with water and agitated to make a froth; when the crushed rock is added, the desired mineral adheres to the bubbles, the gangue sinks, and the froth is skimmed from the surface. The separated mineral is removed and dried, ready for the next stage in its preparation, leaving behind the wet gangue.

Most ores are then heated to a temperature at which the metal melts, and the elements with which it was combined react to form compounds which float above the molten metal and can be removed as 'slag'. This is smelting and it usually proceeds as a series of chemical reactions. In the smelting of iron ore in a blast furnace, for example, the oxide ore is mixed with coke to supply carbon and limestone as a 'flux' that reacts to bind the slag (see Figure 3.19). The carbon is partly oxidized to carbon monoxide (CO), then oxidized further by reducing the iron oxide ($Fe_2O_3 + 3CO \rightarrow 2Fe + 3CO_2$). The impure iron may then be mixed with any of a range of alloying metals and heated again in a 'converter' to make steel.

Some metals are purified by electrolysis. Copper, for example, is obtained by passing an electric current through a copper sulphate solution. The anode (positive electrode) is made of ore, the cathode (negative) of pure copper. Copper ions move from the solution to the cathode and sulphate irons recombine with copper at the anode. Aluminium is also purified by electrolysis because, although it occurs as oxide ores, its affinity for oxygen is so great that heating cannot reduce it without also reducing all its impurities.

At every stage, from cutting the ore from the ground to extracting the metal from its ore mineral, the pollution risk is obvious and high. The pollution is contained by sealing tailings so dust cannot blow from them, or noxious liquors leach from them, removing gases and dust from smelters before they reach the outside air, and by treating liquid effluents.

A completely different technology now exists for extracting some metals. Bacteria that either possess the ability to isolate particular metals from their compounds or can be genetically modified to make them do so allow metals to be obtained with much less environmental disruption. *Thiobacillus ferroxidans*, for example, separates copper into an acid solution containing about 50 parts of copper per million. Sulphuric acid containing the bacteria is sprayed on to the ore rock, the liquor is collected, and the metal is removed, in this case at about one-third the cost of conventional processing, and yielding nickel as a by-product. Uranium can also be 'mined' in this way.

The limits to growth

In 1968 a group of 30 industrialists, economists, scientists, civil servants, and others met in Rome at the invitation of Dr Aurelio Peccei, an industrialist, to discuss 'the predicament of man'. The meeting resulted in the formation of the Club of Rome, which eventually grew to a body with about 70 members drawn from 25 countries. Members met at intervals and collaborated to sponsor studies. Their first, launched in 1970, was the construction of a computer model to trace the consequences of interactions among five factors: population growth; agricultural production; depletion of natural resources; industrial production; and environmental pollution. The model was developed on computers at the Massachusetts Institute of Technology (MIT) by a team of 16 people led by Professor Dennis L. Meadows. Its results were published in 1972 as a non-technical account, called *The Limits to Growth*. The technical information experts needed to evaluate the methods used was published later.

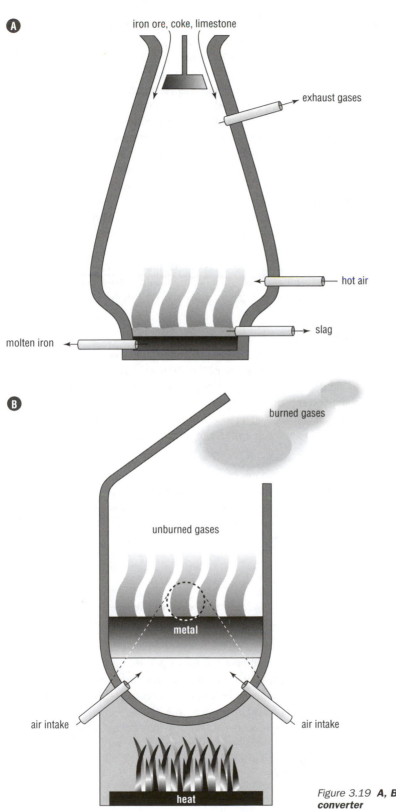

Figure 3.19 *A, Blast furnace. B, Steel converter*

Pollution is only one concern arising from our use of metals; the other centres on fears of their depletion. Such fears were being expressed in the 1960s, but were stated more forcefully in the 1972 study *The Limits to Growth* (MEADOWS *ET AL.*, 1972), which predicted the imminent exhaustion of most of the mineral resources on which we depend. The fear was based on a misunderstanding of the way the amount of a reserve is determined. The concept is economic, not physical. Figures for the reserves of any particular mineral are prepared by the mining industry primarily for its own use. They refer to the amount of the substance that has been identified and that can be profitably extracted under present conditions. They say nothing whatever about the total amount the world possesses, and changing circumstances may bring about an increase or decrease in the amount of reserves. In an apparent paradox, should consumption increase reserves may also increase to meet the demand; it was for this reason that between 1950 and 1970 global reserves of bauxite (aluminium ore) increased by 279 per cent, copper by 179 per cent, chromite (chromium ore) by 675 per cent, and tin by 10 per cent (ALLABY, 1995, pp 176–178). Historically, mineral consumption has always increased faster than the rate of population growth. World population doubled between 1950 and 1990, for example, but during the same period production of aluminium, copper, lead, nickel, tin, and zinc increased eightfold. It is now estimated that when allowance is made for anticipated population growth and economic development, mineral reserves are adequate for the next century and the environmental problems associated with their exploitation can be contained (HODGES, 1995).

Nor did the fears of exhaustion take sufficient account of the extent to which technological change renders resources obsolete. Ceramics, made from clays and sand, are now starting to replace metals for a number of uses that require tolerance of very high temperatures. Glass fibres, essentially made from silica (i.e. sand) are replacing copper cables. Orbiting satellites handle communications that were formerly transmitted by submarine cable. Electronic devices now form the basis of industrial switchgear, formerly based on mercury. Such changes, and there are many, are introduced not because of actual or anticipated shortages of the original material, but because they are superior.

The environmental implications of mining and mineral processing are well known. They can be minimized, although the restoration of mined land is sometimes difficult. In years to come, however, environmental pressures may ease. Technological advances now promise to reduce our dependence on some metals by substituting superior materials that are extremely plentiful and can be processed with much less risk of environmental harm, and by developing new, cheaper, and less disruptive extractive methods.

End of chapter summary

Without water we would die and so we may think of fresh water as the most fundamental of all the resources on which we depend. Water moves between the ocean and land, but we can manipulate the hydrologic cycle where the amount of fresh water is insufficient for our needs.

We also need soil. It is important to understand how soil forms and that it goes through a life cycle, much like a living organism. There are young, mature, old, and even senile soils. The age of a soil is directly relevant to the use that can be made of it. Fears over tropical deforestation, for example, arise partly from the realization that many tropical soils are very ancient and that this makes them inherently infertile, so alternative forms of land use may fail in the long term. Soil can be lost through erosion. The mechanisms involved in this process are well known, as are the management techniques by which they can be avoided.

Water and soil to grow food and fibre are resources essential for all animals. In addition to these, humans need industrial resources. These comprise so-called fossil fuels, rocks for building, and minerals from which metals and a range of chemical compounds are obtained. Obtaining these materials from the ground and their subsequent processing create environmental problems that must be addressed. It is usually better to prevent them by advance planning than to remedy them later.

End of chapter points for discussion

Can shortages of water be remedied by desalination?
What are the differences between young and old soils?
What are the environmental effects of mineral extraction?
What are the principal causes of soil erosion?

See also

Weathering (section 8)
Albedo and heat capacity (section 12)
Greenhouse effect (section 13)
Atmospheric circulation (section 15)
Fresh water (section 22)
Eutrophication (section 23)
Salt water, brackish water, desalination (section 24)
Irrigation, waterlogging, salinization (section 25)
Transport by water and wind (section 27)
Erosion (section 29)
Farming (section 53)
Genetic engineering (section 55)

Further reading

Biology of Freshwater Pollution, 2nd edn. C.F. Mason. 1991. Longman Scientific and Technical, Harlow, Essex. A detailed, somewhat technical explanation of the effects of discharges into rivers and lakes.

Elements: Earth. Michael Allaby. 1993. Facts on File, New York, 1993. A broad, general account of rocks and soils and the uses we make of them.

Fundamentals of Soil Science. H.D. Foth and L.M. Turk. 1972. John Wiley & Sons, New York. A fairly simple textbook on soil science.

The Human Impact on the Natural Environment. Andrew Goudie. 1986. Basil Blackwell, Oxford. Provides a clear overview of the environmental effects of industrial activity.

Only One World. Gerard Piel. 1992. W.H. Freeman, New York. A general account of the human effect on the environment, lucidly written by the former publisher of *Scientific American*.

Soil Conservation. Norman Hudson. 1971. B.T. Batsford, London. A detailed account of the mechanisms of soil erosion and techniques for preventing it.

Soils: An Introduction to Soils and Plant Growth. Roy L. Donahue, Raymond W. Miller, and John C. Shickluna. 1958. Prentice-Hall, Englewood Cliffs, NJ. Simply and clearly written, mainly for students of agriculture.

Waste and Pollution: The Problem for Britain. Kenneth Mellanby. 1992. HarperCollins, London, 1992. An assessment of the environmental effects of industrial and domestic waste disposal by possibly the leading British authority on environmental pollution.

Note

1 Figures from *Britannica Book of the Year 1985, 1995,* and *1999.* Encyclopaedia Britannica, Chicago.

References

Abelson, Philip H. 1995. 'Renewable liquid fuels', *Science,* 268, 995.

Allaby, Michael. 1987. 'Environment' entry in *Britannica Book of the Year 1987.* Encyclopaedia Britannica, Chicago.

—. 1993. *Elements: Fire*. Facts on File, Inc., New York.
—. 1993a. *Elements: Earth*. Facts on File, New York.
—. 1995. *Facing the Future*. Bloomsbury Publishing, London.
—. 1998. *Dangerous Weather: Droughts*. Facts on File, Inc., New York.

Appleby, A. John. 1999. 'The Electrochemical Engine for Vehicles', in *Scientific American*, July 1999, pp. 58–63.

Cruickshank, James G. 1972. *Soil Geography*. David and Charles, Newton Abbot.

Dajoz, R. 1975. *Introduction to Ecology*, 3rd edn. Hodder and Stoughton, London.

Donahue, Roy L., Miller, Raymond W., and Shickluna, John C. 1958. *Soils: An introduction to soils and plant growth*. Prentice-Hall, Eaglewood Cliffs, N.J.

Foth, H.D. and Turk, L.M. 1972. *Fundamentals of Soil Science*. John Wiley and Sons, New York.

Goudie, Andrew. 1986. *The Human Impact on the Natural Environment*. Basil Blackwell, Oxford.

Harvey, J.G. 1976. *Atmosphere and Ocean: Our fluid environments*. Artemis Press, Horsham, Sussex.

Hodges, Carroll Ann. 1995. 'Mineral resources, environmental issues, and land use', *Science*, 268, 1305–1312.

Hudson, Norman. 1971. *Soil Conservation*. B.T. Batsford, London.

Hunt, Charles B. 1972. *Geology of Soils*. W.H. Freeman and Co., San Francisco.

Inhaber, Herbert and Saunders, Harry. 1994. 'Road to nowhere,' *The Sciences*, Nov./Dec. 1994. NYAS, New York. pp. 20–25.

Kupchella, Charles E. and Hyland, Margaret C. 1986. *Environmental Science*. Allyn and Bacon, Needham Heights, Mass.

Kvenvolden, Keith A. 1994. 'Natural gas hydrate occurrence and issues', in Sloan, E. Dendy Jr., Happel, John, and Hnatow, Miguel A. (eds) *Natural Gas Hydrates*, Annals of the NY Acad. of Sciences vol. 715. NYAS, New York. pp 232–246.

Lave, Lester B., Hendrickson, Chris T., and McMichael, Francis Clay. 1995. 'Environmental implications of electric cars,' Science, 268, 993–995.

L'hirondel, Jean-Louis. 1999. 'Are dietary nitrates a threat to human health?', in Morris, Julian and Bate, Roger (eds) *Fearing Food*, Butterworth Heinemann, Oxford. pp. 38–46.

Mason, C.F. 1991. *Biology of Freshwater Pollution*, 2nd edn. Longman Scientific and Technical, Harlow, Essex.

Meadows, Donella H., Meadows, Dennis L., Randers, Jorgen, and Behrens, William W. III. 1972. *The Limits to Growth*. Earth Island, London.

Mellanby, Kenneth. 1992. *Waste and Pollution: The problem for Britain*. HarperCollins, London.

Ministry of Agriculture, Fisheries and Food and Department of Agriculture and Fisheries for Scotland. 1968. *A Century of Agricultural Statistics; Great Britain 1866–1966*. HMSO, London.

Moore, Peter D. 1995. 'Too much of a good thing', *Nature*, 374, 117–118.

Piel, Gerard. 1992. *Only One World*. W.H. Freeman, New York.

Raven, Peter H., Berg, Linda R., and Johnson, George B. 1993. *Environment*. Saunders College Publishing, Orlando, Florida.

Reinhold, Robert. 1992. 'The lingering US drought,' *Britannica Book of the Year 1992*. Encyclopaedia Britannica, Chicago. p. 168.

Royal Commission on Environmental Pollution. 1979. *Seventh Report: Agriculture and Pollution*. HMSO, London.

Smith, George E. 1967. 'Fertilizer nutrients as contaminants in water supplies,' in Brady, Nyle C. (ed.) *Agriculture and the Quality of Our Environment*, AAAS, Washington, DC. p. 177.

Tolba, Mostafa K. and El-Kholy, Osama A. 1992. *The World Environment 1972–1992*. Chapman and Hall, London, on behalf of UNEP.

Winter, E.J. 1974. *Water, Soil and the Plant*. Macmillan Press, London.

4 Biosphere

When you have read this chapter you will have been introduced to:

- the concepts of biosphere, biomes, and biogeography
- nutrient cycles
- respiration and photosynthesis
- feeding relationships, food chains, and food webs
- pyramids of energy, numbers, and biomass
- ecosystems
- succession and climax
- arrested successions
- colonization
- stable and unstable environments, and reproductive strategies
- simplicity and diversity
- homoeostasis and feedback
- limits of tolerance

32 Biosphere, biomes, biogeography

All the organisms living in a particular area can be described collectively as the 'biota'. The noun implies nothing about the size or type of area: it can be applied to the inhabitants of a vast forest, a small puddle of rain water, or the entire Earth. The region of the Earth occupied by its biota is known as the 'biosphere', sometimes called the 'ecosphere' to emphasize the fact that the biota comprises an interacting system which can be studied at the global level.

Most of the biota occurs at or very close to the surface of land or water, and the biosphere forms an extremely thin layer on the Earth. In the soil, organisms inhabit the upper horizons, where they feed on organic material reaching them from above. They, together with their predators and parasites, constitute the soil biota and their domain extends to no more than a few metres below ground level; most of them are confined to the uppermost few centimetres. When soil samples are being taken in order to study the biota, cores to a depth of 10 cm are considered adequate in some soils. Up to 60 cm in deep, peat soils is considered adequate for microorganisms, whereas 10 cm is usually sufficient for sampling larger animals such as worms (PHILLIPSON, 1971, p. 89).

Above ground, birds and insects fly, newly hatched spiders migrate by 'ballooning' attached to lengths of their own silk, and 'aeroplankton', consisting of plant pollen grains, fungal and bacterial spores, and some single-celled plants, is carried aloft by thermal currents and transported by the wind. The domain of these organisms is effectively limited by the air density below which powered flight becomes impossible, the availability of oxygen for respiration, and temperature. It extends to about 6.5 km, although dormant spores, securely sealed within their casings, can reach the tropopause, at an average height of about 9 km, but sometimes reaching 17 km at the equator (TIVY, 1993, 13–14). Tardigrades, which comprise about 700 species of tiny animals sometimes called 'water bears' or 'moss piglets', have been known to cross the Atlantic, carried on winds that bore them through the lower stratosphere, where they survived in temperatures lower than −100 °C (COPLEY, 1999).

Marine life is also concentrated near the surface. It is based on photosynthesizing, single-celled plants, the phytoplankton on which herbivores feed. Plants need light and are consequently confined to the depth of light penetration. In the clearest water, blue-green light can be detected at a depth of almost 1 km, but at least 95 per cent of incident light is absorbed in the uppermost 50 m. There is sufficient light for photosynthesis down to about 150 m in very clear water, but in most sea areas the biologically productive region, called the 'euphotic zone', is much shallower (MARSHALL, 1979, pp. 39–40). At greater depths, organisms rely on organic material descending to them from above, like rain. This sustains life on the ocean floor. In some places, water heated by contact with molten rock below the sea bed and rich in dissolved minerals issues from hydrothermal vents. Depending on their chemical composition the issuing fluids may be black or white; they are known as 'black smokers' and 'white smokers'. A little distance from them, where the temperature is about 40 °C, colonies of bacteria able to synthesize nutrients directly from the vent fluid form the bases of richly productive animal communities. Vent communities are unique in their complete isolation from and independence of communities ultimately deriving their energy from sunlight.

Hydrothermal vents are associated with ocean trenches, which are the deepest part of the ocean, and the Marianas Trench is the deepest of all, its floor lying about 11 km below sea level. In addition, it is now known that there are large bacterial populations living up to a few kilometres below the Earth's surface. They are found in deep aquifers and spaces in rocks of a variety of types and in some places the size of the population increases with depth. The fluid extracted from some North Sea oil reservoirs contains up to 16 kg of bacteria a day (PARKES, 1999). The depth at which these populations are found may not exceed that of the Marianas Trench, however. At its greatest, then, the biosphere extends from about 11 km below sea level to the lower stratosphere at a maximum of about 20 km above sea level. Nowhere is it more than about 31 km deep, a figure that can be compared with the mean radius of the Earth, which is 6371 km. If you pictured the Earth as a large orange, diameter 6.4 cm, the biosphere would fill the pits in the skin and cover its surface to a maximum depth of less than 0.3 mm and in most places to no more than half of that.

Although the biosphere is confined to so thin a layer, the biota contained within it is diverse and its character varies markedly from one region to another. Wolves and conifer forests are typical of northern Canada and Eurasia, tall, dense, broad-leaved evergreen forests of the equatorial regions, grasslands of the interiors of continents, and deserts of the continental subtropics. The consistency of composition of large areas covered by such biotic types allows them to be grouped. If only the vegetation is considered, the groups are 'formation types', the largest unit recognized in a hierarchy of types of plant community and usually identified by the predominant kinds of plant. If the animals are included, the groups are known as 'biomes'. Biomes are defined by all the species within them and the relationships between those species and their environment.

Such a definition describes a type of community, but says little about the detail of its composition. A hot desert, for example, will support plants and animals of a certain type, but the actual species may be local: the Australian, Sahara, and North American deserts support different species of organisms adapted to very similar conditions.

This suggests that biomes are, in fact, climatic regions and that the concept refers mainly to the environment. This is broadly true, and Figure 4.1 shows how major biomes are related to temperature and precipitation. The similarity of type, but with different constituent species, arises through convergent evolution. Natural selection favours those species best adapted to the conditions under which they live, and some climatic regimes impose constraints to which only a limited number of biological adaptations seem possible. Species thriving under such regimes resemble one another, though they are not closely related. Just as biomes appear to be climatically defined, climatic regimes appear to be definable biologically (see section 21).

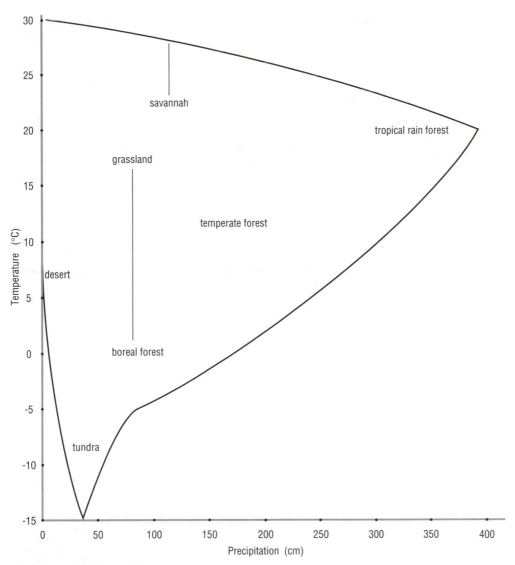

Figure 4.1 **Biomes and climate**

Difficulties with so direct a linkage between biomes and climates begin to appear at the boundaries between biomes. These are never sharp, biomes being separated by transition zones (called 'ecotones') containing species from both adjacent biomes, and some ecotones are broad: that between tundra and coniferous forest ('taiga') is up to 160 km wide (TIVY, 1993, p. 96). It is also possible that the northern edge of the taiga may still be advancing northward following the end of the last glaciation. Further south, some biomes may have resulted from human activities. Repeated forest burning, to drive game or clear land for cultivation, favours the growth of grass, allowing populations of grazing herbivores to increase, leading to the destruction of tree seedlings by nibbling or trampling and the permanent conversion of forest to grassland. Savannah grasslands, prairie, and steppe are all believed to have resulted from such intervention. Not only does this weaken the link between biome and climate, it blurs the natural borders between biomes. If over-grazing exacerbates the spread of deserts during prolonged drought, for example, where should the purely climatic border be located?

With these limitations in mind, the world can be divided into biomes. Between the high-latitude tree line and the deep snow of the polar regions, the tundra comprises lichens, mosses, sedges, flowering herbs, and dwarf shrubs and trees, the animal population swelling by migration during the brief summer. Along its low-latitude borders, the tundra gives way to taiga or boreal forest, comprising mainly coniferous trees and animals associated with them. At still lower latitude, these merge into mixed coniferous and broad-leaved temperate forest. Continental interiors support temperate grassland, such as the prairie of North America, pampas of South America, veld of South Africa, and steppe of Eurasia, or the tropical savannah, which is partly wooded in places. Mediterranean climates support chaparral, a type of grassland with drought-resistant scrub. Deserts occur in the subtropics and humid equatorial regions support tropical rain forest. Biomes closely resembling those of high latitudes also occur on mountains in low latitudes. Climb a high mountain at the equator, and you will move from rain forest to temperate forest and eventually to a type of vegetation similar to tundra before entering the permanently snow-covered heights.

Although they are not called biomes, the oceans can also be divided horizontally into zones, each of which supports a typical biota. There is some migration between zones. Benthic organisms are those which inhabit the sea floor, but pelagic species move vertically between the surface and the floor. These are the minority of animals, however, benthic species accounting for about 98 per cent of the total (MARSHALL, 1979, p. 40). Some pelagic organisms have little or no power of independent movement and drift with tide and current near the surface. These are known as 'plankton', a grouping that includes bacteria, single-celled plants (phytoplankton), and small animals (zooplankton).

Below the surface, marine zones are defined by depth. As Figure 4.2 shows schematically, ocean depth is determined largely by distance from continental coasts. Beyond the shore lies the continental shelf, sloping at a gradient between about 1 : 500 and 1 : 1000 for a distance of several hundred kilometres. At a depth of about 200 m the gradient steepens, to an average of about 1 : 700 but up to 1 : 4 in places, down the continental slope to a depth of about 2000 m. The continental rise, where sediment that has slid down the shelf and slope accumulates, slopes at 1 : 100 to 1 : 700 and continues to the deep ocean floor. The neritic zone corresponds to the level overlying the continental shelf, the bathyal zone to the region above the slope, and the abyssal zone continues

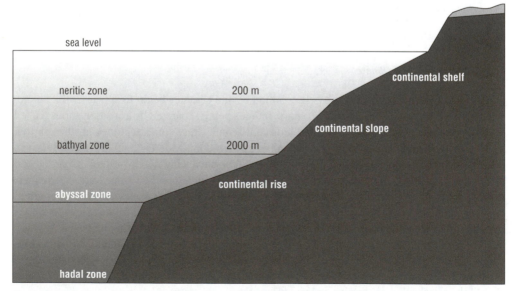

Figure 4.2 **Marine zones and continental margin (not to scale)**

down the rise to the floor. The deepest part of the ocean, in trenches below the general level of the floor, is called the hadal zone.

Close examination of the composition of biomes quickly reveals that although the biome concept is useful as a kind of shorthand to describe a type of environment, species vary widely from one example of a particular biome to another. Temperate forests comprise different trees in different continents, for example, and Old World monkeys are taxonomically distinct from New World monkeys. Biogeography, the study of the geographic distribution of species, is the discipline which addresses such differences.

Biogeographers have divided the world into regions according to the flora and fauna more or less peculiar to them and generally separated by geographic barriers to migration. The number of regions varies from one authority to another, but at least six faunal areas are generally recognized. In the northern hemisphere, the Nearctic covers North America and Greenland, the Palaearctic Eurasia as far south as the Himalayas, North Africa, and part of Arabia. In the southern hemisphere the Neotropical region covers Central and South America, the Ethiopian the remainder of Africa, Arabia, and Madagascar, the Oriental the Indian subcontinent and southern and south-eastern Asia excluding New Guinea and Sulawesi, and the Australasian Sulawesi, New Guinea, Australia, and New Zealand. In addition, some biogeographers recognize an Oceanian region, covering the Pacific islands, and an Antarctic region.(Tivy, 1993, p. 58).

Floral divisions follow similar boundaries, but the Nearctic and Palaearctic are merged and given the name Boreal, and the Ethiopian and Oriental together form the Palaeotropical. Holarctica is a name formerly used to describe the whole of the northern hemisphere, all of which was once joined in the supercontinent of Laurasia.

When these vast areas are studied in more detail, a hierarchy develops, allowing increasingly local designation. Regions can be divided into 'domains' and domains into 'sectors' and 'subsectors'.

At local level, however, further complications emerge. Some plant species are described as 'native' to distinguish them from 'exotics', which are plants introduced by humans. Botanists use this distinction frequently, but what is exotic today may have been native in the past, perhaps during a previous interglacial, and some exotics have been present for a very long time and are fully naturalized, meaning that they are able to grow and reproduce without human assistance. In areas that were once glaciated, most plants arrived as immigrants when the climate warmed and land was exposed by the glacial retreat. Other species may become extinct in part of their range, these days depressingly often.

Plant and animal distribution is constantly changing. It cannot be studied in isolation from the climatic and tectonic history of the Earth and its study is directly relevant to current concerns over the loss in biodiversity (see section 50).

33 Major biomes

Several biome classifications are used, but most agree the world can be divided into about 10. These are usually listed as: polar regions and tundra; temperate forest; tropical rain forest; tropical seasonal and monsoon forest; tropical grassland; temperate grassland; deserts; mountains; wetlands; and oceans.

As their name indicates, the polar regions occupy approximately the area in latitudes higher than 60.5° N and S. These latitudes mark the Arctic and Antarctic Circles. They are defined as the latitudinal limit at which the Sun fails to rise above the horizon on at least one day in each year and fails to sink below the horizon on at least one day.

Ice sheets cover a large area at each Pole. Sea covers most of the Arctic. Some of the sea is frozen throughout the year. The ice sheet proper – ice lying permanently on land – is confined to Greenland. Antarctica is a large continent divided into two parts by the Transantarctic Mountains. There are two Antarctic ice sheets, the East and West, lying to either side of the mountain range. The East Antarctic ice sheet covers about twice the area of the West Antarctic sheet. The West Antarctic ice sheet extends over the sea surface as ice shelves. There are several ice shelves, the largest being the Ronne shelf in the Weddell Sea and the Ross shelf in the Ross Sea.

Polar climates are cold and away from the coasts they are extremely dry. These are also the most extremely seasonal climates.

No plants can grow where the ground is permanently covered by ice and snow. Animals cross the ice and feed in the water. Both Arctic and Antarctic waters are very productive.

Beyond the edge of the permanent ice the ground supports tundra vegetation. There is no tundra vegetation on the continent of Antarctica, because all of the mainland is at too high a latitude. Small areas occur on the Falkland (Malvinas) and other islands. Tundra covers a much larger area of North America and Eurasia. The plants include grasses, sedges, and flowering herbs. Woody plants occur only as low shrubs in places sheltered from the desiccating winds.

Protecting Antarctica

Antarctica is unique in being the only continent with no native human inhabitants, although it has a population of scientists and support workers, some of whom remain through the winter. At one time seven countries claimed territorial rights in Antarctica. These were Argentina, Australia, Chile, France, New Zealand, Norway, and the United Kingdom. In addition, the Soviet Union, United States, Belgium, Germany, Poland, Sweden, Japan, and South Africa engaged in Antarctic exploration, but claimed no territorial rights.

Recognizing the great value of the continent, on December 1, 1959, Argentina, Australia, Belgium, Chile, France, Japan, New Zealand, Norway, South Africa, the Soviet Union, the United Kingdom, and the United States signed the Antarctic Treaty (*www.acda.gov/ treaties/arctic1.htm*). This made the entire continent a demilitarized zone that would be preserved for scientific research. Without commenting on any territorial claims, it forbade signatory nations from establishing military bases on the continent, or conducting manoeuvres, testing weapons, or disposing of radioactive waste. The Treaty binds its members in perpetuity, but allows for a review of its provisions every 30 years. The initial signatories were later joined by Brazil, China, Germany, Finland, India, Italy, Republic of Korea, Peru, Poland, Spain, Sweden, and Uruguay.

On October 4, 1991, a protocol to the Treaty was signed. This banned oil exploration for 50 years and included measures to protect the natural environment.

Temperate forests occur mainly in the Northern Hemisphere. They are bordered to the north by the tundra and extend southward to the Tropic of Cancer (23.5°N). Much smaller areas occur south of the Tropic of Capricorn (23.5°S) in a narrow strip along the western coast of

South America, in the south and west of South Africa, in the south of Australia, and in both islands of New Zealand.

In the north the forests are dominated by coniferous species, especially spruce, pine, fir, larch, and hemlock, in places mixed with birch. In Canada and Alaska this is known as the boreal forest. In Russia it is called the taiga.

Along its southern borders broad-leaved deciduous trees become increasingly common in the coniferous forest, producing mixed forest. This gives way to predominantly broad-leaved deciduous forest across most of Europe north of the Alps and on the eastern half of North America as far south as the Gulf of Mexico. Around the Mediterranean and in parts of California, South Africa, and Australia the forests consist of evergreen trees, some broad-leaved and some coniferous, that are adapted to a climate with a pronounced summer dry season.

Tropical rain forest is found in low-lying areas along a belt to either side of the equator. The temperature is consistently warm. In Libreville, Gabon, for example, the mean monthly daytime temperature varies between 28 °C and 31 °C. Rainfall is heavy, but often variable and with a distinct dry season. Libreville has fairly dry weather from early May to the end of August. In the Amazon Basin, too, the rainfall is less in the summer months than in winter.

Broad-leaved evergreen trees dominate the vegetation. Many grow to a great height and their crowns form a completely closed canopy. Smaller trees and shrubs grow beneath the dominants, producing several horizontal layers. Each layer supports its own populations of epiphytes (plants that grow on the surface of other plants, but without obtaining nutrients from them) and animals. Together with the large number of species contributing to the community, the three-dimensional structure gives tropical rain forest an unparalleled ecological richness.

Monsoon climates are strongly seasonal, with a long dry season in winter and a short wet season in summer. Forests adapted to this type of climate, found mainly in southern Asia, consist of broad-leaved deciduous trees that shed their leaves during the dry season. Monsoon forest contains fewer tree species than tropical rain forest.

On tropical mountainsides, and especially on the eastern slopes of the Andes and southern slopes of the Himalayas, the character of the tropical forest changes with increasing elevation, as the temperature decreases with height (by an average 6.5 °C per kilometre). On the lowest slopes the forest becomes more open. There are gaps in the canopy and the trees, although of the same species as those below, have thicker trunks and smaller leaves. There are more ferns, mosses, and other plants growing on the tree bark. This is known as mountain, or montane forest.

Higher still, moist air forced to rise up the mountainside is chilled and its water vapour condenses to form mist. This produces cloud forest. The ground is wet and often boggy and mosses are common. The trees are small, often stunted, and have tough, leathery leaves.

Above the cloud forest the trees become smaller and more scattered. Ferns and mosses are abundant and the forest has a strange, ethereal look. It is known as elfin woodland and its upper boundary is the tree line, beyond which the temperature is too low for trees to survive.

Saving the tropical forests

Tropical forests are being cleared at a rate of about 2 per cent of the area every 10 years in Asia and about half of that in South America. Montane and monsoon forests are being lost more rapidly than lowland forests, which often lie on very wet, boggy or swampy ground.

They are cleared mainly to provide land for farming. The demand for timber contributes to this because although young trees will grow naturally to restore areas of cleared forest, they are not always given an opportunity to do so. Farmers occupy the cleared land.

Attempts to arrest the loss of tropical forest are based on establishing plantations to supply timber and on finding alternative livelihoods for local people so they no longer need to farm. Throughout the tropics, large areas of plantation forest are being established and the total area of plantation is increasing rapidly. Plantations supply timber more profitably than the natural forest, and adjacent land is used for farming by modern methods that produce higher yields. Tourism brings in a steadily increasing income.

The natural forest is then seen as a resource for research, tourism, and for the harvesting of certain products that cannot be grown on farms.

Grasses thrive in climates that are too dry to support forests and temperate grasslands are found in the interior of continents, far from the nearest coast, where the climate is of the continental type. Winters are cold, summers hot, and precipitation is sparse. These grasslands are known as prairie in North America, steppe in Eurasia, pampas in South America, and veld in South Africa.

They cover an immense area. In Eurasia grassland is the natural vegetation over the area approximately bounded by latitudes 40°N and 50°N and longitudes 20°E and 90°E. The steppes extend from Hungary to Mongolia. In North America the prairie extends from the southern half of the Canadian prairie provinces of Manitoba, Saskatchewan, and Alberta to Texas.

Land that will grow grass will often grow cereal crops, which are also grasses. Consequently most of the prairie and much of the steppe has been cleared for farming.

Tropical grasslands are found in South America, over much of Africa between the Sahara and Kalahari Deserts, on the western side of India, and surrounding the deserts of Australia. The climate is warm and strongly seasonal, with pronounced dry and rainy seasons. Mean temperature ranges between about 18 °C in winter and 32 °C in summer. Winter is the dry season.

In addition to grasses, which are the dominant vegetation, tropical grasslands also include small trees that are adapted to the dry season. Acacias are typical. The extensive pasture supports large herds of grazing mammals, together with the predators that hunt them.

Wetlands occur throughout the world, but they are localized. They are found along most coasts and in river deltas and estuaries. Outside the tropics coastal wetlands form salt marshes, comprising plants that tolerate submergence by salt water at each high tide followed by exposure to the air at low tide. In the tropics they form mangrove swamps, usually with just a few species of mangrove trees.

Inland, wetlands occur as bogs and around freshwater lakes and where large rivers flow across very level ground. The Everglades, Florida, are a freshwater wetland, as is the extensive marshland area around the mouths of the Tigris and Euphrates Rivers, in southern Iraq. Wetlands support a wide variety of wildlife.

People like to live near sea coasts and beside rivers. River flood plains are very fertile, because of the silt deposited on them with each flooding. Consequently, coastal and riverside wetlands are under constant threat of development for housing or farming. Draining them is not usually difficult, although it may be expensive.

In 1971, an international conference at Ramsar, Iran, led to the Ramsar Convention (*www.ramsar.org/index.html*) on Wetlands. By 1999, the governments of 116 countries had signed the Convention. This commits them to designating and then protecting wetlands of international importance, especially as waterfowl habitat, within their own borders. Altogether, 1006 Ramsar sites have been designated, covering a total area of 71.8 million hectares.

Oceans are counted as a single biome because, although they occur in every part of the world, their waters are joined and many of the species living in them migrate over long distances. Together, the oceans cover 70.8 per cent of the surface of the Earth, a surface area of 361 million km^2, to an average depth of 3.7 km. They hold about 1370 million km^3 of water. Phytoplankton, comprising minute, mainly single-celled, plants supply food for marine animals. These include members of the zooplankton, which include the larvae of many larger species, as well as fish, marine mammals, and sea birds (see section 4.32).

Deserts form wherever the potential rate of evaporation over a year exceeds the amount of precipitation. The rate of evaporation varies according to temperature, so the minimum precipitation needed to prevent an area from becoming desert varies with the mean temperature. There is an absolute lower limit, however, of about 250 mm of rain a year, below which desert will develop in any climate.

In a belt stretching across the edges of both tropics desert climates are produced by subsiding air at the edges of the Hadley cells (see section 2.15). This air is warm and extremely dry. It produces high surface pressure, with air flowing outward from it and preventing moister air from entering the region. The Sahara, Arabian, and Thar Deserts are caused by this climatic regime in the Northern Hemisphere and the Kalahari and Gibson Deserts in the Southern.

Deserts also occur in the deep heart of large continents. The Turkestan, Takla Makan, and Gobi Deserts are of this type. Air loses moisture where it is forced to rise as it crosses a mountain range. This can produce a desert on the lee side. The deserts of the southwestern United States are formed in this way, as is the Atacama Desert in South America.

Air also subsides over both Poles, producing the polar deserts. These are covered thickly with ice and snow, but only because such precipitation as does fall is unable to melt or evaporate. In fact, the precipitation is low and the interior of Antarctica is the driest of all deserts, with an average annual precipitation of about 25 mm.

Despite their aridity, many species have adapted to survive in deserts. There are succulent plants, such as the cacti in America and certain euphorbias in Africa, that store water in their leaves and stems. Other plants survive most of the time as seeds, germinating and completing their life cycles in a matter of a few weeks, or even days, whenever rain falls, which very occasionally it does. These plants sustain desert animals and there are many other animal species that venture into deserts without being resident there.

Desertification

Lands bordering deserts have an arid climate. Droughts are common and from time to time prolonged droughts occur. During a drought most small plants disappear. Trees and shrubs may appear as though dead, and may be dead. The land closely resembles the desert nearby. This type of change has been observed in recent years in parts of Ethiopia and in the region known as the Sahel, along the southern border of the Sahara. Eventually, though, the drought ends. When the rains return the vegetation quickly recovers. Despite appearances, drought of this kind does not indicate the permanent expansion of the desert that is implied by the term 'desertification'.

Desertification can occur, however. If marginal land adjacent to a desert is cropped too intensively the fertility of the soil may become exhausted. Crop yields will decline and when drought strikes the vegetation may be unable to recover. Similarly, grazing the land with too many livestock will destroy the vegetation. In both cases plant roots that bind soil particles together are lost. Wind can then lift the dry soil, blowing it away, sometimes to bury any plants or crops that have managed to survive nearby. Tree roots bind soil particles and the trees themselves provide some shelter from the wind. Trees are often cut down for use as fuel or for building materials.

Poor irrigation can sterilize land by rendering it saline (see section 3.25). Plants die and farmers abandon their fields.

Nearly one-quarter of the land area of the Earth is estimated to be at risk from soil degradation leading to desertification. Remedies are based on better land management. Water can be used more efficiently for irrigation. Crop varieties can be genetically modified to grow in saline soils. Trees can be grown to shelter crops.

Mountain habitats change with increasing elevation, due to the fall in temperature with height. This averages 6.5 °C per kilometre and determines the elevation at which the temperature remains below freezing throughout the year. This varies with latitude. In the high Arctic and in Antarctica the summer freezing level is at sea level. In temperate latitudes, where summer sea-level temperatures are typically about 22 °C, it is at about 3400 m above sea level and in the tropics, where the sea-level temperature is about 30 °C snow will lie permanently above 4600 m.

Temperature determines the type of vegetation. In the tropics, rain forest gives way to a transitional type of forest at about 1000 m. Starting at about 2000 m there is montane forest, extending to about 3000 m, where it gives way to elfin woodland. At about 4000 m there is grassland, in some places forming alpine meadows. This extends to the permanent snow line. Mountains are exposed to strong winds and heavy precipitation, but there are many sheltered places. Consequently, the climate varies over quite short distances. These variations are reflected in the type of plants.

Although all mountains support similar types of habitats, the composition of the plant and animal communities varies. The alpine plants of the Andes belong to entirely different species from those of the Himalayas, for example. Regardless of their elevation, plants growing in low latitudes receive more intense sunlight than those in high latitudes. Consequently, species that thrive in a particular temperature in the tropics may fail at the same temperature in northern Europe. Tropical mountains may support a type of tundra, but one that contains many more species than the tundra of the far north.

Outside the conventional biomes, there are organisms known as extremophiles that inhabit environments much too hot, cold, acid, alkaline, or saline for other species. Acidophiles thrive where the pH is below 5.0 and alkaliphiles where the pH is higher than 9.0. Halophiles inhabit very saline environments. Hyperthermophiles live in temperatures of about 105 °C. They can tolerate temperatures as high as 113 °C and some species fail to multiply at temperatures below

90 °C. Thermophiles prefer cooler temperatures, of about 60°C, and psychrophiles prefer temperatures below –15 °C. All extremophiles are single-celled and are classified in the domain Archaea.

34 Nutrient cycles

Plants synthesize carbohydrates directly from gaseous carbon dioxide and water, and use mineral compounds dissolved in soil water to synthesize the other substances they need to construct their own tissues. Animals consume the plants, carnivores consume herbivores, and the metabolic wastes and dead tissue from plants and animals provide food for another range of organisms, which break down complex organic molecules into simpler compounds that dissolve in water to be taken up once more by plant roots.

All the nutrient elements on which life depends are involved in cycles during which they may pass through the soil, water, air, and rock, as well as through living organisms. These biogeochemical cycles operate on a global scale; 'nutrient' cycles are those biogeochemical cycles involving elements necessary for life. Agriculture, manufacturing industry, and transport also move chemical elements between land, air, and water, nowadays in amounts equal to a significant proportion of those moving through the natural cycles. Of all the phosphorus carried by rivers to the sea, for example, human activities contribute an estimated two-thirds (BEGON ET AL., 1990, p. 705), and the carbon emitted by the combustion of fossil fuels is about 5.1 per cent to 7.5 per cent of the amount released through the respiration of the world's biota (BEGON ET AL., 1990, p. 708).

Cycling represents the movement of elements between reservoirs. Rocks, the oceans, the atmosphere, and the biota are reservoirs. The routes by which molecules move among them vary in length. A molecule released from the bedrock by weathering processes that produce soil may be absorbed by a plant root. It has now moved from one reservoir, the rock, to another, the biota. When the plant dies, that molecule will be returned to the soil as part of a much larger organic molecule. Decomposition will break the large molecule into smaller ones and the nutrient molecule may be taken up by the root of another plant. This is a very short cycle; if the plants involved are annuals it may take only a matter of months to complete. Should the molecule be released into the atmosphere rather than almost immediately being taken up by a plant, the cycle will take longer; the molecule will remain airborne for days or weeks until rain water washes it to the ground. If the molecule enters the ground water and moves from there to the sea, its cycle will be longer still.

Molecules may also enter reservoirs in which they remain for many millions of years. Consider, for example, the carbon that once resided in plants and, some 300 million years ago, became part of the organic material that was converted to coal, where it remains to this day. Calcium reaching sea water may find its way into the shells of marine organisms and from there enter sediments that eventually become limestone, where it remains until geologic processes expose it to the air once more and weathering releases it.

Reservoirs also vary in size. Far more nitrogen is in the air than elsewhere in its cycle. The atmosphere is also the principal reservoir for carbon, whereas the hydrogen that plants need for photosynthesis is derived from water, for which the oceans are the biggest reservoir. Rocks are the principal reservoir for many nutrient elements.

Living cells consist predominantly of carbon, hydrogen, and oxygen. Together with nitrogen (N) and phosphorus (P), these account for more than 1 per cent of their total dry weight. In addition to these, the biota as a whole require calcium (Ca), chlorine (Cl), copper (Cu), iron (Fe), magnesium (Mg), potassium (K), sodium (Na), sulphur (S), together with very small traces of aluminium (Al), boron (B), bromine (Br), chromium (Cr), cobalt (Co), fluorine (F), gallium (Ga), iodine (I), manganese (Mn), molybdenum (Mo), selenium (Se), silicon (Si), strontium (Sr), tin (Sn), titanium (Ti), vanadium (V), and zinc (Zn). To give an idea of the proportions in which these nutrients are used, a mature holm oak (*Quercus ilex*) forest was analyzed. The proportion of the above-ground dry weight accounted for by each mineral is shown in Table 4.1 (DAJOZ, 1975, p. 278). These figures refer only to a particular example and would be different for other plant communities.

Table 4.1 **Minerals in an oak forest as a proportion of the total**

Element	% of total
Ca	68
N	13
K	11
P	4
Mg	3
Na	0.6
Mn	0.3
Fe	0.3
Zn	0.1
Cu	0.09

Nitrogen, a constituent of proteins, moves through a fairly complex cycle, illustrated in Figure 4.3. Gaseous nitrogen (N_2) is the main ingredient of our atmosphere, but it is chemically unreactive and plants cannot use it directly. Lightning provides the energy to oxidize some, which then dissolves in rain droplets to form nitric acid (HNO_3), but 96–97 per cent of soil nitrogen is 'fixed' biologically. Soil bacteria, of the genus *Rhizobium*, live in nodules attached to the roots of legumes, the plant family of peas, beans, lupins, clovers, and their relatives. Other genera, including *Azotobacter* and *Clostridium*, live freely in the soil. They are able to use gaseous nitrogen directly, as are some aquatic cyanobacteria such as *Anabaena*. These organisms convert nitrogen into ammonia (NH_3).

Some of the nitrate (NO_3^-) and ammonia is absorbed by plant roots and some leaches from the soil to enter ground and surface water, where it nourishes aquatic plants. Plants are eaten by animals, which utilize their nitrogen in the form of the amino acids from which they synthesize proteins.

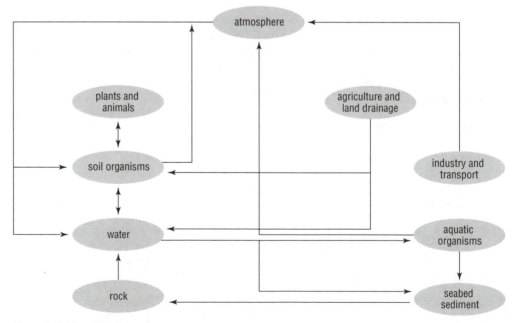

Figure 4.3 **The nitrogen cycle**

Organic wastes return to the soil and water, where other bacteria convert organic nitrogen into ammonia or ammonium (NH_4^+) compounds. Yet more bacteria, including *Nitrosomonas* and *Nitrobacter*, convert these into nitrites (NO_2^-) and then into nitrates (NO_3^-), which can be taken up again by plants.

In the oceans, some nitrogen forms part of the organic material that accumulates as sediment and this may eventually be transformed into sedimentary rock. Weathering reactions can then release the nitrogen into water, to re-enter the cycle.

Ammonia is volatile, and whenever it is released a certain amount evaporates. This takes it back to the atmospheric stage in the cycle, but not for long, because it soon dissolves in rain droplets and returns to the surface.

It is denitrifying bacteria which complete the cycle. Reversing the nitrifying process, various species convert nitrates to nitrites, nitrites to ammonia, and nitrates to gaseous nitrogen or nitrogen oxides.

Humans enter the cycle through the manufacture of nitrogen fertilizers, an industrial process that fixes gaseous nitrogen into a soluble compound. More than 50×10^6 tonnes yr^{-1} of nitrogen fertilizer is now being produced (BEGON *ET AL.*, 1990, p. 706), compared with the 54×10^6 t yr^{-1} fixed biologically (KUPCHELLA AND HYLAND, 1986, p. 55). High-compression internal combustion engines and high-temperature industrial furnaces and incinerators also cause the oxidation of nitrogen and its emission in exhaust gases. Such emissions cause pollution, with consequences that are fairly well understood, but the consequences of doubling the rate of the nitrogen cycle are unknown.

Carbon also resides in the atmosphere, as carbon dioxide, and passes through a similar cycle, illustrated in Figure 4.4. In this case, however, green plants can use the gas directly, the process of photosynthesis incorporating it in carbohydrates. These are consumed by animals and the decomposition of organic wastes involves the oxidation of their carbon, releasing it once more as

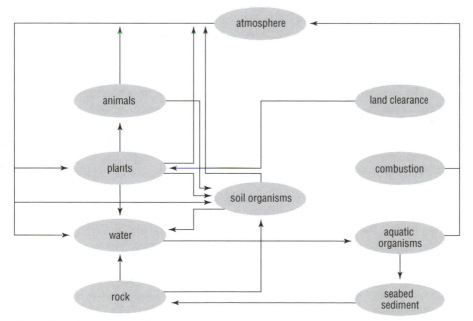

Figure 4.4 **The carbon cycle**

carbon dioxide. This release occurs through respiration, and all aerobic organisms, including all plants and animals, oxidize carbon in this way and excrete carbon dioxide.

Atmospheric carbon dioxide also enters water directly, dissolving to form carbonic acid (H_2CO_3). This dissociates into hydrogen (H^+) and bicarbonate (HCO_3^-) ions, then into more hydrogen and carbonate (CO_3^{2-}) ions. Carbonate ions combine with positively charged ions, such as calcium (Ca^{2+}), forming salts, some of which are insoluble, such as calcium carbonate ($CaCO_3$). In the oceans the formation of calcium carbonate is exploited by animals and some single-celled plants, but whatever the route by which it forms, calcium carbonate tends to accumulate in sediments in shallow water (below about 4 km, at a level called the carbonate compensation depth, the low temperature and high carbon dioxide saturation of sea water cause calcium carbonate to dissolve). Carbonate sediments may be converted into sedimentary rock, effectively removing carbon from the cycle until such time as the rock is exposed to weathering.

'Stored' carbon is returned to the cycle by human activities. The burning of fossil fuels is the best known and most important, many climatologists believing that an increase in the atmospheric concentration of carbon dioxide may perturb climates (see section 13). It is not the only human intervention, however. The kilning (heating) of limestone converts calcium carbonate to calcium oxide, or 'quicklime' (CaO), with the release of carbon dioxide; the subsequent addition of water produces calcium hydroxide ($Ca(OH)_2$), or 'slaked lime'. Calcium oxide is used mainly in the chemical process industries, and both products are used in construction, for example to make mortar, although this recombines the hydroxide to form calcium carbonate. Calcium hydroxide is also used in flue-gas desulphurization processes to remove sulphur dioxide from the gases produced by burning fossil fuel containing sulphur compounds, mainly in power stations. The exhaust gas is passed through an aqueous suspension of calcium hydroxide and the sulphur reacts to form insoluble calcium sulphate, which is precipitated. The technique removes sulphur dioxide, but at the cost of releasing carbon dioxide when the limestone is kilned to convert the carbonate to oxide.

Sedimentary rock may not be returned to the surface at all. Some is subducted at destructive plate margins, where one crustal plate is sinking beneath another. This returns the rock to the Earth's mantle. Nitrogen, carbon, and the other elements of which the rock is composed then form part of the interior of the Earth, returning to the surface, after an indeterminately long time, in volcanic eruptions.

All the nutrient elements are engaged in broadly similar cycles, but the atmosphere is not the only principal reservoir. Phosphorus is derived from rocks and released by weathering. Plants absorb it mainly as orthophosphate (the ions $H_2PO_4^-$, HPO_4^{2-}, and PO_4^{3-}). Some become incorporated in the bones of vertebrates, where it may remain for many years, but eventually the phosphorus is precipitated with ocean sediment and is converted once more into rock. Each year some 13×10^6 tonnes of phosphorus is removed from ocean water by sedimentation (BEGON ET AL., 1990, p. 702).

Sulphur, also an important ingredient of proteins and protein–carbohydrate complexes, is derived from both lithospheric (rock) and atmospheric reservoirs of comparable size. Volcanoes release a fairly small amount of sulphur. Several species of marine phytoplankton excrete dimethyl sulphide (DMS), possibly as a by-product of metabolic processes for osmotic regulation. Much of the DMS breaks down in the water, but some enters the air, where it is oxidized in several steps to sulphur dioxide and then to sulphate aerosol. This contributes about 44×10^6 t S yr^{-1}, and sulphate-reducing bacteria release gases, principally hydrogen sulphide (H_2S). Rock weathering also releases sulphur compounds. The amounts of sulphur reaching surface fresh waters from the atmosphere and the weathering of rocks are approximately equal.

Coal contains an average of 1–5 per cent S and oil 2–3 per cent (BEGON *ET AL.*, 1990, p. 707). When the fuel is burned the sulphur is released as sulphur dioxide unless measures are taken to remove it from exhaust gases. At present, the amount entering the air from fossil-fuel combustion is comparable to that entering the air by natural processes, but with the difference that combustion emissions are concentrated in industrial regions of the world.

Human intervention in the biogeochemical cycles is now occurring on a very large scale. This increases the amounts of nutrients available to plants. In the 1960s, for example, rainfall over most of Britain was delivering more than 12 kg S ha^{-1} annually and sulphur deficiency in farm crops was uncommon (COOKE, 1972, p. 75). Increasing the atmospheric concentration of carbon dioxide increases the rate of growth in many plants and the addition of nitrogen and phosphorus also stimulates growth. This is not necessarily welcome; the addition of phosphorus to lakes and slow-moving rivers can cause eutrophication (see section 23).

At the same time, such emissions and discharges can cause damage to wildlife and, in some cases, to human health. The fertilizing benefit does not justify the pollution cost. Both are indiscriminate and there are more efficient ways of supplying fertilizer to plants which need them.

35 Respiration and photosynthesis

Photosynthesis and respiration are the complementary sides of the process by which we, together with all aerobic organisms, obtain energy. That energy comes, of course, from the Sun. Photosynthesis is the series of chemical reactions in which light energy is used to synthesize carbohydrate; respiration is the series of reactions by which carbohydrate is broken down and oxidized to release energy as cells require it.

Respiration is not the same thing as breathing, although the words are often used synonymously. It is the series of chemical reactions by which cells derive energy from the metabolization of nutrients. Breathing is the mechanical pumping of air to bring it into contact with gill or lung membranes across which gases are exchanged: the process by which oxygen enters the body for the purpose of respiration and carbon dioxide, the by-product of respiration, leaves it and is exhaled.

It seems likely that breathing evolved in early fishes from movements originally linked to obtaining food (YOUNG, 1981, p. 102). In invertebrates, oxygen enters the body passively, by diffusing through openings in the body surface. Whether such animals can be said to breathe is questionable, but they certainly respire, as do all single-celled organisms and plants.

Historically, photosynthesis evolved before aerobic respiration. During the early evolution of life, the atmosphere contained little or no free oxygen, and organisms derived energy from other exothermic (energy-releasing) reactions. Some bacteria, descendants of the original anaerobes, continue to do so. Denitrifying bacteria obtain energy by reducing nitrate to nitrite or to gaseous nitrogen; others reduce sulphate or carbonate, releasing sulphide or methane as by-products. Hydrogen sulphide (H_2S) is produced by the bacterial reduction of calcium sulphate, and methane by the reduction of organic compounds.

Yeasts, which we use in brewing and bread-making, obtain their energy by breaking down glucose, as do aerobic organisms, but the breakdown does not proceed so far and the by-products are ethanol (alcohol) and carbon dioxide (which makes bread rise). The anaerobic reaction is $C_6H_{12}O_6 \rightarrow 2C_2H_5OH + 2CO_2$, the aerobic one $C_6H_{12}O_6 + 6O_2 \rightarrow 6CO_2 + 6H_2O$. Energy can be obtained by oxidizing alcohol; it can be (and is) used as a fuel. Because it does not proceed all the way, to the complete breakdown of glucose into carbon dioxide and water, anaerobic respiration, or fermentation, is less efficient than aerobic respiration.

Significantly, both these types of respiration involve detaching carbon from an organic molecule and oxidizing it to carbon dioxide. They form part of the carbon cycle, and other varieties of anaerobic respiration play an essential role in other biogeochemical cycles (see section 33). Indeed, were it not for them, nitrogen and sulphur would long ago have accumulated as stable compounds from which the elements could not be released and life on the planet might have come to an end.

Some bacteria can live either aerobically or anaerobically, depending on the environment in which they find themselves. Others are obligate anaerobes. Oxygen kills them and consequently they are found only in environments from which gaseous oxygen is rigorously excluded. Hydrogen sulphide, for example, is produced below the surface of waterlogged mud and, as its common name of 'marsh gas' suggests, methane is produced in stagnant water. Methane is also produced by bacteria that inhabit the airless digestive tracts of ruminant mammals and termites. The bacteria produce enzymes that break down cellulose and thus enable their hosts to digest plant material, a task which requires the tough cell walls to be broken and the cell contents released. In return, the bacteria enjoy an environment sheltered from the air, and abundant food. They live in the rumen (the first chamber of the ruminant stomach) in vast numbers: there are 10^{10}–10^{11} of them in every millilitre of rumen contents, as well as 10^5–10^6 protozoa, some of which feed on the bacteria.

Methane absorbs long-wave radiation and nowadays there is concern about its contribution to the greenhouse effect (see section 13), but this arises from an increase in its atmospheric concentration due to human activities, especially the great expansion in cattle and sheep farming and leaks from natural-gas pipelines (natural gas is principally methane). Methane-producing bacteria also inhabit the flooded fields in which rice is grown, and the expansion in rice production is associated with an increase in methane production. Whether steps should be taken to reduce emissions from these sources is a matter for debate, and we are now increasing greatly the amount of nutrients engaged in the global cycles, but the fact remains that the contribution of the anaerobes is crucial. For this reason it may be wise to protect certain of their habitats, such as estuarine mudflats and marshes.

Anaerobic environments are now relatively uncommon, but during the early stages of evolution there was no free oxygen on Earth. All environments were anaerobic and all organisms were adapted to them. This remained the condition of the Earth for a very long time. Living organisms are believed to have been present on Earth nearly 4 billion years ago. By about 3 billion years ago some of them, perhaps resembling present-day algae and cyanobacteria, were exploiting sunlight as a source of energy: they had evolved a method of photosynthesis. This uses carbon dioxide and water as raw materials for the synthesis of carbohydrate and releases oxygen as a by-product.

Oxygen is highly reactive. As it entered the environment in gaseous form, so it combined with substances around it. At that time, the environment was in a reduced state and little by little it became oxidized. Atmospheric methane was oxidized to carbon dioxide and water, carbon monoxide to carbon dioxide, and exposed iron rusted to form the banded iron formations (BIF) that are now among the most important sources of iron ore, in places hundreds of metres thick, extending over vast areas, and containing 40–60 per cent iron. Some sedimentary rocks, especially sandstones, acquired a coat of this haematite rust and became what are now known as the 'red beds'. This gradual oxidation of the environment began some 3.2×10^9 years ago, BIF were completed about 2.2×10^9 years ago, and the red beds formed between about 2.2×10^9 and 6×10^8 years ago (KUPCHELLA AND HYLAND, 1986, pp. 109–110).

Then, with nothing left to oxidize, oxygen began to accumulate in the air and dissolve in water. For organisms living at the time it was extremely poisonous, because what could oxidize minerals could equally well oxidize organic compounds, disrupting cell metabolism. To this day, much of the biochemistry in our bodies is dedicated to rendering surplus oxygen harmless. Should the mechanism falter, free oxygen can cause cancer and is believed to be involved in the processes of ageing (HARMAN, 1992). Most of the single-celled organisms alive while free oxygen was accumulating must have been killed by it in what was the most serious pollution incident the world has ever experienced (JOSEPH, 1990, pp. 99–103). Survivors from that global poisoning retreated to the airless muds where they have lived ever since, some migrating many millions of years later to the equally airless digestive tracts of animals.

Respiration in those days was anaerobic and, in the absence of aerobic respiration to utilize the oxygen released by photosynthesis, the oxygen continued to accumulate. By about 600 million years ago, at the beginning of the Cambrian Period, oxygen accounted for about 0.2 per cent of the atmosphere and it did not reach 2 per cent until 200 million years after that, at the commencement of the Silurian Period. Then the concentration increased steadily to its present 21 per cent, where it has remained fairly constant for the last 400 million years, although there have been excursions. There is evidence that in the late Palaeozoic Era, about 350 million years ago, the oxygen concentration rose, possibly to as much as 35 per cent, then decreased to 15 per cent in the space of 120 million years (GRAHAM ET AL., 1995). This fluctuation may have been due to biological events, but no one knows.

Photosynthesis, the source of the free oxygen, can be described simply: $CO_2 + H_2O \rightarrow CH_2O + O_2 \uparrow$, the energy for the reaction being supplied by light and captured by chlorophyll, the coloured compound contained in chloroplasts, cell organelles that are believed to be descended from free-living organisms. In fact, there are four groups of chlorophylls; those in land plants are known as chlorophylls a and b, those in marine algae c and d. Chlorophyll absorbs light, especially from the blue and red parts of the spectrum, and for each photon absorbed an electron is emitted. Water tends to dissociate spontaneously into hydrogen (H^+) and hydroxyl (OH^-) ions. Within the cell, some electrons from the chlorophyll combine with hydrogen ions, which move to a hydrogen acceptor (nicotinamide adenine dinucleotide phosphate, or NADP); the chlorophyll replaces its electron with one taken from a hydroxyl, and hydroxyls combine into water molecules with a surplus of oxygen ($4(OH) \rightarrow 2H_2O + O_2 \uparrow$). This is the source of the released oxygen. Other excited (energized) electrons emitted by chlorophyll are passed along a chain of donors and acceptors in the course of which they supply the energy to attach phosphate groups to adenosine diphosphate (ADP), converting it to adenosine triphosphate (ATP), the substance that transports energy, releasing it (by losing a phosphate) where energy is required; the process is known as photophosphorylation.

With its attached hydrogen, the NADP engages in a series of reactions, powered by the ATP, in which carbon dioxide is reduced and carbohydrates are synthesized. Because the first stage, directly involving chlorophyll, requires light, it is known as the 'light stage' of photosynthesis. The reduction of carbon dioxide and synthesis of carbohydrate is known as the 'dark stage'. The reactions are summarized, in very simple form, in Figure 4.5. They are described more fully in biology textbooks.

When the first organisms evolved a means of using free oxygen to oxidize carbohydrate they gained a major advantage. Respiration exploits exothermic reactions to produce ATP, the universal biological source of portable energy. A single anaerobic respiratory reaction produces 2 ATP molecules; an aerobic reaction produces 38. Inevitably, once aerobic respiration appeared the organisms equipped to use it prospered. This is why the world today is dominated by aerobes and, because of the energy available to them, probably why some were able to evolve to a much larger size than would be possible for an anaerobe.

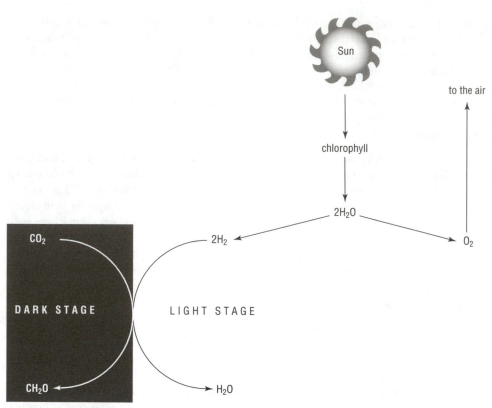

Figure 4.5 **Photosynthesis**

Photosynthesis consumes carbon dioxide and releases oxygen, respiration consumes oxygen and releases carbon dioxide. Essentially, this is the carbon cycle and it balances: the amount of carbon dioxide absorbed by photosynthesis equals the amount released by respiration (including respiration by plants). Energy derived from the burning of fresh plant material ('biomass fuels') does not affect the cycle, because precisely the same amount of carbon dioxide is released by combustion as was absorbed by photosynthesis during the growth of the plants.

There is a final twist to the story, and it concerns photorespiration. This is a version of respiration associated with the light stage of photosynthesis, but one that generates no ATP. It amounts to a waste of carbon dioxide by the plant. There is a group of plants, however, which do not photorespire, at least not to any significant extent. They are known as C_4 plants, because the first step of their photosynthesis yields a four-carbon compound (oxaloacetic acid); the other, C_3, plants produce a three-carbon compound (phosphoglyceric acid). C_4 plants grow faster than C_3 plants, because they synthesize more glucose for each unit of leaf area, they can photosynthesize at higher light intensities than C_3 plants and, because of the efficiency with which they use carbon dioxide, they can grow at lower carbon dioxide concentrations than C_3 plants.

C_4 plants occur naturally in low latitudes. Most are grasses and they are economically important. The group includes maize, sugar cane, and sorghum. Should the atmospheric concentration of carbon dioxide continue to increase, it is uncertain how C_4 plants might respond. C_3 plants, which include wheat, rice, sugar beet, and all trees, would benefit (provided other nutrients were available to them) but possibly at the expense of C_4 plants where the two groups grow in the same area.

Photosynthesis and respiration are the two processes on which all life is based, at least on our planet. They provide the mechanisms by which energy is captured and utilized.

36 Trophic relationships

Rabbits eat grass and foxes eat rabbits. The grass, rabbits, and foxes comprise a 'food chain', and at first glance the relationship appears simple. True, many food chains are longer, but not much longer. The grass → rabbits → foxes chain has three links, or levels, and some chains may have four or, more rarely, five. A tree leaf, for example, may be eaten by a caterpillar, the caterpillar by an insect-eating bird, and the insect-eating bird by a bird of prey. This chain, of leaf → caterpillar → bird (1) → bird (2), has four levels. To add a fifth level it would be necessary to find an animal that feeds on birds of prey. This would be difficult, perhaps impossible, and fifth levels are uncommon, because few carnivores specialize in hunting other carnivores.

It is not too difficult to see why this is so. Rabbits must eat a considerable quantity of grass, which is a not especially nutritious food, even for an animal such as a rabbit with a digestive system adapted to deal with it. The rabbit provides the fox with a substantial meal, but it is only one meal and the fox needs to eat every day. Thus the fox thrives only if there is a population of rabbits to keep it supplied, and the rabbit thrives only if there is a substantial area of grass on which it may graze. Take these together, omit the rabbit, and it is clear that the fox subsists on the produce of a considerable area of grass. Try to add a fifth level and the area of grass must be increased in proportion, to supply sufficient grass to feed a large enough rabbit population to sustain enough foxes to feed whatever monster it is that has taken to eating foxes.

This relationship can be measured and described in several different ways, which are explained in the next chapter. It can also be demonstrated in terms of the land area required to support an individual of a particular species, although this varies according to the size of the individual and the richness of the habitat. Rabbits do not travel far, most grazing within an area of no more than about 60×10^3 m^2 and many in a much smaller area. Rabbits released experimentally 600 m from the place where they had been captured found their way back to their burrows, but rabbits released 1 km from their burrows failed to do so (THOMSON AND WORDEN, 1956, pp. 104–105). Foxes seek food in ranges varying, in mid-Wales, from 2.5 to 15 km^2, females having smaller ranges than males (CORBET AND SOUTHERN, 1977, p. 317). Clearly, the bigger the animal the more space it needs. A European wildcat (*Felis silvestris*) has a range of 600×10^3 to 700×10^3 m^2, a tiger 65–650 km^2. Today we are concerned for the survival of tigers and similar large predators, but they are at risk precisely because they are large predators. Exclusively carnivorous mammals of their size occupy such extensive ranges that they were never common, and any fragmentation of their habitat that breaks up the ranges into smaller 'islands' with barriers the animals cannot cross threatens them with starvation, even if the total area of suitable habitat is little reduced.

The wolves of Isle Royale

Populations are usually regulated by the food supply available to them. In the grass → rabbits → foxes food chain, the number of foxes in an area is ultimately determined by the amount of grass available to the rabbits. While this is usually the case, there are exceptions.

Isle Royale, in Lake Superior, U.S.A., is a national park in which moose (*Alces alces*) feed on balsam fir (*Abies balsamea*) and grey wolves (*Canis lupus*) feed on moose. In winters when there is more snow than usual, moose cannot move so quickly and wolves hunt in larger packs. The wolves kill more moose than they do in years when there is less snow and they hunt in smaller packs. Because of this, winters with heavy snow lead to a reduction in moose numbers. This leads to a reduction in browsing and is followed by an increase in the growth of balsam fir.

Changes in the weather cause changes in the hunting behaviour of wolves, with an effect that cascades through the populations of moose and balsam fir (POST ET AL., 1999).

Feeding relationships, based on grouping species according to their diets, are known as 'trophic', from the Greek *trophe*, meaning 'nourishment'. They can be generalized. At the base of every food chain there are 'autotrophs', organisms that can synthesize organic molecules from inorganic ones. Autotrophs may be 'photoautotrophs', using photosynthesis (see section 34) to combine carbon and hydrogen into carbohydrate, or 'chemoautotrophs', which derive energy from the oxidation of other compounds, but also synthesize organic molecules. Autotrophs are primary producers. Green plants are photoautotrophs and are by far the commonest primary producers in both terrestrial and aquatic food chains.

All other organisms, which cannot synthesize organic compounds from inorganic precursors, are 'heterotrophs' ('other nourishment'). All animals, fungi, and some bacteria are heterotrophs. They are described as consumers, because their food comprises organic material obtained by consuming other organisms, and they form a hierarchy. Herbivores, which feed exclusively on plant material, that is, on the primary producers, are primary consumers. Secondary consumers feed on primary consumers and are, therefore, carnivores. Carnivores which feed on other carnivores are tertiary consumers. Any food chain can thus be represented as: primary producer → primary consumer → secondary consumer → tertiary consumer.

The food chain concept is extremely useful, but only up to a point. It has been used, for example, to trace the fate of certain chemically stable compounds that are consumed by organisms low in a food chain and then concentrated at higher levels. DDT is the best-known example. In the 1950s, large amounts of DDT were used in Illinois to control the beetle (*Scolytus scolytus*) that spreads the fungus (*Ceratocystis ulmi*) which causes Dutch elm disease. The insecticide was ingested by earthworms, presumably from eating dead leaves coated with it. They were not harmed by it, but in spring American robins (*Turdus migratorius*) feed mainly on earthworms. The DDT they ingested with each earthworm accumulated, eventually to a lethal dose. Earthworms were found to have a sublethal concentration of 33–164 mg DDT kg^{-1} (milligrams per kilogram of body weight) in their bodies, and autopsies revealed that robins had an average of 60–70 mg DDT kg^{-1} in their brains. This story was recounted by Rachel Carson in *Silent Spring* (CARSON, 1962, pp. 102–105), but although the effect was very serious, it was also local. Many birds died, but in the United States generally, the overall population of robins actually increased during the period of most intensive spraying (MELLANBY, 1992, pp. 62–63).

Such stories gave rise to the popular and widely accepted view that stable compounds, such as DDT, can accumulate along food chains in a simple fashion. Each consumer stores the dose acquired from each of the organisms it eats and eventually an organism high on the food chain receives a lethal dose. In fact it is far from simple. The substance may be excreted or converted metabolically into harmless compounds, and even when it is stored in the body the amount may peak before it reaches a lethal concentration. In humans, for example, DDT continues to accumulate in the body for many years, but eventually reaches a peak, far below the concentration that could be harmful, after which it is excreted at approximately the same rate as it is being absorbed (MELLANBY, 1992, p. 67).

This suggests that the food-chain concept should be used cautiously, but it is not its principal limitation. In the real world, few organisms are restricted to a single food item. Foxes do eat rabbits, but rabbits are not all that they eat, and rabbits graze plants other than grass. Figure 4.6 illustrates a very simplified trophic structure for a temperate-climate pond. In this diagram, organisms are arranged in five trophic levels, with plants, the primary producers, at level 1 and birds, quaternary consumers, at level 5. Birds, however, do not feed exclusively on the fish that comprise level 4; they also feed on all the organisms at level 3 and on several of those at level 2. The only general rule to be derived from the diagram is that organisms never feed on those at a trophic level higher than their own, and, although generally true, this is more a logical consequence of the way trophic levels are defined than a reflection of what you might observe were you to study pond life. Dragonfly nymphs, for example, capture and eat fish larvae and sometimes small fish. Most animals, in fact, eat a fairly wide variety of food, depending on what is available, and diets often change from season to season. Over the course of the year, the blackbird (*Turdus merula*) has a typically varied diet, summarized in Table 4.2 (HILLSTEAD, 1945, p. 72).

Table 4.2 **Items making up the diet of the blackbird Turdus merula**

Item	% of total
Animal	
Insects (pest species)	22
Insects (beneficial)	3.5
Insects (other)	5.5
Earthworms	4
Snails and slugs	2.5
Miscellaneous	1.5
Vegetable	
Fruit (cultivated)	25
Fruit and seeds (wild)	24.5
Wheat	2.5
Roots	2.5
Miscellaneous	6

When linked, food chains become 'food webs', and even when reduced to their barest essentials these are usually very complex. Figure 4.7 illustrates a food web typical of European heathland (GIMINGHAM, 1975, pp. 55–59). This is also arranged into trophic levels, of which in this case there are four, but as with the pond food web, animals at higher levels feed on organisms from more than one lower level. Wildcats, for example, hunt mice and voles, which are primary consumers (vegetarians) at level 2, but will also eat shrews, which feed on insects and spiders and are consequently secondary consumers at level 3. Frogs, lizards, and meadow pipits also straddle trophic levels by feeding on both herbivorous (level 2) and carnivorous (level 3) insects, and they themselves are preyed upon by carnivores at higher levels.

Except in environments with few species, food-web diagrams must necessarily group species together; to list them all would make the diagrams so intricate as to render them incomprehensible. Some 'insects' feed on leaves, others on shoots, or sap. All are primary consumers, but their feeding preferences have different effects on plants, and although sheep are the only large herbivorous mammals listed, heathlands also support deer and cattle, which feed on different parts of the vegetation. Nor do most food-web diagrams include organisms that feed mainly or exclusively on dead plant material – the decomposers which complete the nutrient cycle.

Illustrations of food webs provide more information than do food-chain diagrams, but their principal value lies in their ability to identify the ecological niche (see section 37) each group occupies. Relationships in almost all natural environments are extremely complex. By expressing them, in effect, as niches, food-web diagrams are useful summaries of that complexity.

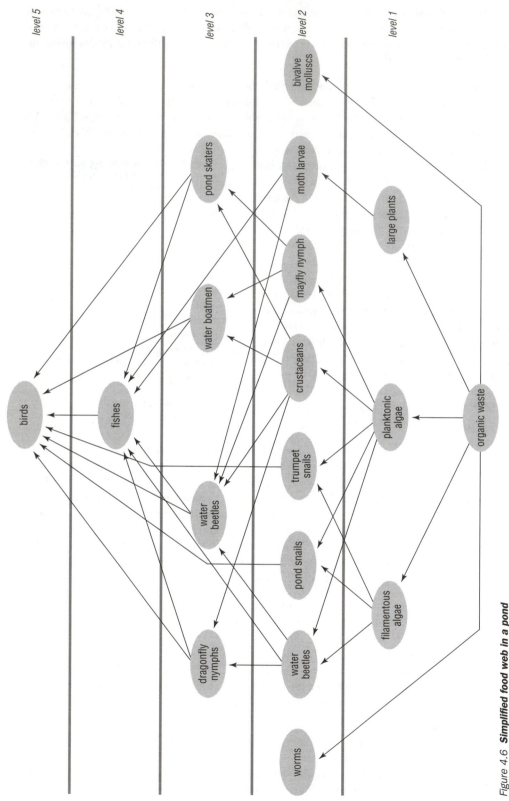

Figure 4.6 *Simplified food web in a pond*
After Dowdeswell, W. H. 1984. Ecology: Principles and practice. *Heinemann Educational Books, Oxford*

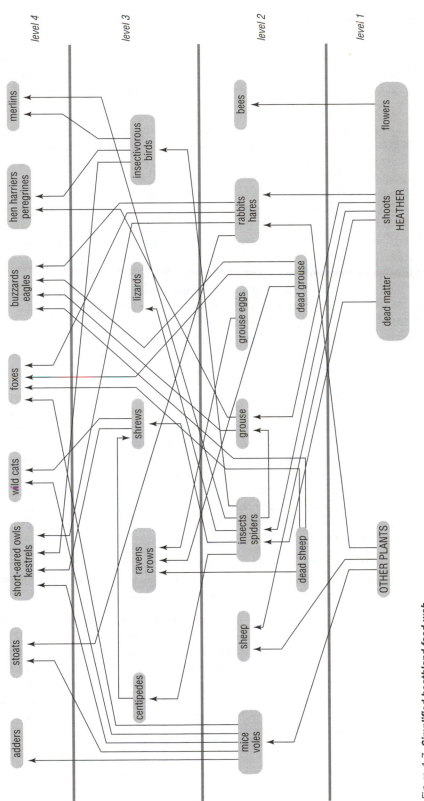

Figure 4.7 *Simplified heathland food web*
After Gimingham, C. H. 1975. An Introduction to Heathland Ecology. Oliver and Boyd, Edinburgh

Biosphere / 159

37 Energy, numbers, biomass

Many plants are required to sustain a herbivorous animal and many herbivores are required to sustain a carnivore. Anything that affects the size of a population at one level must have repercussions at higher levels, but these are not always straightforward, such is the complexity of the food webs of which food chains are only a part.

In the simple grass → rabbits → foxes chain, a reduction in the rabbit population might be expected to result in an increase in the amount of grass (fewer consumers) and a decrease in the fox population (fewer prey). Some years ago, this happened in Britain, when large numbers of rabbits died from myxomatosis. Predator populations fell. In some places, stoat numbers were reduced to one-third of their pre-myxomatosis levels, rabbits being the principal prey of stoats (POLLARD ET AL., 1974, p. 139). No longer nibbled, grasses also grew taller; and that was where the complications began. Being taller, they cast more shade on the ground, which suppressed the growth of many small herbs, including wild thyme (*Thymus drucei*), which until then grew in a few places in Devon and Cornwall. When they first hatch, larvae of the large blue butterfly (*Maculinea arion*) feed on wild thyme and are camouflaged to resemble its flowers. After their third moult, the caterpillars leave the thyme and enter into a symbiotic relationship with ants, living in ant nests until they pupate. As the grasses grew and the thyme disappeared from the sward, adult butterflies were unable to lay their eggs in appropriate places, and on 12 September 1979 the large blue butterfly was formally pronounced extinct in Britain, its last colony, in Devon, having died out (ALLABY, 1981, pp. 142–143). A reduction in the population of primary consumers led to an increase in the producer population, but in this case it also caused the decline of another primary consumer.

Recognizing the need to quantify the organisms at each trophic level, in his book *Animal Ecology*, published in 1927, the British ecologist Charles Elton (1900–91) proposed a method for showing graphically the relationships between levels. If the producer organisms are represented by a rectangle of a size proportional to their number, and similar rectangles are drawn one above another to represent the numbers present through the hierarchy of consumer organisms, the result is a stepped pyramid. Consumer organisms ingest organic matter and convert it into a different kind of organic matter: rabbits convert grass into 'rabbit stuff' and foxes convert 'rabbit stuff' into 'fox stuff'. Consumers, therefore, are also producers from the point of view of those at higher trophic levels, and are sometimes described as secondary or tertiary producers (but not primary producers, of course).

Figure 4.8 shows such a pyramid for four trophic levels comprising the typical population living on temperate grassland. In this form it is known as a 'pyramid of numbers' and it illustrates the way numbers at each level are usually about one-tenth of those at the level below.

In the example chosen, however, the relationship does not hold at the highest level, of tertiary consumers: 10 000 secondary consumers provide sustenance for no more than 10 tertiary consumers, not the 1000 that might be expected. This arises from the organisms themselves. At levels 2 and 3 these are insects, and at level 4 they are insectivorous birds and mammals, which are very much larger than insects. If animals, or plants for that matter, at one level are of markedly different size from those at another, their relationship will not be revealed very clearly merely by counting them. Many more herbaceous plants than large trees are needed to support a herbivore population, and a given number of plants can support many more rabbits than cattle.

Numbers can be misleading, but it should be possible to relate organisms of different sizes by an equation along the lines of: x herbs = 1 tree, or x rabbits = 1 cow. This leads to the concept of 'biomass' or 'standing crop', which is simply the total mass of organisms, usually expressed as

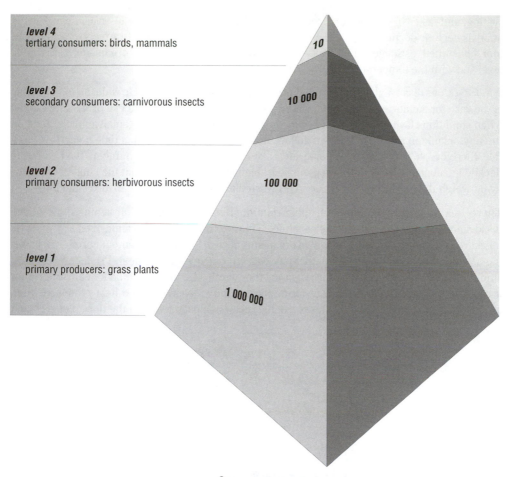

level 4
tertiary consumers: birds, mammals

10

level 3
secondary consumers: carnivorous insects

10 000

level 2
primary consumers: herbivorous insects

100 000

level 1
primary producers: grass plants

1 000 000

Figure 4.8 **Pyramid of numbers per 1000 m² of temperate grassland**

their dry weight, because the amount of water held in tissues varies widely from one species to another. In principle, it is obtained by collecting, drying, and weighing, although in the real world this is rather more easily said than done. Removing the entire root system of a plant can be difficult, some small and very specialized primary consumers are liable to be missed, and some plant parts are temporary, such as leaves that die back before flowering (BREWER, 1988, p. 318).

Nevertheless, biomass has been estimated for various biomes and for the world as a whole. In tropical rain forest, for example, the mean biomass is 45 kg m^{-2} of ground surface, in temperate deciduous forest it is 30 kg m^{-2}, and in tundra it is 0.6 kg m^{-2}. The global biomass, counting both land and marine organisms, is 1841×10^9 t (BEGON ET AL., 1990, p. 652).

Biomass can be estimated for each trophic level and drawn as a stack of rectangles, in the same way as numbers. This produces a second type of pyramid, called a pyramid of biomass. Usually, this slopes less steeply than the pyramid of numbers.

Although the pyramid of biomass is more useful than the pyramid of numbers, unfortunately it does not quite resolve the difficulty. In particular, it takes no account of metabolic rate, which is the rate at which an organism converts food into energy for its own use. This is related to body size, activity, and temperature. We metabolize faster when we are active than we do when resting, and faster in cold weather than in warm weather, but in cold weather plants and many animals become dormant and their metabolism slows. Since metabolic rate affects the amount of food consumed, estimates of biomass can vary with the average size of organisms at a trophic level and with the temperature.

A more serious disadvantage emerges where population sizes are subject to wide fluctuation. This is the case for aquatic environments, where phytoplankton (small plants) can proliferate rapidly when conditions favour them, then decline just as rapidly. Zooplankton (small animals) feed on the phytoplankton and their numbers follow those of the plants, but with a time delay. It is possible, therefore, to take a sample shortly after a decline in phytoplankton numbers but before the decline in zooplankton, and draw a pyramid of biomass that suggests that the consumers heavily outweigh the producers; the pyramid is inverted and this is clearly an absurd situation.

There is a third way to express trophic relationships, however, which avoids most of the disadvantages and is the one ecologists prefer. This is expressed as a pyramid of energy. What is measured is not the number or weight of organisms at each level, but the amount of energy, entering as sunlight and becoming food, that passes from one level to the next. Producers 'capture' sunlight for photosynthesis. The carbohydrate they synthesize provides food for consumers, but it can also be considered as the transfer of energy and measured as the heat produced when organic matter burns. Figure 4.9 shows trophic relationships in terms of the flow of energy and

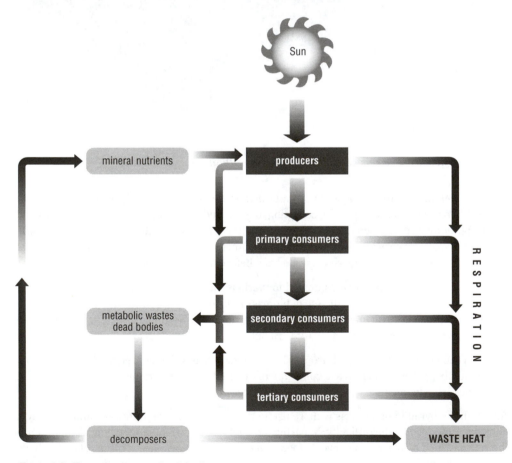

Figure 4.9 **Flow of energy and nutrients**

mineral nutrients. In doing so it explains why numbers and biomass decrease so sharply at higher trophic levels as to produce pyramids. All organisms, including plants, use most of the energy they receive in respiration. Carbohydrates are oxidized to drive the ADP–ATP mechanisms that provide cells with the energy to maintain themselves, and whenever energy is changed from one form to another a proportion is lost as low-grade heat. Unless biomass is increasing, as when a newly planted forest is still growing, for example, eventually all the original solar energy is transformed into waste heat. Biomass then remains constant. In a mature forest, it has been calculated that total photosynthesis delivers 188 MJ m^{-2} yr^{-1} (MJ = megajoule), plant respiration uses 134 MJ m^{-2} yr^{-1} and respiration by all heterotrophs uses 54 MJ m^{-2} yr^{-1}; all incoming energy is used in respiration and ends as waste heat (BREWER, 1988, p. 319).

Of course, the forest may still be growing, or declining, and it may be useful to know this. The technique for estimating overall increase or decrease begins by measuring or calculating the total amount of sunlight used in photosynthesis by plants within a defined area. This produces a figure for gross production (G). The amount of energy used in respiration (R) is then calculated. What remains is the net production (N). It remains distributed among the plants present in the area and indicates an increase in their growth or numbers, its value being calculated as $N = G - R$. If N is positive, growth is occurring, if negative the plant community is in decline.

Unfortunately, the results can be misleading, because the calculation takes no account of the herbivores feeding on the plants. In fact, N also equals $T + C + D$, where T is the increase in plant biomass, C the amount consumed by herbivores, and D the amount lost through the death of plants and the shedding of dead parts, such as leaves and old branches. In the case of agricultural or forestry crops, an ability to calculate T accurately is of obvious value.

Techniques have been developed for acquiring the data needed to make these calculations. Instruments can measure the amount of sunlight to which the plants are exposed over a year. The amount of that energy used by different plant tissues is calculated from measurements of the amount of heat energy they yield when burned. The stomach contents of animals can be measured, as can amounts of dead material reaching the ground surface.

Production and consumption are being measured in terms of energy, and the results of those measurements describe how energy flows from producers to the hierarchy of consumers. Displayed graphically, it forms the third and most useful of the ecological pyramids: a pyramid of energy.

In some well-studied cases energy flow has been measured in great detail, and, in addition to providing information about the plant and animal community as a whole, such studies can also throw light on the efficiency with which energy is used at each trophic level. In particular, they can determine how efficiently herbivores are consuming plant material and how this changes over time, perhaps from season to season. This is of considerable interest to those seeking to grow a commercial plant crop.

'Ecological energetics', the branch of ecology that deals with the flow of energy, represents a boundary zone in which life scientists and physicists meet, but as ecologists. Its existence and great importance in ecological studies demonstrates the interdisciplinary character of all the environmental sciences.

38 Ecosystems

It is all very well to measure the flow of energy through the trophic levels of a community of organisms, but limits must be set to the boundaries of that community. It must be defined in such a way as to distinguish it from other adjacent communities. Such a boundary will imply nothing about the size of the community; it may be as small as the untended corner of a field or as large as a forest. What matters is that in some way it is noticeably different from the communities adjoining it. An area of woodland, for example, may be limited by the boundaries of the cultivated fields surrounding it, a pool by its banks, a marsh by the adjoining drier land where the water table is lower.

There are several ways of approaching the task. In Europe it was undertaken primarily by botanists, who devised ways of classifying plant communities and, from this, a discipline called 'phytosociology', or the sociology of plants. A leading figure in this development was a Swiss botanist, Josias Braun-Blanquet (1884–1980). He began his career in Zürich and in 1930 became the first director of the Station Internationale de Géobotanique Mediterranéenne et Alpine, at Montpellier. The scientists who worked with him from about 1913 became known as the Zürich–Montpellier (or ZM) School and their ideas are still influential.

In their system, sometimes known as 'Sigmatism', from an acronym of the name of the Montpellier Station (BOWLER, 1992, p. 526), plant associations were classified in ways similar to the taxonomy used to classify species. It begins with the concept of the 'minimal area', the smallest area in which a particular plant association can develop fully. For oak woodland this is about 200 m^2, for acid grassland about 9 m^2, and minimal areas have been calculated for every type of vegetation. A stand of plants can be examined only if it covers at least the minimal area for its type. The stand is then sampled within a marked area or quadrat, known as a relevé or Aufnahme, which is usually half the area of the stand and must be larger than the minimal area. All the plants within the relevé are recorded with estimates of the area each covers and the way they grow: solitarily, as clumps, small or large patches, or large colonies. This is their 'sociability'. Relevés are then grouped into classes, called phytocoena, which can be compared.

A somewhat similar scheme was developed at about the same time as the ZM School at Uppsala, Sweden, by ecologists led by J. Rutger Sernander (1866–1944) and later by Gustaf Einar Du Rietz (1895–1967). The Uppsala School called their basic plant unit a 'sociation' and sociations with the same dominant species could be merged, as 'consociations' (MOORE, 1982, pp. 59–62).

During the early years of this century, the foremost British ecologist was Sir Arthur George Tansley (1871–1955), who became a prominent conservationist. Tansley also held that the key to understanding natural communities lay in studying plant associations, but he placed great emphasis on the relationships among plants and between plants and animals, what he called 'biotic factors'. He described his scheme in *Practical Plant Ecology*, published in 1923. He called the basic vegetation unit a biome. An updated and revised version of this book was published in 1946, with the title *Introduction to Plant Ecology*. In it he introduced to a popular readership a word he had coined in 1935, in an article in the journal *Ecology*, which had been absent from the earlier edition. It is a word that has become familiar to us all. 'A wider conception still is to include with the biome all the physical and chemical factors of the biome's environment or habitat – those factors which we have considered under the headings of climate and soil – as parts of one physical *system*, which we may call an *ecosystem*, because it is based on the οικος or home of a particular biome' (TANSLEY, 1946, p. 207).

In the United States (*www.bio.swt.edu/simpson/ecology/ecointro.html*), meanwhile, ecologists were strongly influenced by a Danish botanist, Eugenius Warming (1841–1924). Warming recognized what he called plant 'communities' that were influenced in their development by other organisms, such as parasites. He maintained that a plant has certain physical capabilities which determine where it can grow. Many eminent ecologists adopted the Warming approach, but eventually one group broke away from it. Frederic E. Clements (1874–1926) was one of a team of botanists who embarked on a study of the ecology of the prairie, a task that was considered urgent, because the natural grassland of North America was everywhere threatened by agricultural expansion. The European method, which relied on the skill of an observer to characterize vegetation types, was useless on the prairie. Clements and his colleagues devised a more rigorous method. They marked off measured areas, called 'quadrats', of varying sizes but usually one metre square, and recorded every plant growing within them by species and by number in each species. A number of quadrats sited randomly over a large area allowed the distribution of species to be determined fairly accurately and provided raw data for statistical treatment.

Clements also cleared selected quadrats of all vegetation, then monitored the way species recolonized them. This led him to his theory of succession leading to a climax (see section 38).

The British and American approaches to ecological studies strongly influenced one another. Today they have almost merged, and the ecological methods based on the ecosystem concept that are taught in Britain and North America are very similar. It is important to remember, however, that the tradition of the ZM and Uppsala Schools remains strong elsewhere in Europe, and with the introduction of the National Vegetation Classification (NVC) British ecologists have also embraced the phytosociological approach.

An ecosystem, then, comprises all the organisms living within a definable area and ecosystem studies take account of relationships among them and of the physical and chemical factors affecting them. Figure 4.10 illustrates, very simply, the way an ecosystem may be structured. Its energy is supplied by the Sun and transmitted from one trophic level to another (see section 36). Its

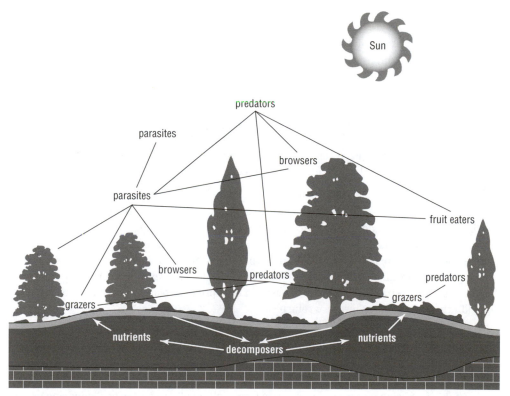

Figure 4.10 *Ecosystem*

chemical nutrients are supplied from the underlying bedrock by means of weathering, to which the soil biota contributes. Green plants, here trees, shrubs, and herbs, are the primary producers. Producers compete with one another for light, water, and nutrients. The herbs are grazed, the leaves of trees and shrubs browsed, and some animals feed on fruits. Fruit-eaters, grazers, and browsers compete with one another for food and for other resources such as nesting sites. All the primary consumers (herbivores and fruit-eaters) are subject to predation, and there are also carnivores, such as birds of prey, that will take small predators; they are tertiary consumers. All the plants and animals are hosts to parasites and these have parasites of their own (known as 'hyperparasites' or, if they kill their hosts, 'parasitoids'). All these are also subject to predation. Metabolic wastes and dead organic material provide food for decomposers, which form another trophic hierarchy (which can be represented by inverted pyramids otherwise similar to the other ecological pyramids).

The illustration allows for parasitism, but makes no mention of mutualistic relationships, which are often important. These occur when two different species live in close association with one another to the benefit of both; if one benefits but not the other, the relationship is known as 'commensalism'. The term 'symbiosis', by which people often mean what ecologists call 'mutualism', is now used to describe all close relationships between organisms of different species; it includes commensalism, mutualism, and parasitism.

As the illustration suggests, ecosystems exist in three dimensions, although one of those dimensions may be small. Animals that live on the surface of water, known collectively as 'pleuston', and the bacteria, microscopic plankton (called 'nanoplankton'), and minute animals that live in the uppermost few centimetres of the sea (called the 'neuston') inhabit an ecosystem as wide as the ocean itself, but only a few centimetres deep (MARSHALL, 1979, pp. 42–43).

Other ecosystems have to be considered in all three dimensions, the most obvious being forests of all types. These are dominated by mature trees, most of them so spaced that their foliage overlaps to form a canopy. Below them there are young trees and shrubs, the young trees unable to grow taller for want of light for photosynthesis. When mature trees die and fall a gap appears in the canopy, sunlight penetrates more strongly to lower levels, and a young tree completes its growth to replace the dead tree; the canopy closes once more. Lower still there are tall herbs and below them small herbs and such plants as mosses. Figure 4.11 shows the arrangement schematically, with the several forest strata labelled. Each of the layers supports its own population of specialized consumers, with their predators and parasites, and so although the forest comprises a total ecosystem, the layers can also be regarded as smaller ecosystems within it.

Ecosystems are alive, of course, but they are also like supermarkets, vast stores of resources offering space, shelter, food, and drink to those who visit them. As in supermarkets, some of those resources are alive; some, such as water, are not. Unlike supermarkets, they require that most of the goods they display be consumed on the premises, although limited exceptions are permitted. Birds, for example, may gather food in one ecosystem but roost in another. In return for this licence, they distribute seeds and so assist in the spread of plants. Flying insects have the same liberty and pollinate flowers in return.

There is another difference. Resources are not advertised as such, but rather presented as raw materials from which customers may make what they will, as though supermarkets displayed not vegetables, but seeds and soil; not butter, but grass and cows. A member of a species entering an ecosystem afresh must explore, seeking the resources it needs. If it finds them it may settle and establish itself, utilizing a space or food source for which no other species has found any use. It makes a home for itself, finds the food it prefers, and is said to have created a 'niche'.

Figure 4.11 **Forest stratification**

Niches do not exist until organisms fill and thus define them, but once a niche is defined it can be vacated. Then another member of the same species, or of a different species, may enter to occupy it. The concept is of function: the niche describes the occupied habitat, what its occupant does, how it feeds, when and for how long it is active, and when and for how long it is present and absent.

Ecosystems must be definable, but their borders are not always sharp. Except where cultivation or other human activities produce sharp boundaries, they more commonly shade into one another. Forests, for example, do not simply stop at their edges. The trees become more widely scattered, different species thrive, often of smaller trees and shrubs, and the forest seems to merge into the different ecosystem adjoining it. Such border regions are called 'ecotones' and are often richer in species than the ecosystems to either side. They support many of the species from both adjacent ecosystems as well as species peculiar to themselves.

Studies of ecosystems reveal how species interact, and they allow an important generalization. No matter how different two ecosystems may appear, as systems they function in very much the same way. This means that ecologists can use the same approach to unravel the complexities of rain forests, grasslands, deserts, or any other type of community, including the artificial ones produced by humans. Agricultural ecosystems are of great interest and even private gardens in towns prove rewarding subjects.

In recent years there has been much popular concern over the supposed fragility of ecosystems. Here it is impossible to generalize; each case is different. It is characteristic of all systems that they maintain themselves in a fairly constant state and ecosystems are no different in this regard, reacting to perturbations in ways that restore their integrity. Not all ecosystems are equally robust, however, and for any ecosystem there is a degree of disturbance that exceeds its tolerable limits, after which it cannot recover.

Fears of damaging ecosystems are based on the sound conservationist principle that we should aim to minimize the disruption we cause, but there is a risk that this principle may be confused with the old idea of a 'balance of nature'. This supposes a perfect order of nature that will seek to maintain itself and that we should not change. It is a romantic, not to say idyllic, notion, but deeply misleading because it supposes a static condition. Ecosystems are dynamic, and although some may endure, apparently unchanged, for periods that are long in comparison with the human life-span, they must and do change eventually. Species come and go, climates change, plant and animal communities adapt to altered circumstances, and when examined in fine detail such adaptation and consequent change can be seen to be taking place constantly. The 'balance of nature' is a myth. Our planet is dynamic, and so are the arrangements by which its inhabitants live together.

39 Succession and climax

When buildings are demolished and the land cleared, after a short time plants appear. Before long they blanket the site, their flowers cheering the desolate landscape, their seeds blowing in the wind. One of the most poignant features of war is the speed with which wild flowers colonize the ruins of bombed and shelled cities. In Britain, the symbol by which we remember the war dead is the common field poppy, the flower that coloured the battlefields of northern Europe.

These plants are opportunists. Their seeds are everywhere and those of some can remain in the soil, dormant but viable, for a very long time. Unable to survive in competition with most other plants, when the land is bare they germinate, grow rapidly, flower, and produce seed before more aggressive species arrive and they lose their advantage. Many have brightly coloured flowers to attract pollinating insects. They are annuals.

Farmers and gardeners regard them as weeds, but ecologists recognize them as members of a pioneer community, the first colonizers to arrive on newly exposed ground. They are tolerant of strong, direct sunlight, they can grow in a soil that contains few nutrients and is too acid or alkaline for most plant species, and neither drought nor temporary waterlogging can destroy them.

Soon, other plants begin to appear, their seeds borne on the wind or dropped by birds, and the composition of the plant community changes. Some of the primary colonizers vanish from the scene. The new plants include biennials and perennials, in which the vegetative parts do not die at the end of a growing season, although the stems and leaves may disappear. Many perennials spread vegetatively as well as producing seed. The annuals produce their seed, but in the following year it is unable to germinate, because all its previous sites are now occupied.

Woody plants, which grow more slowly than herbs, also establish themselves. They grow taller than the herbs and shade them, so the community changes yet again as it becomes dominated by shrubs. Among the woody plants are trees, which grow even more slowly. Eventually, however, they shade out some of the shrubs. The early arrivals tolerate exposure to full sunlight but, as the taller ones shade the ground, conditions develop that are suitable for shade-loving plants.

Leave the bare ground undisturbed and little by little it will be transformed. In most parts of lowland Britain it is likely to develop into woodland. The sequence of distinct vegetation types is called a 'successional series' or 'sere', each type within the series being a 'seral stage', and the process is known as 'succession'. Figure 4.12 illustrates five seral stages in the succession from open ground to mature woodland. It starts with annual herbs, only partly covering the ground,

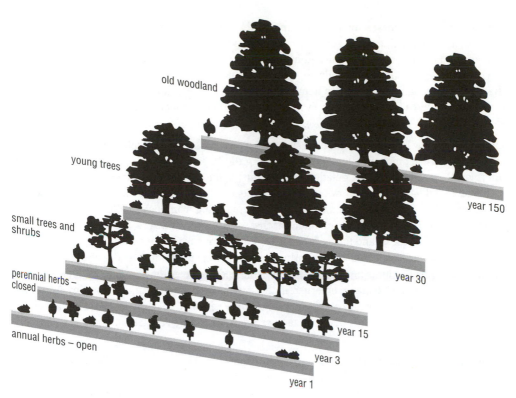

Figure 4.12 **Succession to broad-leaved woodland**

some of which remains bare. These are succeeded by perennials, which establish a complete ground cover, and are followed in turn by small trees and shrubs, young woodland trees and, finally, the mature woodland. The full succession takes about 150 years from the arrival of the first annuals to maturity.

It is not only bare soil that will be colonized in this way. Any open surface may provide a site for a succession provided it offers some physical stability. A newly exposed sand bar will be colonized, as will a lava flow once it has cooled, and a succession does not need to occur on a large area. When a forest tree falls, exposing an area of ground to full sunlight, a 'microsere' ensues as woodland herbs flourish and the dead tree decomposes, the site eventually returning to its original, woodland, condition. This fact even has forensic applications. A study of the organisms present in a decaying corpse provides information about the time elapsed since death, because decomposition is effected by a series of fungi, flies, beetles, and bacteria that appear in a known order, a 'carrion microsere' (BEGON *ET AL*., 1990, p. 383).

A succession that begins on very dry land, such as bare sand or rock, is called a 'xerosere' (from the Greek *xeros*, meaning 'dry'). It can be seen on stabilized sand dunes or in abandoned quarries. If a succession begins in water it is known as a 'hydrosere'.

Aquatic environments sometimes border waterlogged ground with dry ground beyond and in such places it is possible to see all the seral stages in a hydrosere at the same time. Sediment on the lake bottom provides anchorage for plants such as water lilies. Close to the shore, sedges grow in partly decomposed organic matter, sedge peat, undercut by a wedge of lake water. The sedge peat adjoins a band of sphagnum peat lying above rock. Sphagnum peat develops beneath mosses of the *Sphagnum* genus, plants which can absorb and hold large amounts of water. In hollows, *Sphagnum* mosses sometimes form raised bogs supporting a variety of herbs and woody plants that rely on rain for their supply of water and mineral nutrients. Such bogs and peat, lying well above the water table and isolated from it, are known as 'ombrogenous'. Still further from the lake edge, a deep layer of woody peat supports coniferous trees. Birch trees grow on part of the woody peat that is overlain by a humus-rich soil layer. The humus also extends further from the water to where it overlies a typically dry-land soil profile supporting larger trees. This succession is illustrated in Figure 4.13.

Land clearance by humans will initiate a succession on a subsequently undisturbed site, but most successions are set in train quite naturally. During the Pleistocene, the repetition of glacial and interglacial episodes led to vegetation successions, some of which can be traced through the pollen still held in the soil. Following the most recent retreat of the ice sheets, regions that had been covered by them were recolonized by plants migrating from lower latitudes. Arctic–alpine species, such as *Dryas octopetala* (mountain avens), were early colonizers, followed by birch trees, pines and other conifers, and finally broad-leaved deciduous trees (GODWIN, 1975, pp. 451–483). Islands built from lava ejected by submarine volcanoes can appear quite suddenly, and once the surface has cooled, colonizing species start arriving and a succession commences. The island of Surtsey, to the south of Iceland, emerged in this way in 1963, and in June 1995 a tiny, 1-hectare island called Late Iki (*www.volcano.si.edu/gvp/volcano/region04/tonga/metis/var.htm*) appeared near Tonga, in the Pacific. Volcanoes on land can burn and bury vegetation. Such an event, too, is followed by a plant succession.

Seral stages appear to occur in a regular fashion and lead to a plant community, with its associated animals, that remains unchanged until some event disturbs it. This final stage is called a 'climax'. According to Frederic E. Clements, the American grassland ecologist who first proposed the concept in the early years of this century, the climax is the natural vegetation of a region, the type of vegetation that will develop if large enough areas are left free from outside interference. This being so, any vegetation type that differs from the climax must be regarded as immature. The climax, in Clements's view, acquired the status of a single, mature organism. Within it, individual plants develop according to the conditions under which they grow, a successful species being one that can adapt to a variety of different conditions by assuming different forms. This

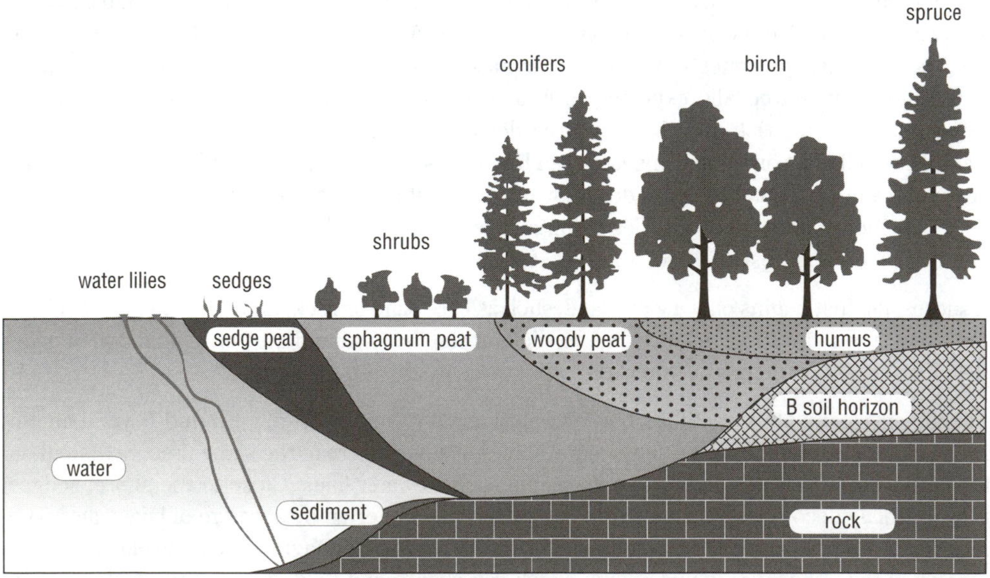

Figure 4.13 **Succession from a lake, through bog, to forest**

holistic view contradicted the Darwinian explanation of evolution by natural selection, but it became very influential and has left an enduring popular impression of the climax as an almost mystical superorganism.

Doubts were soon cast on this interpretation, most notably by another American ecologist, Henry Allan Gleason (1882–1973). He argued that all plants grow wherever they can; if similar plant communities occur in different places it is because conditions in those places are similar (BOWLER, 1992, pp. 521–525). In 1927, Gleason wrote that detailed studies showed very complex successional relationships among plant communities within a region and he rejected entirely the idea that seral stages proceed in a systematic fashion (BREWER, 1988, p. 381).

In fact, seral stages can be explained in Darwinian terms. Species arrive haphazardly, and those tolerant of the conditions survive. In surviving, however, they may alter those conditions in ways that favour other species more strongly than themselves. This is natural selection acting on variation between and within species, and it can produce unexpected but well-documented surprises. When conifer forests are cleared, their regeneration may be delayed for a long time if alder invades, because nitrogen-fixing bacteria live in alder roots, allowing the species to thrive in nitrogen-poor soil. Other plants actively inhibit the establishment of species that would otherwise replace them in the succession (BREWER, 1988, pp. 386–387).

For all their over-simplification and mystical overtones, the concepts of succession and climax remain useful. It is true that any exposed surface will be colonized by plants and that the annual pioneers will usually be replaced by perennials, arriving later. It is also true that the ensuing succession will reach a stage that endures. It is not true, however, that the seral stages can be predicted accurately. Once a succession begins, chance plays too great a part and small variations in conditions from one part of a site to another have too large an effect for its progress to pursue a regular, repeatable course. Clear an oak woodland, for example, and it may regenerate as oak woodland, but then again it may not. What follows may be a 'secondary succession' (where the primary succession was the one leading to oak woodland) involving communities of different composition and leading to a quite different climax. Some ecologists call the resulting climax a 'plagioclimax'; others restrict this term to the climax resulting from a succession interrupted by human activity and use 'biotic climax' for the result of a secondary succession in which humans played no part. Nor is it true that the apparently final stage, the climax, is permanent. It remains dynamic and capable of change, although its rate of change is very much slower than that observed in preceding stages.

With these qualifications, it is possible to use the concepts. When derelict land is restored, for example, the first plants to be sown are pioneers, the opportunist colonizers that can survive the initially harsh conditions. They are followed by others, and an understanding of successions and an idea of the climax most suitable to the intended use of the land provides guidance for subsequent plantings to those responsible for managing the site. Many people maintain that forests should be replanted on land from which original primary forest was cleared by farmers many centuries ago. There are plans in Britain to develop large forests in several parts of the country, mainly for amenity use. This cannot be achieved simply by planting attractive trees. The operation begins with the planting of pioneer species, short-lived but tolerant of wind and full light, that will provide shelter for the longer-lived, but more slow-growing, desired species. In other words, a succession must be initiated and supervised, albeit an abbreviated one.

'Succession' and 'climax' are concepts that should be used with care. In their simplified, popular versions they are misleading. Properly understood, however, more as metaphors than as detailed accounts of processes observable in the real world, they provide valuable guidance to those planning new uses for land and for conservationists seeking to enhance the environmental quality of degraded sites.

40 Arrested successions

If ever you are fortunate enough to visit California, you may have a chance to visit the Sierra Nevada, where you will see in its native habitat one of the most famous trees in the world. Its botanical name is *Sequoiadendron giganteum* and its common names include the Sierra Redwood, Big Tree, Giant Sequoia, Mammoth Tree, and Wellingtonia. It is famous, of course, for the superlatives that describe it, and some individual trees have been given their own names. The General Sherman Tree, for example, is said to be the most massive tree in the world, weighing about 2000 tonnes, and is believed to be more than 3000 years old. It is in the Sequoia National Park. Although *S. giganteum* is not the tallest tree species in the world, it can grow to an impressive 100 m (the tallest is another native of the Californian coastal belt, the Coast Redwood, *Sequoia sempervirens*, which can grow to 120 m).

For those of us who may never visit California, smaller specimens of *S. giganteum* are grown in many gardens and arboretums. There is a fine avenue of them at the Younger Botanic Garden, near Oban in Scotland.

Should you encounter one of these trees, examine its bark. This is so spongy you can punch it quite hard without hurting your hand. The bark protects the trunk of the tree against fire. It burns only slowly and the air entrapped in its loose fibres provides thermal insulation. Not only can the tree survive fires, the heat of a fire is needed to cause its seeds to germinate. Fire clears the ground of dry litter and smaller plants, and before these can recover, *S. giganteum* seedlings will have established themselves.

Sierra Redwoods are not the only trees to rely on fire to maintain their dominance in the plant community. Longleaf or pitch pine (*Pinus palustris*) is fully adapted to environments where fire occurs naturally every 3–10 years. At first its seedlings grow very slowly, producing a dense clump of long needles on a short stem with the growing tip of the tree hidden at their centre. The needles protect the tip against fire, and while it remains small the tree develops a large root system. Its roots established, the seedling grows rapidly and its growing tip is carried upward, beyond the reach of low-level fires. Then the tree develops a thick, fireproof bark. Its cones will not open to shed their seeds unless they are heated strongly, but once released they germinate rapidly; such cones are described as 'serotinous'. This adaptation to fire allows longleaf pines to form large stands along the coastal plain of the south-eastern United States (KENDEIGH, 1974, p. 113). Longleaf pines do not grow in splendid isolation, of course. Other plants share the habitat with them and some of these are similarly adapted to fire. Wire grass (*Aristida stricta*), in particular, flowers profusely following a fire in spring, which is when natural fires are most common, but not after winter fires, although fires at any season encourage its growth (BREWER, 1988, p. 69).

Forest fires may occur at the surface, at ground level, or among the tree crowns. Surface fires clear away accumulated litter and destroy smaller plants, but although they are hot, with temperatures reaching more than 100 °C, conditions remain cool just a few centimetres below ground. Ground fires smoulder below ground, advancing very slowly, destroying roots and consuming litter, but rarely producing flames. They can cause great damage to ecosystems such as temperate heaths and even bogs, which are not adapted to them. Crown fires are commonest in pine forests, where trees retain dead needles for some time before shedding them. They can advance rapidly. Generally, trees will be killed by fire if the actively growing tissue all around the inside of the stem (the cambium) is heated to more than about 64 °C. Thick or spongy bark protects the tree by insulating the cambium. Some tree species survive crown fires, which kill their tops, by sprouting new stems from the base (BREWER, 1988, pp. 66–69). Where there is a pronounced dry season, if many of the trees in a coniferous forest have multiple stems it is often an indication of past fire (in British broad-leaved forest, multiple stems are more likely to indicate past coppicing).

Consider the effect of recurring fires on the seral stages by which plant communities approach a stable climax. As Figure 4.14 shows, the number of species increases as the community develops, but the fire repeatedly destroys this growing diversity. In effect, the fire returns the community to a much earlier seral stage, after which the succession resumes. Should the fires end or be prevented, as in the right-hand curve in the figure, the number of species continues to increase until it reaches a maximum. After that, some species are eliminated by competition, according to the competitive exclusion principle derived mathematically by A.J. Lotka and V. Volterra and first demonstrated experimentally in 1934 by the Russian biologist Georgyi Frantsevich Gause, 1910–86 (and sometimes known as Gause's principle). The principle states that two or more species requiring identical resources cannot coexist in an environment where one or more of those resources are limited, because one species will prove more successful than the others, monopolize the scarce resource(s), and its competitors will fail and disappear. Consequently, the usual pattern for diversity during seral stages shows a slight fall prior to the stable climax, as shown in the figure.

For as long as fires continue to interrupt the succession, however, the final climax cannot be attained. The succession is arrested and then repeated. Such a succession, controlled by abiotic factors, in this case fire, is known as an allogenic succession (one controlled by biological organisms, where those of one seral stage produce conditions favourable to those of the succeeding stage, is known as an autogenic succession) (BEGON ET AL., 1990, pp. 630–633).

Arrested successions clearly complicate the concept of an ecological climax. F.E. Clements, who first proposed climax communities argued for what is now known as the 'monoclimax hypothesis', according to which large regions support a single type of climax community determined by climate. If the succession is repeatedly disturbed, for example by fire, he would maintain that the climax which would be attained were it not for disturbance is the true climax. Most ecologists consider this unsatisfactory, because it fails to describe the situation they observe in the field.

A possible alternative introduces the concept of the 'polyclimax', in which a region supports a number of different climaxes, each identified according to the factor controlling it, of which climate is but one. A particular soil may produce an 'edaphic' climax, for example, and repeated fire produces a fire climax (or pyroclimax).

Attempts to escape the difficulties inherent in the monoclimax hypothesis lead to further complications, not to mention a rapid proliferation of terms. Is a forest mainly composed of mature trees a climax community? It may seem so, but trees grow slowly compared to a human lifespan and within the forest some species may be increasing in number at the expense of others. Is a fire climax really a climax, or merely an arrested succession on its way to a climax it never reaches? Add to these sources of confusion the fact that few supposed climaxes remain for long in a true equilibrium and it is hardly surprising that many modern ecologists prefer to avoid using the climax concept at all (BREWER, 1988, pp. 400–401).

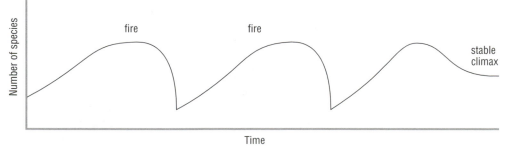

Figure 4.14 **The effect of fire on species diversity**

Fire provides an obvious example of an abiotic factor disturbing successions, but by no means is it the only one. Very fierce sea storms, of the kind that occur only occasionally at intervals of some years, generate winds and waves that can cause similar disturbance in coastal environments. Hurricanes in tropical regions and even storm-force winds (103–120 km h^{-1}), which rarely occur inland in temperate regions, but are capable of uprooting trees when they do, will arrest a succession. Nor can climate be regarded as constant. Glacial and interglacial episodes cause profound ecological change and if the word 'climax' is held to imply permanence, then such major climatic changes, and some less dramatic ones, appear to qualify as disturbances.

Human activities also cause disturbance, of course, when natural vegetation is cleared and the land put to economic use. Farmers exploit the natural process by removing all plants from their fields, then introducing crop plants to take the place of the primary colonizers that would otherwise soon appear, their crops taking opportunistic advantage of the absence of competition for light, water, and nutrients to achieve rapid growth.

Fire in Yellowstone

In 1988 fire swept out of control through Yellowstone National Park in the United States. In all, some 400 000 ha was burned, amounting to 45 per cent of the total park area (of 890 000 ha) and about 8000 ha was said to have been virtually destroyed. Despite all efforts to extinguish them, the fires ended only with the first substantial falls of snow in the autumn.

Yellowstone is naturally prone to fire and its species are adapted to it. Fire removes surface litter and old, woody plants, leaving ash on the surface and stimulating fresh growth. Official policy over the years had aimed to curb natural fires, most of which had been extinguished successfully. The policy arose in part from public concern about the effect of fire on wildlife, which it was presumed to harm, an impression strongly conveyed by the Disney movie *Bambi*. The result of the policy was that surface litter accumulated until, when eventually it did ignite, the fire proved unquenchable. In fact, controlled burning at intervals of a few years is ecologically beneficial. By the spring of 1989 plants were recovering and a secondary succession had begun, and as the park recovered its biological diversity increased.

Since 1992 a management scheme has been adopted that divides the park into three zones. In the first, comprising about 10 large areas, fires will be suppressed. In the second, covering most of the park, fire will be allowed to burn but will be controlled. In the third, a 2.4 km strip just inside the park boundary, fires will be permitted but under even more strictly controlled conditions. Even in zone 1, fires will be started if too much dry, flammable material accumulates.

It can also happen that humans, other species, and abiotic factors combine to produce an arrested succession. Suppose, for example, that an area of temperate forest is cleared. Humans may have felled the trees or fired them. Unless the forest is fire-adapted and already at a fire climax, the removal of shade will cause grasses and herbs to germinate and flourish. Ordinarily, after a time tree seeds will start to germinate, seedlings will appear and grow and eventually the forest will return. The first two diagrams in Figure 4.15 illustrate this. It may be, however, that the humans cleared the forest precisely in order to encourage the growth of grass to feed their livestock. Grass growth is encouraged by grazing, because the blades grow from ground level and are unharmed, but as they graze the herbivores also destroy tree seedlings. This delays the regeneration of forest, but does not halt it altogether and after a few years scrub vegetation begins to shade out the grass. At this point, the humans remove their animals, wait for dry weather, and fire the pasture, destroying both grass and scrub, but encouraging the grass by once more returning the succession to an earlier stage. Repeated firings and intensive grazing gradually eliminate most of the woody plants by destroying them each time seeds lying dormant in the soil germinate and before the new plants produce seed. In this way forest can be converted into grassland, as is shown in the third illustration in the figure. In North America, prairie vegetation can be maintained only by setting fire to it every 1–3 years (BREWER, 1988, p. 75) to rid it of the shrubs and seedlings that would otherwise come to dominate it as it developed towards forest. African pastoralists also maintain their savannah pastures by periodic burning. It may well be that at least some of the great grassland biomes were formed by this type of human intervention, although natural fires occur readily enough in such regions and sweep rapidly across the plains.

It is attractive to think of biological communities increasing in diversity as they proceed through seral stages to a climax. Undoubtedly this does occur, but successions are often arrested

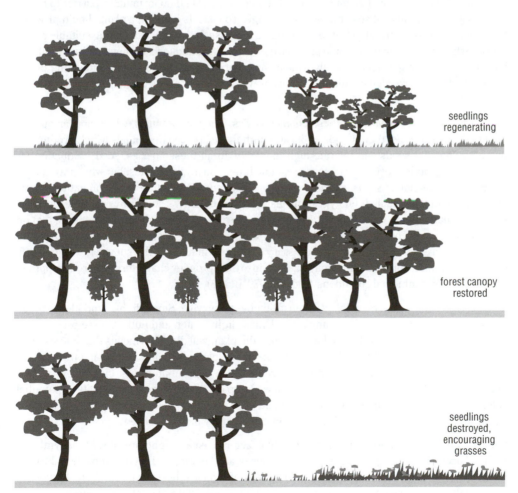

seedlings
regenerating

forest canopy
restored

seedlings
destroyed,
encouraging
grasses

Figure 4.15 **Effect of grazing on succession**

naturally or by human activities. Most of the landscapes we see around us, from town gardens and parks to the farmed and forested countryside, are artefacts, made by humans, and the biological communities they support are those of successions arrested at some stage short of their climax. Too great a concentration on the concept of the climax may obscure the importance and considerable biological value of such arrested successions.

41 Colonization

When buildings are demolished, exposing an area of bare ground, in most of Europe the first plant to arrive is often the rosebay willowherb (*Chamaenerion angustifolium*). Indeed, the speed with which it appears after a major fire has earned it its other common name of 'fireweed'. It is a tall, colourful plant that quickly forms large stands and its prominence tends to hide the smaller plants growing with it. Groundsel (*Senecio vulgaris*) and coltsfoot (*Tussilago farfara*), for example, are other early arrivals. Why do these plants appear first so predictably and other common wild plants only later?

If a species is to establish itself in a new habitat, it must be able to reach the site. The first species to arrive will be those that had only a short distance to travel and no formidable barrier to cross, such as a wide expanse of sea or a mountain range. To ease its journey, a plant should produce light seeds that are distributed by the wind and, to increase their chance of finding suitable conditions in which to germinate, it should produce them in great abundance. Having reached a suitable location, they must germinate rapidly. This means that many of the seedlings and young plants will die in the overcrowded conditions that occur when so many seeds germinate, so mortality is very high.

The early colonizers have all of these characteristics. They are annuals, which appear quickly only to disappear when their season ends, although there are also perennials which produce an overabundance of seeds and tolerate high mortality. In all these species, seed production far exceeds the quantity necessary to ensure establishment on available sites, but what may appear to be over-production of seeds probably serves a useful purpose. Natural variation in the genetic composition of individuals (genotypes) means there are countless small physical differences in the plants that begin to grow (known as phenotypic variation). These differences ensure that on any vacant site a proportion of individual plants will possess the characteristics that allow them to thrive, whereas those lacking the necessary features will die, but the survivors will be sufficiently numerous to form a viable colony, passing on to their progeny characters that are appropriate to the local environmental conditions (GRIME, 1979, p. 118).

Once arrived, the would-be colonizers must be able to survive. Small, light, wind-blown seeds contain little sustenance for the young plant. Unless light, water, and nutrients are present and available from the moment the seed germinates, the plant will die. This means that initial colonization is often patchy. Within a bare-ground site, some places will be wetter than others, more or less shaded, or offer a more or less balanced suite of mineral nutrients. Figure 4.16 shows, diagrammatically, how this can produce a dense clump of a species, where mortality of potential colonizers is low, surrounded by a thinner stand where conditions are less favourable, within a larger area which the species finds inhospitable and from which it is absent.

On open ground, the most critical environmental factor is usually light intensity. Photosynthesis is the process by which green plants obtain energy and its rate is directly proportional to the intensity of light: the brighter the light, the faster plants can synthesize carbohydrate. Plants use much of their captured energy for respiration and there is a point, called the compensation point,

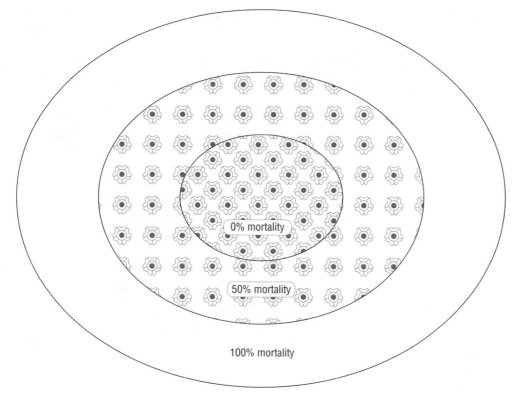

0% mortality

50% mortality

100% mortality

Figure 4.16 **Establishment of colonizers in an area of habitat**

at which the amount of energy being captured (or carbon dioxide absorbed, which is the same thing) by photosynthesis is equal to the amount being used for (or carbon dioxide emitted as a by-product of) respiration. At the compensation point, a plant can survive but not grow; below the compensation point energy output exceeds input and it will fail; but above the compensation point there is a surplus of energy available for growth and reproduction.

It might seem, therefore, that most plants should prosper on open ground, where they are exposed to full sunlight, but it is not so simple. The compensation point varies widely from one species to another. Shade-loving plants can grow well under much lower light intensities than sun-loving plants, because their compensation point is lower. If the light is too intense, however, photosynthesis is inhibited by other chemical reactions in cells, which destroy chlorophyll and inactivate other substances. This adverse effect of high light intensity is called solarization, and susceptibility to it also varies according to species.

The ideal colonizer of bare ground, therefore, will be a sun-loving plant. Should seeds of a sun plant be blown on to shaded ground they may germinate, but the plants will not form a true colony. Rosebay willowherb can be found in woodland, growing in shade, but although it can grow, the dim lighting provides insufficient energy for it to flower. At the end of the growing season the plants die without having reproduced; if they reappear in the following season it is because fresh seed has entered the site from outside (DOWDESWELL, 1984, p. 7).

Ground may be bare not because of disaster or demolition, but for reasons of its chemistry: it may be poisonous to most plants. Even poisoned land may be colonized, however, because the strategy that allows colonizers to adapt rapidly to the conditions they find embraces tolerance for at least some toxic substances. This phenomenon has been most widely observed on heaps of mine waste contaminated with high levels of copper. An early colonizer is bent grass (*Agrostis tenuis*). It is generally intolerant of copper, but among the countless seeds blown on to the heaps a few give rise to tolerant individuals and it is these that reproduce, eventually to produce a tolerant population that in time is joined by other tolerant species (BEGON *ET AL.*, 1990, pp. 74–75).

At one time, the manufacture of sodium carbonate (washing soda), a very important industrial chemical, caused severe air pollution and left behind an extremely alkaline waste, with a pH of nearly 14, which was simply dumped, especially in Lancashire, England. In some places this material covered several hectares to a depth of several metres. Amid the alkaline material were small patches of very acidic soil, where ash from boilers was dumped on top of the soda waste and the two wastes did not mix. Rain, slightly acid because of the dissolved carbon dioxide it contained, and acid air pollutants such as sulphur dioxide reduced the extreme alkalinity at the surface, although 70 years later at depths of more than 55 cm the pH was still more than 12.0. Yet the alkaline waste sites were colonized by lime-loving plants and now support a diverse community, including most notably a wide variety of orchids, and the acid sites have also been colonized, though by fewer species (MELLANBY, 1992a, pp. 64–66).

Much has been learned about colonization from studies comparing the natural populations of islands with those of the mainland from which they originated. In 1963, the ecologists Robert H. MacArthur and Edward O. Wilson proposed that the number of species present on an island is determined by a balance struck between immigration and extinction. At first, we may assume that a newly formed island is bare. Species begin to arrive, so the immigration rate is high. The extinction rate is low, because so far there are few species to become extinct. As time passes, the immigration rate will fall, because an increasing proportion of incoming individuals belong to species already present. Simultaneously, the extinction rate will increase as a result of intensifying competition for resources, predation, and parasitism (BREWER, 1988, pp. 302–306). Figure 4.17 describes the situation as a graph in which the straight-line curve for decreasing immigration intersects that for increasing extinction, the point of intersection indicating the number of species.

Islands vary, of course. Some are larger than others, some further away from the mainland, and these factors affect colonization and the eventual number of species. Immigration to a remote

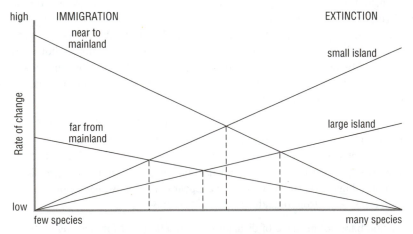

Figure 4.17 **Island colonization as a ratio of immigration to extinction**
After Brewer, Richard. 1988. The Science of Ecology. **Saunders College Publishing, Fort Worth, TX**

island will begin at a lower rate than that to a nearby island, because migrants must travel further and more will perish on the way. The rate of immigration decreases and the rate of extinction increases regardless of the starting value, and so the intersection point, or final number of species, is likely to be lower on a remote island than on one nearer the continental coast. The size of the island also affects the diversity of its established population. Size does not affect the rate of immigration, but small size does increase the rate of extinction, because a small island offers fewer resources than a large one. As common sense would suggest, a large island finally supports a more diverse population than a small one.

The composition of an island population will change constantly as immigrants continue to arrive, albeit at a reduced rate, and residents become extinct. It has been confirmed many times by observation that once equilibrium has been reached, the number of species on an island remains fairly constant regardless of when or how often it is sampled, although the species themselves change.

MacArthur and Wilson supported their theory experimentally, by counting all the arthropods on four small (about 15 m across) islands on the Florida Keys, then employing a professional exterminator to kill all of them by fumigating the islands with methyl bromide, after which they monitored their recolonization. Wilson has pointed out that an island of 1000 km^2 supports, on average, about 50 species of land birds, whereas an island of $10\ 000 \text{ km}^2$ supports about 100 species (WILSON, 1992, pp. 208–214). This is known as the area effect and it is described mathematically by the species–area equation: $S = CA^z$, where S is the number of species, A the area, C a constant, and z a parameter related to the types of organisms (such as grasses, arthropods, reptiles, mammals, birds, etc.) and distance from source areas. It is only fair to mention, however, that the species–area concept has been criticized as an over-simplification, and critics have pointed out that the equation utilizes the constant C and a value, z, that are somewhat arbitrary.

Understanding how colonization proceeds is of great environmental importance, because areas of habitat isolated amid much larger areas of land managed for forestry, agriculture, industry, or urban development are ecologically somewhat similar to islands. Consider, for example, town parks surrounded on all sides by streets and houses. As we seek to restore land damaged by past human activities we need to know what kinds of species to expect the habitat to support and the approximate order in which we may anticipate their arrival. Armed with a knowledge of the processes involved we will be in a better position to predict the resources each arriving group of species will require and ensure they are available. In this way we will be more likely to redress past despoliation effectively. We may also be better placed to prevent future despoliation, because the detailed information about the species composition of different types of community at each stage in their development will make it easier to estimate the consequences of interference. Finally, we may more accurately identify areas, such as islands, that may be especially vulnerable to ecological disturbance.

42 Stability, instability, and reproductive strategies

Every so often, according to popular legend, vast armies of lemmings gather together and then stampede into the sea in a spectacular mass suicide (HUCK, 1995). Curious though it may seem, the popular legend is true. Mass migrations of Norway lemmings (*Lemmus lemmus*) occur every 3 or 4 years, although the huge mass migrations are much less common. Should the migrants meet a barrier, the migration is halted temporarily until the lemmings seem to panic, perhaps because they are so crowded, and rush the obstruction. They swim well, so rivers present no serious obstacle to them, and if they reach the coast, which sometimes they do, they simply set out to swim across the sea as though it were a river, drowning in the attempt. During their flight, the ordinarily pugnacious lemmings become even more than usually aggressive to one another, and if they cross vegetated land they will devastate it, eating everything.

Thousands of miles from Scandinavia, desert and migratory locusts (*Schistocerca gregaria* and *Locusta migratoria*) (*www.fao.org/WAICENT/FAOINFO/AGRICULT/AGP/AGPP/Locusts/ index.htm*) behave in a very similar way. Flightless, immature insects, called 'hoppers', change colour, then begin to move together across the landscape. As they march they also mature and when their wings are developed they take to the air, drifting with the wind in swarms that can exceed 250 km^2 in area (BARON, 1972, pp. 1–2 and 39–52). We think of locust plagues as phenomena confined to Africa and low-latitude Asia, but in the past they have also affected southern Europe and North America; the last serious outbreak in the United States and Canada occurred in 1938 (HUTCHINS, 1966, pp. 86–89).

Norway lemmings and locusts live in environments rendered unstable by climatic extremes, differing from the other inhabitants of such regions only in the spectacular migrations that bring them to our attention. Other species increase in population size and then suffer a collapse in their numbers. When snowy owls (*Nyctea scandiaca*) are seen in Scotland, and occasionally as far south as France and central Europe, it is a sign that populations of the small rodents and hares which form their usual diet have collapsed. Such irruptions of snowy owls from the Arctic occur at intervals of 4 years or so.

Environments are unstable if the living conditions they afford vary widely and unpredictably. Conditions in many parts of the world vary according to season. Winters may be very cold, one season dry, another wet, and temperature and the availability of water will influence plant growth and therefore the amount of food for animals. Being seasonal, however, such changes are predictable, and plants and animals can and do adapt to them. Seasonal climates do not confer instability on environments. In deserts and in the Arctic, on the other hand, there are sequences of favourable and unfavourable years. Food may be abundant for a time and then disappear when temperatures remain depressed during the growing season or when no rain falls for several successive years. These environments are inherently unstable, but organisms adapt to them nevertheless. As always, their adaptations ensure the continuance of the species.

During the good times, species reproduce prolifically, taking full advantage of the abundance of resources, and mortality is relatively low. Numbers increase steadily. Individuals are consequently crowded more closely together, but this does not halt population growth. It is halted by an environmental change wholly unconnected with the reproductive behaviour of species. The usual effect of the environmental change is to reduce the food supply drastically, but it may be the water supply that fails or the temperature may exceed tolerable limits. Most individuals die, but prior to the collapse they were so numerous that natural variation makes it highly probable that just a few among them will survive the period of environmental deterioration to become the parents of the first generation to be produced when conditions improve.

Lemmings, for example, become so numerous that either they quite literally eat themselves out of house and home, or they pollute what food remains, thus rendering it unpalatable, or, more interestingly perhaps, their intensive grazing induces increased production of defensive chemicals by plants, again rendering them inedible (BREWER, 1988, p. 193). At this point the lemmings begin to die in large numbers or, in the case of the Norway lemming, commence their migration. Many lemming migrants also die, of course, but a few find new sources of food and found a new population. In Figure 4.18, the situation is described by curve 2, showing a gentle rate of increase in population, starting from an already high level, followed by a sudden rapid decrease.

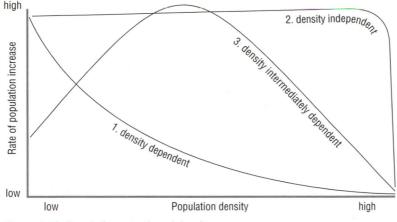

Figure 4.18 **Population growth and density**

Population size is always regulated. Were it not, numbers could increase indefinitely, which is absurd. The question, therefore, centres on the mechanisms by which they are regulated. Lemming populations are regulated primarily by factors external to the animals themselves and independent of the density of the population. The obvious alternative to density-independent regulation is regulation by density-dependent factors, with effects illustrated by the first curve in Figure 4.18. As numbers increase, the pressure on the food supply may increase, causing an increase in mortality, especially among young animals, leading to reduced reproduction. At the same time, crowding may facilitate the transmission of disease, and large numbers of individuals within a confined area may attract more predators. Beyond a certain point, as the population density increases the rate of population growth decreases.

Density-dependent factors affect all species and are usually associated with stable environments. They may also act in conjunction with density-independent factors in unstable environments, however. At high densities lemmings may alter their behaviour in ways that make them more quarrelsome and less inclined to mate (BREWER, 1988, p. 193). Locust migrations appear to be triggered by sexually immature females under crowded conditions: when reared in isolation the insects exhibit their *solitaria* (non-migratory) markings and behaviour; when reared in large numbers they are of the *gregaria* (migratory) type (JOHNSON, 1963). Migration and reduced reproduction are most likely to occur immediately prior to the exhaustion of resources, but in an unstable environment the resources themselves are unreliable. Were they not exploited while they are abundant, leading to population densities high enough to trigger behavioural and physiological responses, they would shortly disappear anyway. The immediate responses may be density-dependent, but the underlying regulation is density-independent.

Rapid and violent fluctuations in the population of one species have consequences for other species. These are most notable among Arctic animals, such as snowy owls, lynxes, and foxes, the secondary consumers which feed on such primary consumers as lemmings, hares, and voles. Predator populations fluctuate, with a time lag, in response to changes in prey populations and also in response to changes in the populations of competing predators. In Canada, for example, populations of snowshoe hares (*Lepus americanus*, also known as the varying hare and snowshoe rabbit), lynxes, and red foxes fluctuate over a roughly 11-year cycle; that of the Arctic fox varies over 3 years, its population increasing when prey is abundant but rival predators are less numerous (DOWDESWELL, 1984, p. 31).

Curve 3 in Figure 4.18 illustrates a third way in which population size may respond to the environment. In stable environments, colonial species are stimulated to breed by the proximity of other members of their own species and do so until they are so crowded as to occupy all the available nest sites. This intermediately dependent regulation of population is most clearly seen in sea birds that nest in large colonies.

Species inhabiting unstable environments are opportunists. They must seize resources while they are there and to achieve this they must be able to reproduce rapidly. Producing large numbers of progeny, they can afford only the minimum investment in each. In plants, small seeds containing little nutrient are released in huge numbers. Among animals, large numbers of eggs are laid and abandoned to their fate or, if they are mammals, large litters are produced several times a year. Survival rates for the progeny are low, but the resource is overwhelmed and competitors excluded by sheer weight of numbers. As a reproductive strategy it is highly successful and is known as *r*-selection ('selection' because it results not from some deliberate scheme devised by the organisms or their ancestors, but by natural selection, which favours it). It is a typical adaptation to unstable environments, shared by primary colonizers of newly cleared sites.

In stable environments, a different reproductive strategy, called *K*-selection, is more successful. Species adopting it produce few young. Offspring develop slowly, often with considerable parental care, and many of them survive. Plants falling into this category produce few seeds, but with large nutrient stores. Animals tend their young. This strategy allows a species to maximize the efficiency with which it exploits reliable resources. Humans are a *K*-selection species. Parents produce very few young, care for them during a long period of development, and most young survive. We know from this that our species is adapted to stable environments.

These reproductive strategies produce different patterns of population growth when species first arrive in a new habitat or when additional resources become available. Opportunistic, *r*-selection species reproduce rapidly, so their numbers increase dramatically. On a graph, this results in a J-shaped curve and that is what it is called, although its resemblance to the letter J may be less than precise. Figure 4.19 shows a J-shaped growth curve and, as the figure indicates, growth is limited when the number of individuals exceeds the carrying capacity of the environment. During the growth phase the rate of increase is so rapid that the failure of resources when the carrying capacity is exceeded produces a dramatic decline in numbers: an 'overshoot-and-collapse'. *K*-selection species may also increase in number rapidly at first, and if they are judged by data for population change alone it can be difficult to distinguish them from *r*-selection species during the

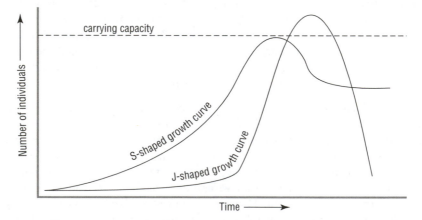

Figure 4.19 **J- and S-shaped population growth curves**

early stage. In fact, though, their rate of increase is not so rapid as for *r*-selection species and they overshoot carrying capacity only slightly. This causes a decrease in their numbers, but only a small one followed by further even smaller increases and decreases until the population adjusts to the carrying capacity and stabilizes. This produces an S-shaped growth curve, also illustrated in Figure 4.19.

In some species the reproductive strategy changes according to circumstances. Under unstable conditions they adopt an *r*-selection strategy and under stable conditions a *K*-selection strategy, thereby optimizing their chance of survival. In these species, *r*-selection can be altered to *K*-selection by stabilizing their environment.

This is relevant to the human situation. Because we know from their reproductive strategy that humans are a *K*-selection species, we can be reasonably confident that our present population increase is located on an S-shaped growth curve, not a J-shaped one. Despite the fears of many people, it is very unlikely that human numbers will overshoot the carrying capacity of the planet by a margin large enough to generate a population collapse.

43 Simplicity and diversity

Farmers devote a great deal of time, effort, and expense to eliminating pests and weeds from their crops. Those crops comprise stands of just one plant species occupying a large area. In effect this is an ecosystem containing only one species (called a 'monoculture') and it appears to be unstable, because it is very vulnerable to invasion. At the other extreme, an equivalent area of tropical rain forest contains many species and it is highly resistant to invasion. With the possible exception of logging companies, it is difficult to imagine how a pest or weed species could establish itself in a tropical rain forest at the expense of the existing community: such forests are not vulnerable to pest or weed infestations.

Many people draw from this example the simple lesson that the more species an ecosystem supports the more stable it will be. Diversity confers stability, simplicity confers instability, and ecological successions lead to increasing diversity. It is easy to see why this should be so. An individual species can occupy only one ecological niche, defined by its physiological requirements. In a monoculture, other niches remain unoccupied and species entering the area by chance will find their needs met and will be able to establish themselves. If the ecosystem supports a wide variety of species, on the other hand, all niches will be occupied and, according to the competitive exclusion principle, two or more species with identical requirements cannot occupy the same area of habitat. Would-be invaders cannot establish themselves, therefore, and the complex system is stable because it resists invasion. A complex system is also more resistant than a simpler one to other forms of disturbance, because it is more likely to contain species, or individuals that can survive.

A problem arises immediately over the definition of 'diversity'. It is one that returns to haunt those concerned about the maintenance of 'biodiversity'. Counting the number of individual members of particular species is of little help, because of the huge differences in the sizes of organisms: 100 bacteria and 100 fieldmice (never mind elephants!) are not at all equal. If the diversity of plant species is being considered, not all plants are easily identified as individuals. A grove of trees may comprise plants grown from individual seeds, but it may also comprise trees growing as suckers from a single root system and, therefore, technically clones and genetically a single plant. Ecologists have devised various techniques for calculating species diversity, some based on biomass rather than numbers of individuals, but the task is not simple (see box).

Measuring diversity

In 1949, E.H. Simpson devised a diversity index named after him. This is given by the formula:

$$D = 1 - \Sigma p_i^2$$

where D is the diversity index, Σ means 'the sum of', and p_i is the proportion of individuals belonging to a particular species.

The Shannon–Wiener index of diversity (H), devised by Claude E. Shannon and Norbert Wiener, also in 1949, is given by:

$$H = -\Sigma p_i \log p_i$$

In 1980, S.F. Wratten and G.L.A. Fry devised the sequential comparison index (SCI). This is calculated as the number of runs (i.e. the number of times a sequence of individuals all of the same species occurs in a sample) divided by the total number of individuals in the sample. A run is counted by calling the first individual 1, the next individual of a different species 2, and so on. Thus, if all the individuals in the sample are of different species the SCI will be 1; if they are all of the same species the SCI will be 1 divided by the number of individuals.

'Diversity equals stability' seems so obvious that those who question it may be thought to fly in the face of common sense. Indeed, so many eminent ecologists once supported the idea that for many years it was the conventional view. Even then, though, ecologists recognized difficulties with the concept.

In the first place, there are examples of ecosystems that comprise relatively few species yet remain highly stable. Throughout the world, for example, salt marshes support plants belonging to a small number of cosmopolitan genera, including *Salicornia* (glassworts), *Spartina* (cord grasses), *Juncus* (rushes), *Plantago* (plantains), and *Limonium* (sea lavenders) (LONG AND MASON, 1983, p. 39). Mangrove forests often contain just one or two species of mangrove and even the most extensive contain only 5–25 species. Salt marshes and coastal mangrove forests occupy variably saline conditions few species have adapted to tolerate, so they may be thought exceptional, but there are large tropical freshwater swamp forests containing only one tree species (*Shorea alba*) and tropical montane forests with no more than five species and sometimes fewer (JANZEN, 1975, p. 45). In temperate regions, bracken covers large areas of upland, where it forms seemingly stable monocultures.

Even the vulnerability of agricultural monocultures does not necessarily establish a link between diversity and stability. When farmers plough the land they remove all vegetation, and in ecological terms the crops they plant are primary colonizers and we would expect them to be replaced by later colonizers in the ordinary course of succession. Certainly the succession will lead to greater diversity, but this does not imply that the agricultural monoculture is inherently unstable because of its simplicity.

It has also been suggested that the stability of natural communities may result partly from long periods of coevolution they, but not agricultural communities, have experienced. It is true that some, although not all, tropical rain forests are ecologically complex but, despite this, populations of insect species within them fluctuate in size just as widely as those in ecologically much simpler high-latitude forests (BEGON ET AL., 1990, p. 795).

Diversity, it seems, is not related to stability in any simple way. Could it be, then, that the opposite is true and that diversity actually reduces stability? As we all know, simple devices (such as bicycles) are less likely to break down than complicated devices (such as cars).

In the early 1970s ecologists began building mathematical models of ecosystems to test the relationship between diversity and stability, and that is precisely what they discovered. The models revealed that increasing the number of species, the fraction of those species interacting with one another directly (called the 'connectance'), and the strength of those interactions made ecosystems less likely to return to equilibrium following disturbance.

It would be a mistake to substitute one sweeping assertion for another by replacing 'diversity equals stability' with 'diversity equals instability'. Different ecosystems respond to disturbance in different ways and probably no general proposition can be derived from the work that has been done so far. The truth is more likely to lie between the two extremes, and in his *Fundamentals of Ecology* (ODUM, 1984) the American ecologist E.P. Odum suggested that the confusion may be to some extent semantic. What, after all, do we mean by 'stability'?

Stability clearly concerns the response of an ecosystem to disturbance. If the system resists being disturbed its ability to preserve its essential features will be proportional to the strength of its resistance. This is one meaning of 'stability', known as 'resistance stability'. An ecosystem would exhibit high resistance stability if it offered no opportunities for invading species, tolerated parasites, and withstood prolonged periods of extreme weather with little or no change in its composition.

Other ecosystems are more easily disrupted, commonly by fire, but they recover quickly, resuming their former composition. This is an alternative type of stability, known as 'resilience stability', and these two stabilities may be mutually exclusive. Should an ecosystem with high resistance stability suffer serious disruption it may never recover fully, and high resilience may preclude high resistance.

Consider a ball lying on a surface. If the ball is free to roll, moving it is easy, because it offers little resistance. Glue the ball to the surface and its resistance to disturbance increases dramatically. Now consider the situation when the ball is moved. If the surface is level, the ball will move when it is pushed and remain in the position it reaches when no more energy is applied to it. It shows no sign of returning to its former position and so, if the ball represents a perturbed system, it exhibits no resilience. If the surface is curved, however, like the inside of a bowl, and the ball is pushed away from the lowest point, as soon as it is released it will roll back to its original position, showing high resilience. The situation is illustrated in the upper part of Figure 4.20, with arrows indicating the direction of movement.

Inherent stability can also be demonstrated using a ball lying on a surface, as in the lower part of Figure 4.20. In the diagram on the left, the ball is in a metastable condition. This means the system will remain stable, with the ball balanced on the small 'hill', for as long as it is not perturbed. The slightest perturbation will dislodge it, however, sending the ball to a lower and more stable position. Once in its new position it can be returned to the top of the 'hill' only by expending much more energy than was needed to dislodge it. If the ball represents an ecosystem, dislodging it from a metastable state may prove irreversible. The system to its right is extremely stable, although superficially it may appear fairly similar. Highly resilient, it will recover rapidly from disturbance. The metastable system on the left is fragile; the resilient system on the right is robust. Stability may also be described as 'local', if the system recovers from small perturbations, and 'global' if it recovers from any perturbation, no matter how severe.

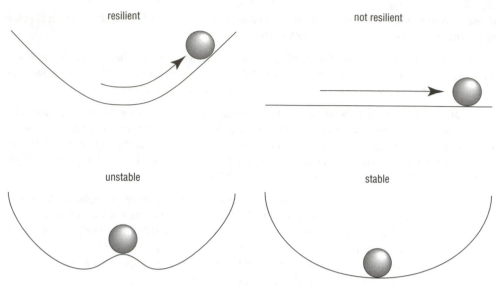

Figure 4.20 **Resilience and stability**

With stability defined more clearly, theoretical studies and a limited number of field studies have found that adding more species generally reduces overall stability, but with some important qualifications. Increasing the number of trophic levels reduces both resistance and resilience, but increasing the interactions (connectance) among species increases them. If species are removed from a system, in most cases its stability is reduced (PIMM, 1984).

Recent research (SANKARAN AND MCNAUGHTON, 1999) in the savannah grasslands of southern India suggests that the stability of an ecosystem is more closely linked to its ecological history than to the diversity of species present in it. If the removal of species reduces the stability of an ecosystem, it does not follow that an ecosystem with little diversity is therefore inherently unstable. The species may have evolved together to withstand disturbances that occur naturally. High diversity does not necessarily imply ecological stability.

Even the concept of diversity should be treated with some caution. Ecosystems are studied as though they were discrete and isolated, but in the real world they are not. Each merges into the next and the 'edge effect' can generate an intermediate but distinct ecosystem between them. Figure 4.21 shows two ecosystems that are adjacent and overlapping. Each contains 3 species, indicated by shading, none of which is present in both systems: those in one system are indicated by circles, those in the other by squares. All 6 species occur in the intermediate zone, along with 3 more species that are confined exclusively to that zone (indicated by triangles). The intermediate zone thus contains 9 species, those to either side 3 each, yet the greater species-richness, or diversity, of the intermediate ecosystem is wholly dependent on the ecosystems bordering it. Were either of them to suffer a perturbation large enough to disrupt them, the intermediate system would necessarily be affected. In the most extreme case, the total destruction of one ecosystem would cause the intermediate system to be absorbed into the survivor and its relative diversity could not prevent its disappearance, eventually followed, of course, by the emergence of a new intermediate system as the lost system was replaced by a new community of species.

One of the few field experiments to test experimentally the relationship between ecosystem diversity and stability was conducted from 1982 to 1992. David Tilman of the University of Minnesota and John A. Downing of the Université de Montréal established 207 control and experimental

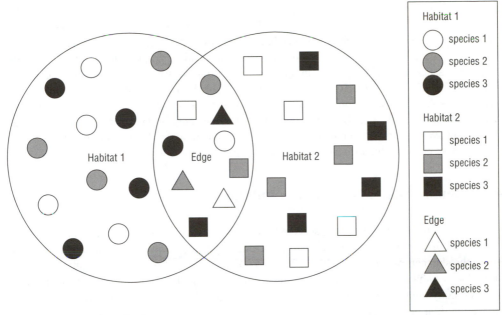

Figure 4.21 *The edge effect*

plots in 4 fields of native grassland in Minnesota. In 1987–88, the region suffered the most severe drought for 50 years, allowing the ecologists to monitor the response of the plots to this major perturbation, which they measured as change in plant community biomass. They found that species-richness increased both drought resistance and resilience; for the 4 years following the drought, the species-poor plots remained further from their pre-drought condition than the species-rich plots (TILMAN AND DOWNING, 1994). Experiments by British scientists have reached similar conclusions using an 'Ecotron' (NAEEM *ET AL.*, 1994) (see box).

The Ectron

The Ecotron is a series of closed chambers in which environmental conditions can be closely controlled and monitored. In 1993, scientists at the NERC (Natural Environment Research Council) Centre for Population Biology, in Berkshire, England, used one containing 14 microcosms (miniature ecosystems), each 1 m^2, and varying in the number of species they contained. The experimenters found that plant productivity was higher in the more complex ecosystems, probably because the greater number of species produced a more complete canopy of leaves and so intercepted light more efficiently. Interestingly for those concerned about possible climate change, they also found the richer ecosystems absorbed more carbon dioxide for most of the time.

Theoretically, then, increasing the diversity of an ecosystem appears to reduce its stability, but the few experiments to test this suggest the opposite. They offer some support to the earlier view that diversity confers stability.

Doubtless it will be some time before the relationship between ecological diversity and stability is finally resolved, but the issue is of considerable importance and ecologists have a keen interest in the outcome. Many human activities reduce the diversity of ecosystems, by removing species from them or removing ecosystems altogether and substituting simpler ones. This is aesthetically displeasing: we find areas rich in species more attractive than those with few species. It would be splendid, therefore, if more tangible reasons could be found for valuing ecological diversity. There is also widespread concern over the rate at which species are being brought to extinction. Again, it would be pleasing to be able to confirm that biodiversity is of real, practical importance. We should avoid jumping to conclusions, however. Unfortunately, this is another area of environmental science in which we will have to be patient while we wait for clear answers.

44 Homoeostasis, feedback, regulation

When we are hot we sweat. Water, secreted through our skin, evaporates, obtaining the latent heat of evaporation from the skin surface and so cooling it. When we are cold, blood vessels just below the skin contract, reducing blood flow and the loss of heat through the skin; if we become still colder, we shiver, the muscular activity generating warmth. These are among the ways humans regulate body temperature. Our bodies, and those of all other organisms, also have means for regulating their internal chemistry.

The human (or non-human) body can be described as a system, an arrangement of interacting components which combine as a discrete, total entity equipped with means for regulating its own internal operating conditions. This tendency to maintain a constant state, returning to it without outside intervention in response to perturbations, is called 'homoeostasis'. It is characteristic of individual living organisms and biological communities, but there are also many abiotic examples.

Machines are also systems. They can function efficiently only within certain tolerances and engineers have devised ways of automatically regulating their operation to prevent those tolerances being exceeded. Mechanical homoeostasis is analogous to biological homoeostasis, but much simpler, so it can provide a clear example of the way all homoeostasis works.

In the days when factory machines were all powered by a single large steam engine, it was important that the steam engine ran at constant speed. This was achieved by means of a governor, illustrated diagrammatically in Figure 4.22 and still used in all constant-speed engines, although most no longer look like the one in the diagram. Essentially, the governor comprises a set of arms (usually four, but only two are shown in the figure) weighted at the ends and pivoted. The governor spins, being driven directly by the engine. As engine speed increases, the governor spins faster and the weights are thrown outward. Because they are pivoted, as the outer, weighted ends of the arms rise, the inner ends descend, pushing down a valve that reduces engine power. If the engine slows, the opposite happens. The weights spin more slowly and fall inward, raising the inner ends of the arms and opening the valve to increase engine power. Without any attention from the human machine operator, the governor regulates the speed of the engine.

It does so by 'negative feedback'. Feedback occurs when one component in a system alters its behaviour in response to the behaviour or condition of another component. If the effect is to counter departures from an optimal state, thus restoring equilibrium, the feedback is said to be negative, as in the case of the governor. Feedback can also be positive, acting to accelerate departure from a former state. Suppose, for example, that during an unusually cool summer some of the winter snow failed to thaw. The white snow would reflect solar radiation, preventing warming of the ground below, and the following winter would commence with the ground already cold,

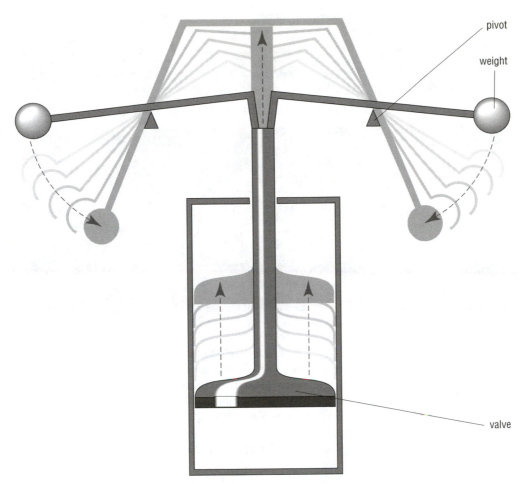

Figure 4.22 **Speed governor of a steam engine**

so early snow would settle at once, rather than melting. This would reflect more radiation, increasing the cooling. Some climatologists believe this is how ice ages begin, in a 'snowblitz' during which snow accumulates rapidly, year after year, by strongly positive feedback.

Ecologists who have studied particular mature ecosystems over prolonged periods have found that the populations within them fluctuate erratically from year to year, but within quite narrow limits. Overall, the composition remains fairly constant provided the system is not severely disrupted from outside, as it might be by a major pollution incident, for example, or a disease such as myxomatosis or Dutch elm disease. This, of course, is what is meant by the stability of mature ecosystems.

Populations do not plan to control their numbers. Indeed, since natural selection favours those individuals which reproduce most successfully, homoeostasis is an entirely automatic consequence of the attempts by all organisms to maximize their reproduction. Figure 4.23 illustrates one way populations may regulate one another by feedback relationships. Allow that for some reason plant growth (primary production) increases; perhaps because of very favourable weather during the growing season. This increases the amount of food available for herbivores (primary consumers). More of their young survive and more are born, so their population increases. Since

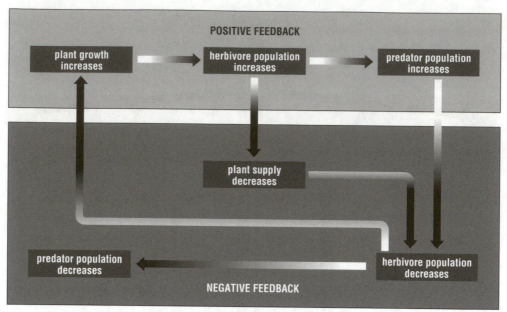

Figure 4.23 *Feedback regulation of a population*

there are now more herbivores, more food is available for carnivores (secondary consumers) and, after a delay, their population also increases. These responses represent positive feedback: the first increase stimulates further increases. Overall, however, the system is governed by negative feedback, which soon becomes evident. Increased consumption of plants causes a decrease in the food supply to the herbivores, which are also suffering increased predation. The herbivore population decreases and, again after a delay, the carnivore population also decreases. Pressure on the plants being relieved, primary production may now start increasing again. Through cycle after cycle, populations rise a little and then fall a little, but the changes are quite minor because they are regulating one another.

The figure greatly over-simplifies the situation in a real ecosystem, of course, particularly in its omission of parasitism and the diseases associated with it, which can exert a large influence on population sizes, but it reflects the principle by which ecosystems achieve homoeostasis. It also demonstrates that herbivore numbers are controlled only partly by predation, but predator numbers are controlled to a much larger extent by herbivores. The over-simplification becomes evident when, for example, a predator species suffers from disease that reduces its population to below the level determined by the size of the herbivore population. This may encourage another predator species to enter the ecosystem and, if it can establish itself securely, to remain there after the original predators have recovered from disease and limit their population by competing successfully for a share of the food supply.

Primary production → primary consumer → secondary consumer relationships are density-dependent. That is to say, the increase in the density of population at each trophic level is what permits the proportional increase at the next higher level, and when population density at one level falls, so do those at the others.

Ecosystem homoeostasis is entirely density-dependent, but trophic relationships form only one aspect of it; it also directly affects breeding in many species. It has been observed that if the population of great tits (*Parus major*) is 1 pair per hectare, for example, the average clutch size is 14 eggs, but if there are 18 pairs of great tits per hectare, average clutch size is only 8 eggs. African

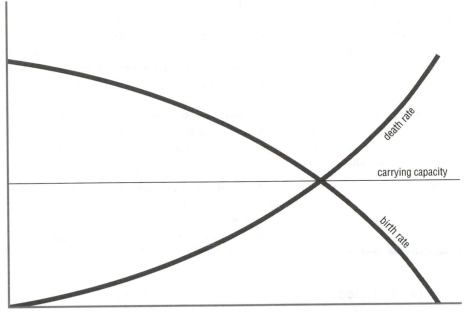

Figure 4.24 *Density-dependent feedback regulation*

elephants reach sexual maturity at an earlier age and give birth more frequently when the population density is low than when it is high. Similar responses have been observed in many species (DAJOZ, 1975, pp. 212–213). Kept in crowded cages, mice will resorb foetuses even when the food supply is abundant.

Eventually, the usual consequence is that for each population the number born each year is approximately equal to the number dying. Birth rate and mortality strike a balance quite close to the carrying capacity of the environment for that population, as is shown in Figure 4.24. It is exceedingly difficult, and sometimes impossible, to calculate a precise value for carrying capacity and so populations are able to fluctuate somewhat in size without exceeding it or falling below it.

Density-dependent feedback maintains homoeostasis only provided it is not over-sensitive. In this respect it is a little like steering a car. When you first learn to drive there is a tendency to overcompensate when the car starts to drift across the road. Unless corrected, the result is a series of swerves from side to side that are liable to increase in magnitude. This tendency to overcompensate is even more marked among trainee pilots; because aircraft move in three dimensions and air offers much less friction than a road, over-compensation quickly sets an aeroplane bucking crazily all over the sky. Similarly in ecosystems, density-dependent over-compensation can lead to chaotic swings in numbers, producing a situation far removed from stability (BEGON *ET AL.*, 1990, pp. 222–223). Such apparently chaotic effects have been observed when systems are perturbed; although eventually they regain their stability, for a time they exhibit huge, erratic peaks and troughs (GLEICK, 1988, pp. 78–79).

Feedback mechanisms have also been proposed as the means by which living organisms regulate important aspects of their own abiotic environments, an idea that has been asserted most forcefully by James Lovelock in his Gaia hypothesis. This states that on any planet supporting life, the physical and chemical conditions necessary for life are maintained through feedback responses by the organisms themselves. If conditions begin to depart from the optimum, biological responses to those departures have the secondary effect of cancelling them and restoring equilibrium. Lovelock developed what he has called 'Daisyworld' computer models to demonstrate this type of homoeostatic regulation. In a world where the only living things are daisies, some white and some black, fluctuations in solar radiation favour one group or the other. When temperature rises, white daisies reflect more heat and so remain comfortably cool; when temperature falls, black daisies absorb more heat and so remain comfortably warm. With each change, the surface area covered by white or black daisies also changes as one type expands and the other contracts. This alters the planetary albedo, white daisies reflecting more radiation when radiation increases and the area they cover expands, black daisies absorbing more radiation when radiation decreases and they expand their area. These changes in albedo then raise or lower the surface temperature and hence the air temperature, so preventing the climate from becoming so warm or cool that neither type of daisy can survive (LOVELOCK, 1988, pp. 35–41).

Perhaps the highly abstract 'Daisyworld' story carries the efficacy of homoeostatic regulation too far. In our own, real world, however, it is a potent mechanism for maintaining ecological stability. Interference with an ecosystem will produce feedbacks. If the perturbation is small, the system will probably recover, but if it is large, the long-term consequences may be difficult and sometimes impossible to predict.

45 Limits of tolerance

Along the Pacific coast of North America, prevailing winds drive the cool California Current, flowing south with many upwellings of deep, cold water that bring nutrients close to the surface, where they feed a wide variety of marine organisms. The average temperature of the surface water, and of the breakers enjoyed by Californian surfers, used to be 13 °C. Since around 1950, however, the current has warmed by 1.2–1.6 °C. Too small a change to do much for the comfort of surfers, this slight warming is believed to account for an observed 80 per cent reduction in zooplankton (minute animals that drift in the surface waters) over the past forty years (HILL, 1995).

Clearly, even apparently robust ecosystems can be disrupted by quite minor changes in certain physical conditions over which they have no control. To produce such an effect, the change must affect something species need, or something that can harm them if it is present in a concentration higher than some critical value. Such a 'something' is known as a 'limiting factor'.

For any limiting factor there is a minimum and a maximum value, below or above which conditions are intolerable. These boundaries are known as 'limits of tolerance'. They apply directly to species and to ecosystems indirectly, through the effects on their constituent species. Somewhere between the two extremes, of 'too little' and 'too much', a limiting factor occurs at highly favourable levels. Measure the rate of growth and reproduction of a species in response to changing values of a limiting factor, and a graph displaying the resulting data will have a bell-shaped curve, as illustrated in Figure 4.25. Boundaries of the highly favourable level, where growth and reproduction reach a maximum, mark the optimum range for that limiting factor for that species.

The concept of limits of tolerance was first proposed in 1911 by the American ecologist Victor Ernest Shelford (1877–1968). In what is now known as Shelford's law of tolerance he said that the presence and success of any organism depend on the degree to which a complete set of conditions are satisfied and that, once each of these conditions has been identified, particular organisms can be encouraged by altering those conditions which approach the limits of tolerance for those organisms. Shelford found, for example, that tiger beetles cannot reproduce unless they find suitable sites to lay their eggs, and their requirements for egg-laying and the survival of larvae are

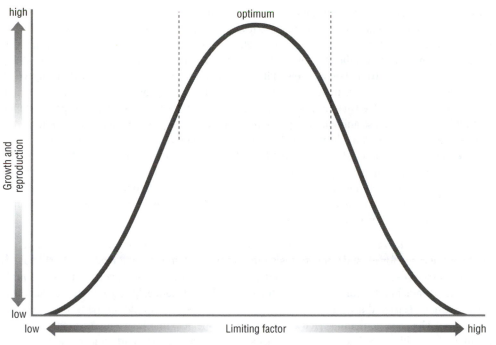

Figure 4.25 **Limits of tolerance and optimum conditions**

quite precise. They must find sandy soil with little humus and just the right temperature and moisture, and stones to provide shade from the light. If these conditions are not present, either the
females will not lay eggs or, if they do, the larvae will die (DAJOZ, 1975, p. 9).

Nutrients, of course, are the most obvious of all limiting factors, and it was from the realization
of this fact, and the 'law of the minimum', that Shelford's law of tolerance was developed.

In 1840, the eminent German chemist Baron Justus von Liebig (1803–73), also perhaps the best
teacher of chemistry of his day, published *Die organische Chemie in ihrer Anwendung auf
Agrikulturchemie und Physiologie* ('Organic chemistry in its application to agricultural chemistry
and physiology'), in which he argued that plants do not feed directly on humus, as was then generally believed, but subsisted on simple, inorganic chemicals taken directly from the air and soil.
By analyzing the chemical composition of plants he was able to make a list of these substances,
and he showed that plant growth ceases if the availability of just one essential nutrient falls below
a certain minimum, regardless of the abundance of all other nutrients. In other words, it is the
availability of the scarcest nutrient, not the most abundant, that determines success or failure for
plants: if the soil is deficient, say in boron, no amount of phosphorus will remedy the lack and
stimulate vigorous plant growth. This is Liebig's law of the minimum. Its acceptance by agricultural scientists led to the development of the fertilizer industry but, more generally, Liebig had
identified the first limiting factor and consequently his work was of relevance to ecologists as well
as agriculturists.

Plants take up soil nutrients in aqueous solution and also need water to provide their tissues with
rigidity (or turgor). Water, therefore, is also a limiting factor in arid and semi-arid climates and
during periods of drought in temperate climates. Deserts support sparse vegetation because for
most of the time the availability of water falls below the limit of tolerance for plants not adapted
to survive in arid conditions.

Some plants (called 'xerophytes') and animals do inhabit deserts, of course. Others thrive in habitats too saline for most and still others tolerate levels of copper that would prove lethal were it not for their adaptation to them. Copper is an essential trace nutrient, however, and in North Ronaldsay, the northernmost island of the Orkneys, off the north coast of Scotland, there is a breed of sheep that has adapted to extremely low levels of it. The sheep are confined to the foreshore by a wall surrounding the arable fields in the interior of the island, built to protect the crops from grazing. Having no choice, they feed mainly on seaweed, a diet rich in copper but also containing a substance that inhibits its uptake, so animals eating too much seaweed are prone to copper deficiency. Over the generations, North Ronaldsay sheep have evolved the ability to metabolize copper four times more efficiently than other sheep, allowing them to obtain the amount they need from an apparently deficient diet, but also exposing them to the risk of copper poisoning if they are moved on to grass. Their seaweed diet also supplies the sheep with huge amounts of iodine, their milk having been found to contain 550 times more iodine than is normal for other sheep, apparently causing them no inconvenience (ALDERSON, 1978, pp. 76–79).

The ability of certain species, or varieties within species, to thrive in circumstances beyond the limits of tolerance for most demonstrates an important evolutionary and ecological principle. If resources are present but in some respect deficient or excessive, natural selection will strongly favour those individual organisms that are tolerant of the imbalance, because their tolerance will allow them unrestricted access to the remaining resources. We should not be too surprised to learn of species, and sometimes rich communities of species, that thrive in very extreme environmental conditions. Saline lakes, which lack any outflowing river and so lose water only by evaporation, are as common in the world as a whole as freshwater lakes, and they are usually very fertile. Deep-sea hydrothermal vents support complex ecosystems thriving at temperatures of up to 40 °C or even higher on a diet of sulphides and heavy metals. The topmost metre of sediment in an area of the North Pacific sea bed, 600 m × 1200 m, has been found to contain 1–1.5 tonnes of uranium, 4000 tonnes of magnesium, and high concentrations of 13 other metals. Cadmium commonly occurs around vents, and some sediments are rich in hydrocarbons. Despite this, or perhaps because of it, ecosystems flourish that comprise organisms adapted to these conditions. In 1995, Greenpeace prevented the oil storage platform *Brent Spar* from being sunk to the sea bed in the deep Atlantic. Had they failed, the platform and its contents might well have nourished the ecosystem receiving it and in the opinion of many marine biologists could have caused little, if any, harm (NISBET AND FOWLER, 1995).

Temperature is an important limiting factor. Food for animals is scarce in cold weather, because at temperatures close to freezing biochemical reactions proceed so slowly that there is little or no plant growth. As Figure 4.26 shows, growth increases rapidly as the temperature rises, peaking at around 25 °C and then falling rapidly. At temperatures approaching 40 °C the rate of respiration exceeds that of photosynthesis and at above about 45 °C many plants die.

Despite the apparently absolute chemical limitation imposed by low temperature, there are algae of the genus *Chlamydomonas* that grow on the surface of snow and die if the temperature rises above +4 °C (DAJOZ, 1975, p. 69), and polar waters support diverse populations of animals. On the islands in the Arctic basin, known as the High Arctic, there are flowering plants that thrive in a climate where the soil thaws for no more than three or four weeks each year and the summer temperature seldom exceeds 4 °C. Many of the plants are brown, their dark colour absorbing heat, and females of the arctic willow (*Salix arctica*) are covered in down that traps warmth and can raise their leaf temperatures to 11 °C higher than the surrounding air temperature. The plants are helped by the 24 hours of summer daylight that is available to them for photosynthesis. The High Arctic ecosystem is well adapted to a cold climate, but how might it fare if the climate should become warmer? Studies have shown that the climate is already warmer than it used to be in high

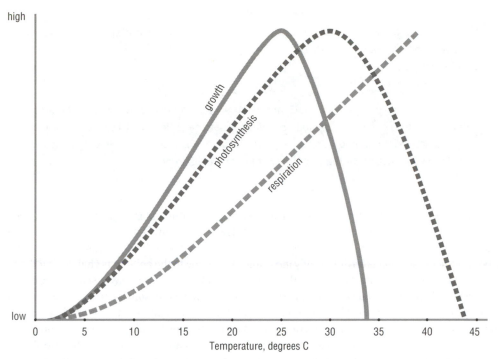

Figure 4.26 **Plant response to temperature**

latitudes and seedlings are appearing in areas for the first time free from permanent snow and ice. At the same time, species have adapted to quite wide variation in types of habitat and seem well equipped to flourish under warmer conditions (CRAWFORD, 1995).

Brent Spar

In the spring of 1995, Shell began to tow the *Brent Spar* oil-storage platform from the North Sea to the Atlantic. There it was to be sunk in water 2000 m deep in the North Feni Ridge, which is part of the Rockall Trough, well beyond the edge of the continental shelf. This was thought to be the most environmentally satisfactory method for disposing of this particular installation, which had reached the end of its useful life.

The Greenpeace vessel *Adair* shadowed the platform as it moved north of the Shetland Islands, and on June 16 two activists boarded it by helicopter. *Solo*, a large Greenpeace tug with a helicopter landing pad, also entered the area. Greenpeace occupied the platform for 23 days, orchestrating world-wide media coverage of their protest against deep-sea disposal. Environmentalists picketed Shell petrol stations in several countries. Shots were fired at a station outside Frankfurt and a Hamburg station was firebombed. Shell backed down and the platform was towed to Erfjord, a deep inlet on the Norwegian coast while an alternative method of disposal was devised.

Greenpeace maintained that the platform contained about 5000 tons of crude oil mixed with radioactive waste and other pollutants. In fact it contained about 150 tons of oil. In September, Greenpeace admitted its mistake and apologized to Shell.

It was decided to dismantle the platform on land and to use the materials to construct a quay near Stavanger. Work began in November 1998 and was completed by September 1999.

The conversion had cost £41m (€62.2m) compared with the original estimate of £21.5m (€32.25m) and an estimated £26.25m (€39.4m) for disposal at sea. Instead of yielding a net energy gain of 48 000 gigajoules it led to a net energy loss of 115 000 gigajoules. This was approximately double the energy cost of disposal at sea. Most scientists agreed that deep-sea disposal of the platform would have caused no harm.

Our planet provides few habitats to which at least a few, and often many, species have not been able to adapt by expanding their limits of tolerance. These are natural habitats, however, in which natural selection has had ample time to operate. Environmental pollution presents a different situation, not because species are inherently incapable of adapting to the pollutants challenging them, but because of the rapidity with which they are exposed to them. This may disrupt established ecosystems leaving areas severely impoverished. Eventually, it is reasonable to suppose they will be recolonized, for they are seldom more inhospitable than some habitats that support rich communities, but it will take time, and this is a valid reason for minimizing the disruption we are prepared to permit. That said, most ecological effects of pollution are greatly exaggerated. While we should certainly take care to prevent avoidable pollution incidents, we should not frighten ourselves into supposing industrial or other pollution could bring life on Earth to an end. Organisms are more robust than some people believe and better able to stretch their limits of tolerance.

End of chapter summary

Physical and chemical resources are used by plants and animals, including ourselves, and any adverse effect that arises from their exploitation and use is felt by living organisms. Indeed, it is difficult to imagine what 'environmental harm' might mean if it does not mean harm to living organisms.

Before we can measure, monitor, and predict the environmental consequences of events we need to understand how living organisms function and how they relate to each other and to their abiotic surroundings. This involves acquiring a grasp of the way nutrients move through cycles and how the basic processes of respiration and photosynthesis work.

Relationships among populations provide the subject matter of ecology. Those relationships are based on the transfer of nutrients or energy from group to group. In this context the concepts of nutrients and energy may be considered interchangeable. Ecological systems are complex, but research has revealed a limited number of processes that underlie them. From an environmental point of view, possibly the most important ecological discovery is that there are definable limits to the conditions each species can tolerate.

End of chapter points for discussion

What is the relationship between photosynthesis and respiration?
What are the advantages and disadvantages of ecological pyramids?
Does greater species diversity imply greater ecological stability?
Is the climax concept always valid?

See also

Greenhouse effect (section 13)
Climatic regions and floristic regions (section 21)
Eutrophication (section 23)
Nutrient cycles (section 33)
Respiration and photosynthesis (section 34)
Trophic relationships (section 35)
Energy, numbers, biomass (section 36)
Ecosystems (section 37)
Succession and climax (section 38)
Biodiversity (section 50)
Human populations and demographic change (section 54)

Further reading

The Ages of Gaia. James Lovelock. 1988. Oxford University Press, Oxford. A description of the Gaian hypothesis by its originator.

Biogeography: A Study of Plants in the Ecosphere, 3rd edn. Joy Tivy. 1993. Longman Scientific & Technical, Harlow, Essex. An excellent introduction to biogeography and the subjects related to it.

Ecology: Individuals, Populations and Communities, 2nd edn. Michael Begon, John L. Harper, and Colin R. Townsend. 1990. Blackwell Science, Oxford. A standard textbook on its subject, and an excellent one.

Gaia: The Growth of an Idea. Lawrence E. Joseph. 1990. Arkana, London. Explains the Gaian hypothesis clearly and also the criticisms it has encountered.

Methods of Study in Quantitative Soil Ecology: Population, Production and Energy Flow. J. Phillipson. 1971. International Biological Programme, London, and Blackwell Scientific Publications, Oxford. Provides detailed information on sampling soil organisms.

Waste and Pollution: The Problem for Britain. Kenneth Mellanby. 1992. HarperCollins, London. Describes simply the character and effects of waste. The late Professor Mellanby was one of the foremost authorities on environmental pollution.

References

Alderson, Lawrence. 1978. *The Chance to Survive*. Cameron and Tayleur in association with David and Charles, Newton Abbot.

Allaby, Michael. 1981. *A Year in the Life of a Field*. David and Charles, Newton Abbot.

Baron, Stanley. 1972. *The Desert Locust*. Eyre Methuen, London.

Begon, Michael, Harper, John L., and Townsend, Colin R. 1990. *Ecology: Individuals, populations and communities*, 2nd edn. Blackwell Science, Oxford.

Bowler, Peter J. 1992. *The Fontana History of the Environmental Sciences*. Fontana Press, London.

Brewer, Richard. 1988. *The Science of Ecology*, 2nd edn. Saunders College Publishing, Fort Worth, Texas.

Carson, Rachel. 1962. *Silent Spring*. Penguin Books, Harmondsworth.

Cooke, G.W. 1972. *Fertilizing for Maximum Yield*. Crosby Lockwood, London.

Copley, Jon. 1999. 'Indestructible', in *New Scientist*, 23 Oct., pp. 45–46.

Corbet, G.B. and Southern, H.N. 1977. *The Handbook of British Mammals*, 2nd edn. Blackwell Scientific Publications, Oxford.

Crawford, R.M.M. 1995. 'Plant survival in the High Arctic', *Biologist*, 42, 3, 101–105. Institute of Biology, London.

Dajoz, R. 1975. *Introduction to Ecology*, 3rd edn. Hodder and Stoughton, London.

Dowdeswell, W.H. 1984. *Ecology: Principles and practice*. Heinemann Educational Books, Oxford.

Gimingham, C.H. 1975. *An Introduction to Heathland Ecology*. Oliver and Boyd, Edinburgh.

Gleick, James. 1988. *Chaos: Making a new science*. Heinemann, London.

Godwin, Sir Harry. 1975. *History of the British Flora: A factual basis for phytogeography*. 2nd edn. Cambridge University Press, Cambridge.

Graham, Jeffrey B., Dudley, Robert, Agullar, Nancy M., and Gans, Carl. 1995. 'Implications of the late Palaeozoic oxygen pulse for physiology and evolution'. *Nature*, 375, 117–120.

Grime, J.P. 1979. *Plant Strategies and Vegetation Processes*. John Wiley and Sons, Chichester.

Harman, Denham, 1992. 'Role of free radicals in aging and disease', in Fabris, Nicola, Harman, Denham, Knook, Dick L., Steinhagen-Thiessen, Elizabeth, and Zs.-Nagy, Imre, eds. *Physiopathological Processes of Aging*, Annals of the NY Acad. of Sciences, Vol. 673. NY Adac. of Sciences, New York. pp. 126–141.

Hill, David K. 1995. 'Pacific warming unsettles ecosystems', *Science*, 267, 1911–1912.

Hillstead, A.F.C. 1945. *The Blackbird*. Faber and Faber, London.

Huck, U. William. 1995. 'Lemming migrations', in Macdonald, David (ed.) *The Encyclopedia of Mammals*, Andromeda Oxford, Abingdon, Oxon., pp. 656–657.

Hutchins, Ross E. 1966. *Insects*. Prentice-Hall, Englewood Cliffs, New Jersey.

Janzen, Daniel H. 1975. *Ecology of Plants in the Tropics*. Studies in Biology No. 58 (Institute of Biology). Edward Arnold, London.

Johnson, C.G. 1963. 'The aerial migration of insects', in Eisner, Thomas and Wilson, Edward O. (eds) *The Insects*. W.H. Freeman & Co., San Francisco.

Joseph, Lawrence E. 1990. *Gaia: The Growth of an Idea*. Arkana, London.

Kendeigh, S. Charles, 1974. *Ecology with Special Reference to Animals and Man*. Prentice-Hall, Inc., Englewood Cliffs, New Jersey.

Kupchella, Charles E. and Hyland, Margaret C. 1986. *Environmental Science*, 2nd edn. Allyn and Bacon, Needham Heights, Mass.

Long, Steven P. and Mason, Christopher F. 1983. *Saltmarsh Ecology*. Blackie, Glasgow and London.

Lovelock, James. 1988. *The Ages of Gaia*. Oxford University Press, Oxford.

Marshall, N.B. 1979. *Developments of Deep-Sea Biology*. Blandford Press, Poole, Dorset.

Mellanby, Kenneth, 1992. *The DDT Story*. British Crop Protection Council, Farnham.

—. 1992a. *Waste and Pollution: The problem for Britain*. HarperCollins, London.

Moore, David M. 1982. *Green Planet*. Cambridge University Press, Cambridge.

Naeem, Shahid, Thompson, Lindsey J., Lawler, Sharon P., Lawton, John H., and Woodfin, Richard M. 1994. 'Declining biodiversity can alter the performance of ecosystems', *Nature*, 368, 734–737.

Nisbet, E.G. and Fowler, C.M.R. 1995. 'Is metal disposal toxic to deep oceans?', *Nature*, 375, 715.

Odum, E.P. 1984. *Fundamentals of Ecology*. W.B. Saunders, Philadelphia.

Parkes, R. John. 1999. 'Oiling the wheels of controversy', book review in *Nature*, 401, 644.

Phillipson, J. 1971. *Methods of Study in Quantitative Soil Ecology: Population, production and energy flow*. International Biological Programme, London, and Blackwell Scientific Publications, Oxford. p. 89

Pimm, Stuart L. 1984. 'The complexity and stability of ecosystems', *Nature*, 307, 321–326.

Pollard, E., Hooper, M.D., and Moore, N.W. 1974. *Hedges*. Collins New Naturalist, London.

Post, Eric, Peterson, Rolf O., Stenseth, Nils Chr., and McLaren, Brian. 1999. 'Ecosystem consequences of wolf behavioural response to climate', *Nature*, 401, 905–907.

Sankaran, Mahesh and McNaughton, S.J. 1999. 'Determinants of biodiversity regulate compositional stability of communities', *Nature*, 401, 691–693.

Tansley, A.G. 1946. *Introduction to Plant Ecology*. George Allen and Unwin, London.

Thompson, H.V. and Worden, A.N. 1956. *The Rabbit*. Collins New Naturalist, London.

Tilman, David and Downing, John A. 1994. 'Biodiversity and stability in grasslands', *Nature*, 367, 363–365.

Tivy, Joy. 1993. *Biogeography: A study of plants in the ecosphere*, 3rd edn. Longman Scientific and Technical, Harlow, Essex.

Wilson, Edward O. 1992. *The Diversity of Life*. Penguin Books, London.

Young, J.Z. 1981. *The Life of Vertebrates*, 3rd edn. Clarendon Press, Oxford.

Biological Resources

When you have read this chapter you will have been introduced to:

- evolution
- evolutionary strategies and game theory
- adaptation
- dispersal mechanisms
- species and habitats
- biodiversity
- fisheries
- forests
- agriculture
- human populations and demographic change
- genetic engineering

46 Evolution

Evolution is the formation of new species from pre-existing species. That is all the word means to biologists and it implies nothing with regard to the steps that might be involved. The word is often linked to the name of Darwin, giving the impression that in some sense he invented the concept. This is quite untrue. The evolution of species from pre-existing species was widely (although not universally) accepted by the time Darwin began to think about it seriously and today it is not in the least controversial. The evolution of species is a fact, well documented and observed.

Darwin contributed to the concept the proposal of a mechanism by which the evolutionary process may occur. He called it 'natural selection', and after his death its merging with the growing body of knowledge about genetics led some people to rename 'Darwinism' 'neo-Darwinism' or 'the modern synthesis'. Nevertheless, it remains fundamentally the explanation Darwin proposed.

Today the great majority of biologists accept Darwinism as a valid explanation for evolution in general. There is argument about details and particular instances, but these tend to strengthen the Darwinian proposition rather than weaken it. When scientists talk of the 'theory of evolution', it is the Darwinian theory, of evolution by means of natural selection, to which they refer and they give 'theory' its scientific meaning of an explanation for observed phenomena. Never do they seek to imply that evolution itself is no more than a vague, albeit attractive, idea. To misuse the word 'theory' in this way, and to conflate the Darwinian theory with the observed fact of evolution, betokens ignorance or intellectual dishonesty.

Evolution proceeds from natural selection and at the centre of this concept lies the idea of 'adaptedness'. This is the degree to which a species is suited to the conditions under which it lives. Those conditions vary from place to place and time to time, and the degree of adaptedness varies from one individual to another. These variations provide the 'raw material' on which evolution operates and its operation leads to 'speciation', which is the dividing of one species into two that in principle (but not always in fact) are unable to interbreed.

Consider the plight of the Red Queen. She explained to Alice that: 'Now, *here*, you see, it takes all the running *you* can do, to keep in the same place. If you want to get somewhere else, you must run at least twice as fast as that!' Species adapt to the environmental conditions to which they are exposed, but those conditions include other species with which they interact. Among a prey species, for example, those individuals that run fastest may be more likely to escape capture. They survive to breed and so the species as a whole comes to comprise animals that run faster than did their ancestors. Their predators, however, are also likely to evolve means of countering this development. Perhaps they, too, will come to run faster or perhaps they will acquire new hunting strategies. Thus natural selection can place species in a situation closely resembling that of the Red Queen in *Through the Looking Glass*, running, or adapting, as fast as they can merely to remain in the same place. In 1973, L. Van Valen called this the Red Queen effect and that is the name by which it is now known (COCKBURN, 1991, p. 125).

Environments are often exceedingly complex, however, and the image of large predators hunting grazing herbivores across the plains of Africa is not typical. It is better to picture an environment, and the relationships producing natural selection within it, as something at once smaller and richer in detail, perhaps in the way Charles Darwin described it:

> It is interesting to contemplate an entangled bank, clothed with many plants of many kinds, with birds singing on the bushes, with various insects flitting about, and with worms crawling through the damp earth, and to reflect that these elaborately constructed forms, so different from each other, and dependent on each other in so complex a manner, have all been produced by laws acting around us (DARWIN, 1859).

This generates a quite different picture and suggests quite different strategies. Although they have not been investigated so thoroughly as the Red Queen effect, the Tangled Bank hypotheses, proposed by G. Bell in 1982, suggest that while a species may be well adapted to its immediate surroundings, its offspring may not inherit that advantage. Offspring may increase their chances of survival, therefore, if they disperse randomly into neighbouring habitats where they will encounter slightly different conditions that will favour at least some of them, because of the natural variation between one individual and another. If true, this implies that sexually reproducing organisms are better equipped to enter and exploit new habitats than those which reproduce asexually and therefore produce offspring that are genetically identical to one another and to their parent (i.e. that are clones). It also implies that competition among siblings is reduced in sexually reproducing species.

Evolution is believed to proceed essentially in the manner proposed by Darwin (see box).

Natural selection

Darwin proposed a theory that can be expressed very simply. It proceeds by the operation of natural selection on the variation within populations.

1 Within any population of a species, individuals are not identical in every respect.

2 Populations generally produce more offspring than are required to replace their parents.

It was the English philosopher Herbert Spencer (1820–1903) who described Darwin's theory as the 'survival of the fittest', a phrase Darwin disliked, although he used it more or less as a synonym for natural selection in later editions of *On the Origin of Species by Means of Natural Selection*, and Alfred Russel Wallace (1823–1913), the co-discoverer of the theory of evolution by natural selection, used it unreservedly (OLDROYD, 1980, pp. 107 and 117). Critics, however, pointed out an apparent weakness. 'Survival' means remaining alive and, it seems, the fittest can be identified by the fact that they survive, so perhaps 'survival of the fittest' can be rephrased as 'survival of the survivors', which is tautological and leads to a circular argument: we identify the fittest (i.e. the survivors) by the fact of their survival (RIDLEY, 1985, pp. 29–30).

In the form in which he presented it, Darwin's theory does, indeed, present difficulties. He believed, for example, that the characteristics of parents are blended in their offspring (a discredited theory called 'blending inheritance') and that, at least to a minor extent, in adapting to environmental pressures, individuals may develop physiologically or behaviourally in ways that are inherited by their offspring (the 'inheritance of acquired characters', another discredited theory). Difficulties there were, but the theory was not tautological.

It avoids tautology by invoking the concept of selection within a changing environment. Consider the well-known case of the development by insects of resistance to insecticides. Within the initial insect population, some individuals will be especially susceptible to a particular insecticide. Most will be susceptible to it, but at a higher dose. There will be just a few, however, that can tolerate the highest dose applied. After one application, all the most susceptible and most of the moderately susceptible insects will die, but the tolerant ones will survive. After several applications, the majority of insects will tolerate the insecticide and the population as a whole will have become resistant to it. Selection, in this case not really natural, of course, because the insecticide is applied by humans, drives adaptation, and also explains it.

This kind of phenomenon was demonstrated very dramatically by H.B.D. Kettlewell in 1973 (FORD, 1981, pp. 88–92) (see box) in the case of the peppered moth, although no speciation occurred because the morphs were not isolated from one another and continued to interbreed. It is a precondition for speciation that, for whatever reason, two populations of the same species cease to interbreed. When this happens each can evolve in its own way and that is how one species becomes two.

The peppered moth

The peppered moth (*Biston betularia*) rests on tree trunks and wooden fences and is hunted by birds, which seek their prey visually. The moth is polymorphic. That is to say, it exists in several forms. Some are pale, others progressively darker. In 1848, the first dark moth (*carbonaria*) was reported near Manchester; by 1895, 98 per cent of the moths were of the *carbonaria* morph. Between 1848 and 1895, soot from factory chimneys had blackened trees and fences around Manchester. H.B.D. Kettlewell bred pale and dark moths and released them, watching with binoculars to see how they fared. When pale moths alighted on clean, lichen-covered trees they were almost invisible, but when they alighted on blackened trees they were clearly seen by their predators. As air pollution decreased, the trees and fences became cleaner, making the *carbonaria* moths more visible, and the pale morphs became commoner, intermediate morphs doing well in the areas through which the amount of pollution was decreasing.

Should the environmental pressure causing selection continue for long enough, members of the adapted population may become sufficiently different from those of the ancestral population and neighbouring populations not subjected to the pressure as to be unable to interbreed with them. At this point, the reproductively isolated population is classified as a new species. It follows, therefore, that natural selection acting on natural variation also explains the evolution of species.

In these cases, natural selection drives variation in a particular direction. This is called 'directional selection' and it is one of the three ways natural selection can influence evolution. Most individuals in a polymorphic population will be 'average', the numbers of variants decreasing as their variation becomes increasingly extreme. Presented graphically, this produces the bell-shaped curve shown as the solid line in the first graph in Figure 5.1. If selection favours one of the variants, after several generations these will become the new 'average' and the population will have changed in the direction of that variant, shown by the broken curve.

Suppose that then the environment remains constant. Selection now favours the average individual at the expense of variants. The bell-shaped curve remains where it is, but becomes taller and narrower (the broken line) as the extreme variants disappear from the population. This is called 'stabilizing selection' and is illustrated by the second graph in Figure 5.1. By favouring average individuals, this can reduce the number of variant types and, by depleting the evolutionary 'raw material', make speciation less likely.

It is possible, however, for two extreme forms to be selected, as shown by the third graph. Again, the selected variants become the average, but now there are two average types and two bell-shaped curves (broken line). This is 'disruptive selection' and may lead to the isolation of two distinct morphs that eventually evolve into distinct species.

Darwin knew from observation that variation exists among individuals within any population, but he had no explanation for the cause of that variation or for the way variations were inherited. This was a major weakness in his theory, of which he was well aware, and he died without learning that it had been resolved in 1866 by an obscure Austrian monk in a paper published in the *Transactions of the Brünn Natural History Society*. If, as Darwin believed, offspring inherited a blend of their parents' characteristics, variation within populations would gradually even out through random matings, leaving nothing on which natural selection could act. What the paper showed was that offspring do not inherit a blend of characters; that is not how heredity operates.

The monk, Gregor Johann Mendel (1822–84), spent eight years growing peas in the garden at St Thomas Monastery in Brünn (now Brno, in the Czech Republic), his experiments ending when

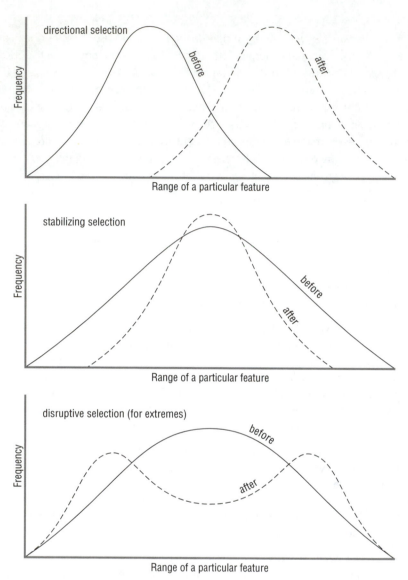

Figure 5.1 *Effects of natural selection*

his election as abbot, in 1868, left him too little time to continue them. He discovered that heritable characters are controlled by 'factors' (now called genes) which individuals possess in pairs (now called alleles). When gametes (egg and sperm cells) form, each contains only one version (allele). This is now known as Mendel's first law, or the law of segregation. Mendel's second law, or the law of independent assortment, states that when gametes combine at fertilization, the alleles behave independently, combining randomly with corresponding alleles. This is now known to apply only to genes that are not closely linked on the same chromosome; these tend to remain together.

Mendel worked with tall and short peas. Expressed in modern terms, his peas possessed two tallness alleles. One, call it T, is dominant: any pea inheriting T will be tall. The other, call it t, is recessive. The consequences of this arrangement are shown in Figure 5.2. If TT is crossed with tt, all offspring will be Tt, and thus tall. If Tt is crossed with another Tt, 75 per cent of the offspring will be Tt and thus tall, and 25 per cent tt and short. Mendel died in obscurity, and it was not until 1900 that his work was rediscovered independently by three botanists studying the past literature in connection with their own work.

Variation within populations results from the mutation in the genes of gametes (the cells which fuse at fertilization and develop to form a new individual) and by the shuffling of genes. DNA (deoxyribonucleic acid) is the substance by which heritable characters are transmitted from one generation

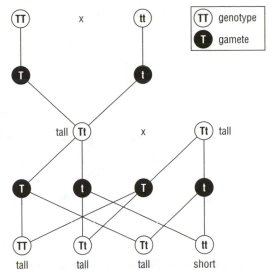

Figure 5.2 **Mendelian inheritance**

to the next, and changes in the units of which it is composed, the order in which they are arranged, or the amount of genetic material (as in polyploidy, where the number of chromosomes increases) alter the effects they produce. Such changes, called mutations, occur randomly from a wide variety of causes. Many mutations cause the death of the cell in which they occur or of the entire organism, but some have no great effect. Eventually these become established within a population, bringing about a gradual change in its genetic composition, called 'genetic drift'. Should the environment change, natural selection acts on mutant organisms.

Natural selection acts upon individual organisms, but evolution occurs at the molecular level; it acts on genes. The genetic constitution of an organism is known as its 'genotype', and its physical characteristics, or the expression of its genotype, are its 'phenotype'.

Genetics combined with evolutionary ecology can explain much about the way organisms adapt to their environments and the links between adaptation and evolution. Not all organisms evolve at the same rate, however. It has been suggested, for example, that birds have evolved very rapidly, with all modern types having appeared during the last 5–10 million years (STOCK, 1995). There is also some evidence that the rate of evolution has slowed in humans (GIBBONS, 1995).

The concept of a species is convenient, but it is no longer so secure as it used to be. Many genes occur in all species, so it makes little sense to think of a 'pig' gene or a 'tomato' gene; there are simply genes. Genetic comparisons within and between species have revealed a great deal of overlap, such that the genetic difference between two individuals of the same species may be greater than that between members of different species. This blurring of what were formerly thought to be sharply defined boundaries between species is highly pertinent to the debate over genetically modified organisms.

Environments are changing constantly and have been doing so since life first appeared on our planet. Were organisms unable to adapt to change, life would have perished long ago, and were natural selection not to lead to the evolution of new species, countless ecological niches would have remained undefined. Environmental change is inevitable and certainly is not to be feared. It is difficult, probably impossible, to imagine any change capable of destroying all life on Earth, short of the eventual and inevitable expansion of the Sun into a red giant.

47 Evolutionary strategies and game theory

Imagine both you and your best friend are criminals. Together, you perform a heinous crime for which you are both arrested some time later. The police place you in separate cells, so you cannot confer, and a detective begins to question you. He admits that although he knows perfectly well that you are both guilty he cannot prove it, so he needs an admission. At this point he makes you an offer and tells you that his colleague is making an identical offer to your friend.

Your offence carries a maximum sentence of 5 years in prison.

If you will swear in court that your friend committed the offence you will go free, but your friend will receive the maximum sentence.

If both of you refuse to implicate the other, you will be convicted of a lesser offence, for which you will go to prison for 2 years.

If both of you implicate the other, both of you will go to prison for 4 years.

What should you do? If you betray your friend you may avoid prison, but what will your friend do? If you remain loyal, but are betrayed, you could go to prison for 5 years. If both of you remain loyal you will still go to prison, but for only 2 years. The trouble is, can you trust your friend?

This conundrum is known as The Prisoner's Dilemma and it is easier to understand if, instead of prison terms, the consequence of each choice is represented as a score, as in a game, and the words 'loyalty' and 'betrayal' are replaced by the more neutral 'cooperate' and 'defect'. In this case we might award scores based on the number of years removed from the sentence, from 0 to 5. The possible options and their scores are shown in Figure 5.3.

Work it out and you may find the result surprising. For each prisoner, or player, the best option is to defect regardless of what the other does. If both cooperate, they each receive 3 points; this is a higher score than they receive if both defect, but carries the risk of a one-sided defection and a zero score. The likelihood, therefore, is that both players will defect (NOWAK ET AL., 1993).

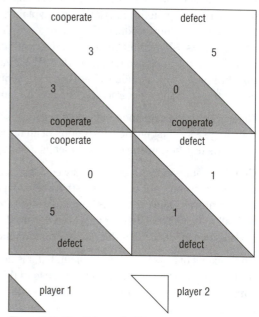

Figure 5.3 *The Prisoner's Dilemma*

Abstract though it may seem, The Prisoner's Dilemma raises an important question. Since, in any transaction, it pays to cheat, why is it that we feel this is wrong and how is it that in nature we see so many examples of cooperation? Remember that non-humans are not constrained by morality and are impelled by a purely Darwinian urge to maximize their reproductive opportunities.

We are accustomed to thinking of games, like soccer or baseball, that are played once and in which one team or player wins and the other loses. These are known as 'zero-sum' games, because if the winner is given a score of +1 and the loser of −1 the sum of the scores is 0.

Life is not really like that. The games of life, or transactions between organisms, are played many times and usually end only with the deaths of participants. This alters the situation, because the strategy that works well in a one-round, zero-sum game may not succeed over an indeterminate number of rounds. Played many times, The Prisoner's Dilemma becomes what is usually called The Iterated Prisoner's Dilemma.

Some years ago, the British biologist J. Maynard Smith and his colleagues G.R. Price and G.A. Parker borrowed ideas from game theory, a branch of mathematics devised originally to help plan military strategy, and applied them to the evolution of behaviour. Their aim was to discover behavioural strategies that could not be defeated if most members of a population adopted them. Because such a strategy would endure, it was called an 'evolutionarily stable strategy' (ESS). As an apparently simple example, Maynard Smith proposed a population consisting only of hawks and doves. When two individuals meet, hawks always fight as hard as they can, doves never fight; if a dove meets a hawk it runs away and if two doves meet they posture at one another until one retreats, but their contests never come to blows. So, if two hawks meet, one of them is badly wounded or killed; if a hawk and dove meet, the hawk wins but neither is hurt because the dove runs away; if two doves meet neither is hurt, but they waste a good deal of time posturing. There is no way to tell until an encounter whether an individual is a hawk or a dove. The ESS for the population as a whole will be achieved by some ratio of hawks to doves.

If the population consists only of doves, they will prosper, but there is a risk that a hawk will suddenly emerge either by invading or by mutation. A single hawk will have an immense advantage. Hawk numbers will increase until a point is reached at which there is a high probability that hawks will encounter other hawks, rather than doves. This places the hawks at a serious disadvantage. When scores are allotted for the outcomes of encounters, with penalties for injuries and wasting time, it emerges that the population will stabilize at around 42 per cent doves and 58 per cent hawks (DAWKINS, 1978, pp. 75–77).

Natural populations are not composed of individuals that invariably react to encounters in one of two extreme ways, but we need not suppose non-humans capable of thinking through the consequences of their actions to understand how an ESS can evolve. Natural selection acts on behavioural strategies just as it does on physiology, and the behaviour that optimizes reproduction will prevail. Some twenty years ago, Robert Axelrod, a political scientist, challenged computer programmers to devise an undefeatable strategy for The Iterated Prisoner's Dilemma, then played the 63 contesting programs against each other repeatedly. The winning strategy, devised by a game theorist, Anatol Rapoport, and called 'Tit-for-Tat' (TFT), is extremely simple. The dilemma, you will recall, is to decide whether to cooperate or defect, but this time in a game proceeding over an indeterminate number of rounds. In TFT, you cooperate in the first round and in every subsequent round you repeat the behaviour of your partner in the last round. If your partner defects, you defect next time; if your partner cooperates, you cooperate.

More recently, biologists have applied game strategies to models that allow an element of chance by introducing a 'mutation' in behaviour once in every hundred generations. This reflects the situation in real biological communities more accurately. TFT still succeeds, but only if a small number of TFT players are present at the start, and it leads to even greater cooperation (NOWAK AND SIGMUND, 1992), but with dangers. From time to time there are phases during which almost all members of the population cooperate or almost all defect.

Still more realistic modelling led to a variant of TFT called 'Pavlov', which corrects mistakes and allows the exploitation of unconditional cooperators. In Pavlov, players repeat their own last move if it brought a reward: if both players cooperated and were rewarded, then cooperate; if you defected, your partner cooperated, and so you were rewarded, then defect; if you both defected and received no reward, then cooperate. The game was repeated over 10^7 rounds, with 10^5 mutant strategies introduced. Pavlov also generated prolonged periods of cooperation and defection, switching from one to the other quite rapidly, but with a clear trend toward increasing cooperation. After 10^4 rounds, only 27.5 per cent of rounds exhibited cooperation, but after 10^7 rounds 90 per cent of them did (NOWAK AND SIGMUND, 1993).

Game theory based on The Prisoner's Dilemma provides powerful insights into the evolution of cooperation in a wide variety of contexts. Mutualism, for example, in which members of two different species perform services for one another, had long puzzled biologists. Why does a large fish not swallow the cleaner fish that moves about inside its mouth picking food from its teeth? The mathematics of the relationship demonstrate that cooperation is an ESS and mutualism is not subverted by occasional cheating (HAMMERSTEIN AND HOEKSTRA, 1995).

Persuasive though it is, the model remains somewhat controversial, and although examples have been found of behaviour that supports it, there are also some that seem to refute it. Lionesses, which cooperate in pairs to repel strangers seeking to invade their territory, may be brave or cowardly. Two brave lionesses will advance together, sharing equally the not inconsiderable risk of injury when the invader is encountered. If one of the pair is a coward, she will hang back. The brave lioness will advance more slowly, glancing behind her to see what her companion is doing, but apparently tolerates this cheating, because in subsequent forays to repel intruders the brave individual does not hang back herself and makes no attempt to punish her cowardly companion in a 'tit-for-tat' way. It is possible that relationships among lionesses are complex, involving much more than the shared defence of territory, and cowards contribute to the welfare of the group in other ways that warrant toleration of them, but in this case at least it seems the model strategy is not being applied (MORELL, 1995).

Cooperation is only one aspect of behaviour that can now be modelled mathematically to discover evolutionarily stable strategies. Natural selection favours those individuals that use their time and resources most efficiently. The dawn chorus of birds, for example, occurs because at first light the birds cannot see well enough to allow them to forage for food, so they can afford to spend time declaring their territories.

Later, when they start foraging, they are likely to adopt an optimum foraging strategy, which can also be calculated. In Figure 5.4, the heavy curve shows the amount of food, as energy, that the forager accumulates during the time spent foraging. The diagonal straight lines connect the time at which the forager starts to travel from one foraging patch to the next with the point on the

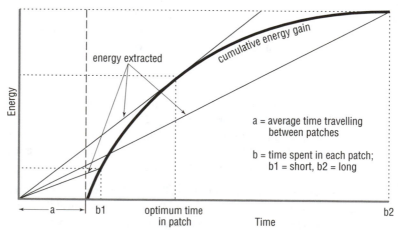

Figure 5.4 **Optimum foraging strategy**

heavy curve corresponding to time at which it leaves that patch. The intersections of the diagonal lines with the heavy curve indicate the amount of food energy actually gained during the time spent in a patch. It is important to remember that the figure implies nothing about the physical size of patches: they are all assumed to be of the same size and quality. The steeper the angle at which the diagonal lines rise, the more rapidly the forager obtains food energy. Obviously, the more time the forager spends in a patch the more energy it gains in total, but there is very little difference in the rate of gain between spending a very short and very long time in a patch. The most efficient use of foraging patches requires the forager to spend an intermediate length of time in each. This optimizes its foraging strategy by providing the most rapid acquisition of food. It is not too difficult to see why this should be so. When the animal first enters a patch the most palatable and nutritious food items are relatively abundant. It eats well, but with each item it consumes the total number of items is reduced and so it must spend more time looking for them. Its rate of energy acquisition slows. If it spends a very short time in a patch it will not be there long enough to acquire very much food, but if it stays in the patch a very long time an increasing proportion of time will be spent searching for items, and as the more nutritious items are consumed the nutritional quality of the patch as a whole will diminish. The optimal strategy, therefore, is to take the most palatable items quickly and when they are gone move to another patch. Such optimal foraging behaviour has often been observed. Similar optimal strategies can be devised for many behaviours (COCKBURN, 1991, pp. 88–94).

Behaviour is subject to natural selection. This applies to human behaviour as much as to the behaviour of any other species, but with a risk and an important difference. In non-human species variable amounts of behaviour are inherited. Web-spinning spiders do not learn to build webs, they inherit that ability, just as cleaner fish inherit their habit of foraging inside the mouths of particular large fish which they recognize as their 'customers', waiting in groups for them at 'cleaning stations'. Being inherited, such behaviour must be transmitted genetically and, therefore, behavioural genes must exist. Indeed, it was the idea that behaviour is to some degree determined genetically that gave rise to the scientific discipline of sociobiology. The risk is of extrapolating from this obvious link to the supposition that all behaviour results simply from 'programmed' instructions carried on genes. For spiders, worms, and other invertebrates this may be largely true, since they behave in highly stereotyped ways. Vertebrates show much more flexibility, however, and their behavioural responses to particular stimuli are not always the same. It is better to think of the genetic 'program' as supplying the capability for a range of behaviours. The resulting flexibility benefits the animals possessing it and is, therefore, favoured by natural selection, but the fact that it has evolved should not tempt us into the fallacy of extreme determinism.

The difference concerns humans. It is quite easy to demonstrate that our behaviour is also subject to genetic influence. Genes code for the synthesis of proteins and there are drugs that affect our moods or behaviour and are direct gene products (proteins) or the products of enzymatic reactions, and enzymes are proteins. Human behaviour is also flexible, and much more so than that of any other species because of our unique ability to contemplate the consequences of our actions, including their consequences for others. We can choose how we behave (TUDGE, 1993, pp. 100–105).

In recent years, unfortunately, press stories about 'homosexual', 'criminal', 'depressive', 'violent', 'alcoholic', and other genes have fostered a popular but misplaced belief in genetic determinism derived from neuroscientific research that has been reported out of context (ROSE, 1995). The truth is that very little is known about the link between genetic constitution and even physical differences between individual humans, let alone behavioural ones. Genetic determinism allows us to blame victims: people are poor because they are genetically disposed to idleness and fecklessness or have inherited a low IQ. This leads to repressive political and social responses and, of course, to racism and gender discrimination.

48 Adaptation

Darwin based much of his evolutionary theory on his observations of domestic animals. He had a great fondness for pigeons and described the great variety of forms, all of which had been bred by pigeon-fanciers from the rock dove (*Columba livia*). Domestic dogs provide an even more dramatic example of the powers of selective breeding. From the Great Dane to the chihuahua, all are descended from a single species, the wolf (*Canis lupus*), and the entire process has taken no more than 12 000 years (CLUTTON-BROCK, 1981, p. 34).

		Darwin's finches				
Genus	Species	Tree-feeding	Ground-feeding	Seed-eater	Insect-eater	Cactus-eater
Geospiza	magnirostris		X	X		
	fortis		X	X		
	fuliginosa		X	X		
	difficilis		X	X		X
	scandens		X	X		X
	conirostris		X	X		X
Platyspiza	crassirostris	X			X	
Camarhynchus	psittacula	X			X	
	pauper	X			X	
	parvulus	X			X	
Cactospiza	pallidus	X			X	
	heliobates	X			X	
Certhidea	olivacea				X	
Pinaroloxias	inornata				X	

Natural selection takes longer, but produces similar results, albeit for different reasons. In his visit to the Galápagos Islands, a volcanic group some 800 km off the coast of Ecuador, Darwin noted a group of birds, clearly of similar general type, in which each species had a bill seemingly adapted for a particular diet. Known now as 'Darwin's finches' (see box), the group comprises 14 species (some biologists count 13, others 15, depending on the taxonomic criteria used), all of them descended from a single species that migrated to the islands from mainland South America. Figure 5.5 shows how they have diverged from the ancestral species. Some, with long, slender bills, feed on insects and nectar. Others eat seeds and have short, stout bills, the size and strength of the bill varying according to the particular seeds the bird consumes and the proportion of birds with large and small bills varying with the relative abundance of the foods to which they are best suited. The woodpecker finch, with a slender bill, does not hammer at wood and lacks both a long tongue and a very long bill with which it might extract grubs from beneath the bark. Instead, it uses twigs or cactus spines as tools (WILSON, 1992, pp. 93–95).

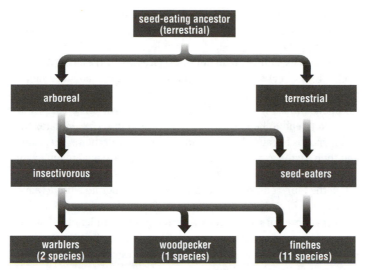

Figure 5.5 **Adaptive radiation of Darwin's finches**

Darwin's finches are adapted to the ecological niches they fill. 'Adaptedness' is the condition of being adapted, 'adaptation' the process by which adaptedness is acquired. In the case of the finches, adaptation occurred fairly rapidly when descendants of the ancestral population found a variety of food sources awaiting exploitation. Individuals with slightly thicker bills found seeds more suitable, those with slender bills fed preferentially on insects, and so the groups began to diverge. Rapid speciation of this kind, based on the colonization of under-exploited habitats, is called 'adaptive radiation'. It may occur at any taxonomic level. In the Galápagos Islands, there was an adaptive radiation of species; during the early Cenozoic Era, following the mass extinction marking the end of the Cretaceous Period 65 million years ago, there was an adaptive radiation of mammalian orders. An explosion of adaptive radiation at the species level even greater than that of the Galápagos occurred in Hawaii, another group of volcanic islands that was once devoid of life. There, a single species of finch produced 47 species and one or two groups of fruit fly (*Drosophila*) evolved into more than 600 species, one-quarter of all *Drosophila* species. Eventually, of some 21 000 species of algae, protists, fungi, plants, and animals on the islands, more than 8500 occur nowhere else (MLOT, 1995).

Adaptation is driven by natural selection, because those individuals most suited to the conditions under which they live will produce most offspring, and these will inherit the characters that favoured their parents. Adaptedness, then, is a product of natural selection. This can lead to confusion between the terms 'adaptedness' and 'fitness'. Adaptedness is the ability to live and reproduce in a particular environment. It is absolute, in that a species either possesses or lacks it. Fitness, in the Darwinian sense, is a measure of the degree to which the particular genetic constitution of an individual (a genotype) contributes genetically to the succeeding generation. The term applies to genotypes and it is relative, because it can be described only by comparison with the fitness of another genotype (PATTERSON, 1978, pp. 57–58). Some biologists prefer the term 'adaptive value' to 'fitness'.

Related organisms have adapted evolutionarily to varying environmental conditions in ways that now make them very distinct. In the last century this led zoologists to propose general rules describing differences that may be ascribed to temperature adaptation. Bergmann's rule (proposed in 1847 by C. Bergmann) states that animals in cold regions are larger than related forms in warm regions. This may be due to simple geometry: the larger the animal the smaller its surface area in relation to its volume and, therefore, the more efficiently it conserves heat.[1] The wing span of puffins (*Fratercula arctica*) reflects this rule; it averages 14 cm in the Balearic Islands, 16.5 cm in the British Isles, and 19 cm in northern Greenland (BERRY, 1977, p. 30). Gloger's rule (proposed by C.W.L. Gloger) states that animals in warm climates are more darkly coloured than those in cold climates, perhaps because they produce more of the dark pigment melanin. Allen's rule (proposed in 1876 by J.A. Allen) states that projecting parts of the body, such as ears, muzzle, and tail, tend to be longer in animals living in warm climates than in those living in cold climates, perhaps because heat is readily lost through them. There are exceptions to all these rules and their status as rules is dubious, but there are also many examples of the adaptations they describe. Arctic foxes (*Alopex lagopus*) have small ears and average 57 cm in body length (excluding the tail); red foxes (*Vulpes vulpes*) of temperate latitude have markedly larger ears and average 66 cm in body length; and fennec foxes (*Vulpes zerda*) of North Africa and the Near East have very large ears and a body averaging 39 cm in length. The biggest of all bears is the polar bear (*Thalarctos maritimus*), up to 2.5 m in body length; the American black bear (*Ursus americanus*) grows up to 1.8 m. Clearly, animals do adapt to climate physiologically, giving some support at least to Bergmann's rule.

Plants also adapt. Mangroves grow in, and also trap, salty, anaerobic mud along tropical coasts. Their roots require air and three genera have evolved root systems to cope with different levels of tidal flooding, illustrated in Figure 5.6. *Sonneratia*, which grows below the low-tide line, has

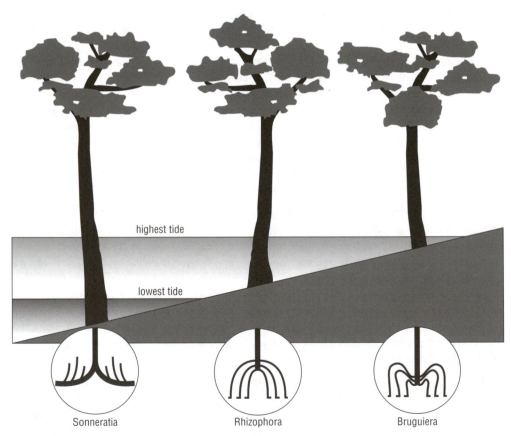

highest tide

lowest tide

Sonneratia   Rhizophora   Bruguiera

Figure 5.6 **Adaptation by mangroves to different levels of flooding**

'peg' roots; the main root lies just below the surface and has vertical extensions that project high enough to be clear of the water except at high tide. *Rhizophora*, which grows in the intertidal zone, has 'stilt' roots that are exposed to the air for much of the time. *Bruguiera*, growing in mud that is flooded only at high tide, has 'knee' roots: horizontal, underground roots with loops that project upward well clear of the surface.

It is not only climate and physical aspects of the environment to which species must adapt. Environments include other species and they also adapt to those. Perhaps the most convincing demonstrations of such adaptations are provided by mimics. It was the English naturalist Henry Walter Bates (1825–92) who, in 1862, first drew attention to species of moths that are palatable to birds, but which closely resemble other, unpalatable species and are avoided by predators. This is now known as Batesian mimicry and its best-known exponent is the viceroy butterfly (*Limenitis archippus*). It is edible, but is virtually indistinguishable from the monarch (*Danaus plexippus*), although it is somewhat smaller. The monarch concentrates in its body poisons produced by the milkweed plants on which its larvae feed. These induce vomiting within a few minutes, so any bird eating a monarch learns to avoid butterflies of its appearance. That appearance also protects the viceroy, but it works only as long as monarchs heavily outnumber viceroys. Were viceroys to become common, it would take birds longer to encounter a monarch, during which time they would consume viceroys, and they might not learn avoidance. Monarchs and viceroys are not unique. Many butterfly species are protected by accumulating poisons from the poisonous plants on which they feed and also have Batesian mimics.

Poisonous or otherwise dangerous species can also benefit by resembling one another. This type of mimicry is called Müllerian, after the German zoologist Fritz Müller (1831–97) who was the first to describe it, in 1879. Predators learn to avoid all species resembling the first distasteful one they encounter.

Mimicry has offensive as well as defensive uses. Some spiders of the genus *Myrmecium* are almost indistinguishable from the ants on which they prey. Spiders have eight legs, of course, and insects six, but the ant-mimics have the first pair of legs greatly lengthened and hold them so they look like antennae. One Red Sea species of blenny exploits the Batesian mimicry of two others. *Meiacanthus* has fang-like teeth that inject venom and predators learn to avoid it. *Ecsenius* is harmless, but is avoided because of its close resemblance to *Meiacanthus*. *Plagiotremus* resembles the other two and is also harmless, but feeds on the skin of fish that allow it to approach, supposing it to be both harmless and inedible (BURTON AND BURTON, 1977. p. 73).

This is called 'aggressive mimicry', and predators and parasites that are aggressive mimics may resemble their prey or hosts. There are species of fireflies, for example, which mimic the flashes, different for each species, by which males locate females and on reaching the females, eat them. The sabre-toothed blenny (*Aspidontus taeniatus*) closely resembles the cleaner wrasse (*Labroides dimidiatus*); when wrasse customers approach it to be cleaned it bites pieces from their fins. Many predatory mantids mimic the flowers on which they rest and capture insects attracted to the 'flower'.

Camouflage is not classed as mimicry, although camouflaged species often resemble the background against which they rest. It is not difficult to see how natural selection must favour well-camouflaged organisms, because camouflage will protect prey species from predators and increase the efficiency with which predators capture prey. It is not surprising that camouflage is very common. Not all species use coloration for concealment, however. Bright colours may also deter predators if they are associated with danger or unpalatability and it is this use of coloration that allows Batesian mimicry to evolve.

Natural selection provides a wholly adequate explanation for the vast range of shapes, colours, and behaviours we see around us. Heritable characteristics have adaptive value (increase fitness) if they enable certain individuals to withstand better than others the physical extremes of their environment, or provide them with shapes or colours that improve their chances of eating while reducing the chance of being eaten. Those individuals will produce more offspring and so their small, advantageous differences will be emphasized and become more widespread. This is how adaptation proceeds and how adaptedness is acquired.

49 Dispersal mechanisms

Opportunist plants that spring up within weeks when ground is cleared are able to arrive so quickly because their abundant seeds are light and carried on the wind. They are a familiar sight, floating past on their 'parachutes' of fine, white hairs. Maples produce one-winged seeds (samaras) that whirl like helicopter rotors, their spinning motion carrying them further than they would travel were they simply to fall. There are winged seeds that glide, and bladder senna (*Colutea arborescens*), a Mediterranean shrub, produces small seeds inside hugely inflated pods which float on the air like balloons. Tumbleweeds break off at ground level when their seeds have ripened and roll with the wind, sometimes for long distances, scattering seeds as they go.

These seeds are visible. Others are so small as to need no assistance in remaining airborne. Orchids produce microscopically small seeds comprising only a few cells, but some species produce a million of them at a time. Fungi release clouds of tiny spores.

There are plants that drop their seeds into flowing water and many produce seeds that are distributed by animals. Burrs are inedible fruits covered with small hooks by which they cling to the fur or clothing of any animal brushing against the plant. By the time they fall off and the fruit decays to release the seed, they may be a long way from their parent plant. The carrying of seeds on the outside of an animal is called 'epizoochory'. 'Endozoochory', the transport of seeds inside animals, is achieved by concealing a tough, indigestible seed inside a delicious fruit. When an animal eats the fruit it is rewarded with food, but the seed passes unaffected through its digestive system to be deposited in faeces, which also supply nutrient for the young plant. Fruits such as grapes, cherries, and dates that have thin skins, no smell, and are attached fairly firmly to the plant have evolved to be eaten by birds. Those adapted to attract other kinds of animal drop to the ground and often have thick skins concealing the edible parts and a strong, sometimes rancid, smell.

Many animals assist plants unintentionally by collecting seeds and storing them for future consumption. This is called 'synzoochory' and it works because although some of the seeds are eaten, many more are buried and never recovered. In effect, the hoarders sow them.

All organisms must disperse their offspring. A seed of a perennial plant deposited very close to the plant that produced it might germinate, but it would be unlikely to develop into a healthy plant, because of competition from its well-established parent. The young must leave home, and most plants, as well as fungi and many animals, achieve dispersal by broadcasting seed, spores, or fertilized eggs. These are known collectively as 'disseminules', because they are disseminated. This is passive dispersal, a random and very wasteful process. During the whole of their lives, a pair of sexually reproducing organisms need produce only two viable offspring to ensure their own replacement, yet broadcasters release vast numbers of disseminules, each of which cost energy and materials to produce. Production has to be prolific, because the parent has no control over the fate of its disseminules and most will fail to develop.

Dispersion is not the same thing as migration. 'Migration' generally describes the mass movement of organisms from one place to another, as when birds migrate between winter and summer locations. Dispersion, or dispersal, is the movement of individuals away from one another.

Passive dispersion is not necessarily truly random, because prevailing winds or the flow of water will tend to transport disseminules in predictable directions. Few passively dispersed disseminules travel far, their distribution forming the kind of pattern shown in Figure 5.7. This is what you might expect, because they begin falling to the ground from the moment they are released, but there are exceptions. Once airborne, disseminules and adults can be carried to considerable height and transported a long way.

Some species of spiders, in particular, exploit this possibility. They disperse by 'ballooning'. A young spiderling climbs to the top of a plant or fence, turns to face the wind, then raises its abdomen and releases a stream of silk from its spinnerets. The silk rises, the spider releases its grip, and the wind carries it where it will. The spiderling may need to produce several strands of silk before one lifts it into the air and the abandoned strands left by thousands of spiderlings can cover vegetation as 'gossamer'. If, on a warm summer evening, you find tiny spiders in your hair or on your clothes they are likely to be ballooning spiderlings. Most of the spiderlings drift at heights below about 70 m, but they have been found in samples of 'aerial plankton' taken at 3000 m and they can sometimes be carried for hundreds or even thousands of kilometres (PRESTON-MAFHAM AND PRESTON-MAFHAM, 1984, pp. 99–101).

As you will appreciate if you have watched swallows or bats feeding on airborne insects (and spiderlings), the air can be quite densely populated. Years ago, a study in England counted 12.5 million insects an hour drifting through a rectangle 91 m high and 1.61 km long (KENDEIGH, 1974, p. 278). Hurricanes can carry mainland organisms to remote islands, and tornadoes have been known to raise aquatic animals, including fish, and drop them somewhere else. Most die, of course, but occasionally some may fall into water and so colonize a new lake or river.

Larger animals also disperse, and although these animals are independently motile their movements are usually random. They simply move away from the area occupied by their parents in search of a patch of habitat where they will experience less competition. Their travels seldom take them beyond the borders of the range their population inhabits and those which do cross the border may find themselves in an inhospitable environment where they cannot survive.

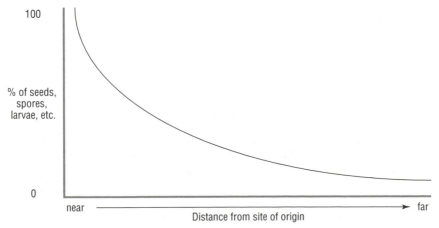

Figure 5.7 **Common pattern for passive dispersal**

There are exceptions. Some marine invertebrates, for example, which spend their adult lives anchored to a particular spot and feed by filtering particles from the surrounding water produce large numbers of motile offspring. These live as juveniles among the plankton, but actively seek sites that are likely to provide them with a substrate on which they can settle securely and where food is likely to be abundant. Their dispersal is certainly not random. Nor is that of some of the species described in the next section, headed 'Wildlife species and habitats'.

It may happen, however, that a disseminule or young animal wanders beyond its ordinary range and into a site, previously unoccupied by members of its species, where it finds the resources it needs. This allows it to expand the range of its species by diffusing into an adjacent area. Daisies quickly establish themselves in new lawns by this type of range expansion.

Species can expand their ranges by diffusion only if their original ranges are adjacent to favourable areas, and eventually their expansion will be halted by a barrier. This may be physical, such as sea for terrestrial species or land for marine species, a desert or mountain range, or biological, such as an entirely different type of vegetation. Barriers isolate populations, but occasionally they disappear and sometimes they can be circumvented. Geologic processes can cause sea levels to change, creating or removing barriers. At one time, species could move between the Old and New Worlds across the land bridge linking Alaska and Siberia across the Bering Strait, and before the Atlantic Ocean opened North America and Eurasia were part of the same landmass. Less than about 8000 years ago Britain was joined to continental Europe by a bridge between what are now Hull and the Dutch archipelago and Esbjerg, Denmark, and across what is now the English Channel (SIMMONS ET AL., 1981, pp. 86–88).

Species sometimes cross or circumvent barriers. Air currents may carry them across mountains as seeds, and small animals occasionally drift to sea on rafts of vegetation washed away by storms. Even more rarely, they make landfall on an island they can colonize. Insects, spiders, birds, and bats can be blown long distances. Such long-distance travellers seldom survive for long. It is fairly common for North American birds to be blown across the Atlantic, but they arrive in Europe singly and so cannot breed, and in any case find themselves in well-populated habitats with few niches awaiting definition, a fate they share with most dispersing individuals that stray beyond the borders of their established range.

When exceptions occur the result can be spectacular and 'jump dispersal' (BREWER, 1988, p. 37) can be highly successful. It most commonly results from the deliberate or accidental introduction of species by humans into regions where food is abundant and predators and parasites are left behind. 'Tramp' species are carried inadvertently; they 'hitch rides'. That is how the brown rat (*Rattus norvegicus*) and house mouse (*Mus musculus*) have travelled the world and it is how the water hyacinth (*Eichhornia crassipes*) first established itself in Louisiana (see box).

Introduced species

Species introduced to a region may not survive, but if they do their numbers may increase until they present serious economic or ecological problems. They proliferate because the environment in which they establish themselves does not include the predators and parasites that restricted their numbers in the environment from which they came.

Once established, exotic species may become fully naturalized to their new environment and, eventually, so familiar as to be often regarded as native. They may still make nuisances of themselves, however. The house mouse (*Mus musculus*), originally a native of the Middle East, was possibly the first mammal to be introduced to Britain by humans, in pre-Roman times. The black rat (*Rattus rattus*) probably arrived in Britain in the eleventh century and the first brown rat (*Rattus norvegicus*) arrived early in the eighteenth century. The domestic cat (*Felis catus*) was present in Britain by the Middle Ages and has been introduced deliberately to places from which it was previously absent, usually to control rodents. Many have escaped or been driven from human homes and established a large feral population that has a marked effect on populations of the small mammals and birds on which cats prey.

The rabbit, introduced to Australia in 1787 and 1791, bred more rapidly than native marsupial animals and adapted equally to semi-arid and humid regions until its consumption of grass and crops caused serious economic loss to farmers. Foxes, introduced as predators, preyed on native marsupials and birds rather than rabbits, devastating their populations.

Exhibitors travelling from Venezuela to the Cotton Exposition held in New Orleans in 1884 took with them specimens of an attractive aquatic plant, the water hyacinth (*Eichhornia crassipes*). It escaped, established itself, and is now a serious weed of freshwater aquatic systems throughout the tropics. It forms huge, floating mats and can double its numbers every 8–10 days, choking navigable waterways.

Others are introduced deliberately. The European house sparrow (*Passer domesticus*) was introduced to North America in 1852–53 and is now widespread. As Figure 5.8 illustrates, the starling (*Sturnus vulgaris*), introduced to North America in 1890–91, expanded its range to more than 4 million km^2 within half a century (KENDEIGH, 1974, pp. 276–277).

Britain, too, has considerable experience of introduction. If you see a squirrel in Britain it is most likely to be a grey one (*Sciurus carolinensis*), which is much more numerous and widely spread than the red squirrel (*Sciurus vulgaris*). Indeed, the scientific name *vulgaris* is something of a misnomer, because there are now no red squirrels over large areas of southern England and it is certainly not common. As its name, *carolinensis*, indicates, the grey is not native to Britain. There may have been a few in North Wales early in the last century, but grey squirrels were introduced to about 35 sites in England, Wales, Scotland, and Ireland between about 1876 and 1929 (CORBET AND SOUTHERN, 1977, pp. 166–167). At first they spread rapidly and continue to do so more slowly.

Rabbits (*Oryctolagus cuniculus*), now so common throughout Britain, were also introduced. The earliest unambiguous reference to them dates from the thirteenth century and they were abundant in England by the sixteenth, but it was not until the eighteenth that they spread throughout Wales and the nineteenth that they reached most of Scotland (THOMSON AND WORDEN, 1956, pp. 12–13).

As the spread of the rabbit shows, a species may establish itself but still take a considerable time before managing to expand its range by diffusion. The grey squirrel was fortunate in that its principal competitor, the red squirrel, is subject to wide fluctuations in numbers, and as it declined the grey was able to occupy and retain its habitat. The expansion of the rabbit, originally raised for food in guarded warrens, may have been due to agricultural changes that involved enclosing open land with hedges that provided shelter.

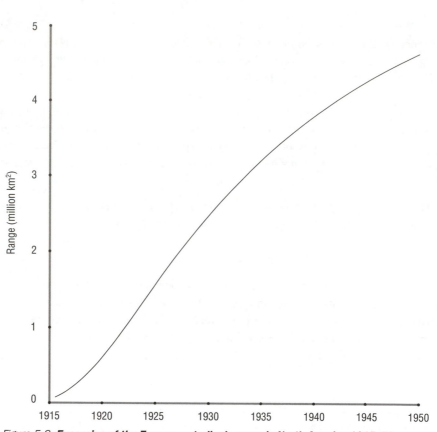

Figure 5.8 **Expansion of the European starling's range in North America 1915–50**
Source: Kendeigh, S. Charles. 1974. Ecology with Special Reference to Animals and Man. *Prentice-Hall, Englewood Cliffs, NJ*

Ecologists usually oppose the introduction of species to new regions, because of the havoc such introductions may cause should the species succeed in its new habitat. The competitive exclusion principle, according to which no two species (or groups of species) with identical resource requirements can survive together in the same ecosystem, strongly suggests that an introduced species can succeed only at the expense of native species. This is not inevitable, but there are enough examples to justify caution.

50 Wildlife species and habitats

If you were looking for badgers, there are certain kinds of places it would be worth visiting and others you could ignore. You would be wasting your time searching in a low-lying, wet field. Badgers do not live in such places. The best place to look for signs of a set might be in deciduous woodland, especially if there were fields nearby used for pasture and arable crops. Together, the woodland and fields would provide badgers with abundant food and a varied, balanced diet. You might call it a good badger habitat.

A habitat is a place, an area in which an organism lives, feeds, and breeds. It is sometimes described as the organism's address. Within its habitat, the organism lives in a certain way peculiar to its species. This is its niche. If the habitat is its address, the niche is its job.

Turn the concept around, and an area of land or water can be examined for the different habitats it provides. Over a large region of the world, the general type of vegetation, such as temperate grassland, tundra, or tropical rain forest, defines a biome, but clearly a biome contains many habitats. In most school atlases, the Amazonian rain forest is coloured a uniform shade of green, suggesting that a traveller traversing it would see little variety in flora and fauna over its entire $4 \times 10^6 \text{ km}^2$. Closer inspection of a $5 \times 10^5 \text{ km}^2$ area has revealed that far from exhibiting a tedious uniformity, the forest comprises a patchwork of 50 or so types of forest (CULOTTA, 1995a), visible in satellite images as patches of different colours identifiable on the ground as distinct vegetative formations (an observation atlas publishers may care to note). This discovery has profound implications for conservation in the region, because it shows that current efforts, based on identifying and protecting areas of significant biodiversity, rely on quite inadequate survey data (TUOMISTO, ET AL., 1995). Scientists do not use school atlases, of course! Their studies of changes in vegetation patterns over very large areas are based on satellite images (PARKINSON, 1997, Chapter 7) that reveal a considerable amount of detail, but require careful interpretation (TOWNSHEND *ET AL.*, 1993). Habitats also have histories and some of their characteristics may have originated long ago. The great variety in the Amazon basin may have developed during the middle Miocene, about 10^6 years ago, when the interior of tropical South America was partly an inland sea, open to the Atlantic in the north, east (through what is now the Amazon River), and south. The Amazon Sea divided the land into a mountainous (Andes) western peninsula and two large islands, the Guayanan and Brazilian shields (WEBB, 1995).

Indeed, habitats can exist on a very small scale. A precisely defined portion of a larger habitat, where a particular species or community is most likely to be found, is called a 'microhabitat'. It is possible, therefore, to consider a main habitat as a mosaic of microhabitats, but with some species moving freely among them.

A pond may provide a habitat for various fish and aquatic birds. Ducks may feed on it and nest close to its banks, and herons may visit it to hunt for fish and other small animals. In more detail, however, the number of microhabitats it provides depends on the size of the organisms inhabiting them. The pond in Figure 5.9 is fairly typical. It has an island, where ducks and other birds probably roost and breed. Around the island the surface water provides a habitat for duckweed and all the small animals that feed on it. Further away, the bottom mud and water depth are suitable for waterlilies, so they have colonized that area. In several places near the banks, where the water is shallow, reeds have established themselves and other birds shelter among them. Dead branches and leaves floating on the surface provide habitats for other small animals, and gravel washed in from the stream feeding the pond provides conditions suitable for a quite different population. Crevices between the stones near one bank provide a habitat for insect larvae. Microorganisms also have 'addresses'. Their habitats are very small, of course, and the pond and its inhabitants provide an even wider variety of those.

Badgers often live in a woodland habitat, but within the woodland each individual tree is also not one habitat, but many. There are insect larvae that feed on its leaves and other larvae that feed on or beneath its bark. Birds feed on the larvae and roost and nest in its branches. In many countries, but not Britain, there are tree frogs that spend their entire lives in their arboreal habitats.

When plant seeds disperse, chance rules their fate. Those which land in favourable spots thrive, those falling on ground that is too dry, too wet, too warm, too cool, too exposed, or too crowded fail to germinate or, if they do germinate, soon die. Plants occupy habitats, they have 'addresses', but they cannot go looking for them. Most animals, on the other hand, do choose their habitats. An animal tours an area, being especially attracted to certain places by their physical appearance. Visits to those places allow it to examine the finer details of the accommodation and

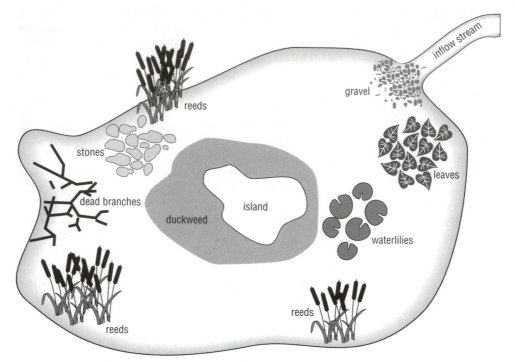

Figure 5.9 **Habitats in a pond**

neighbourhood. If it finds food, suitable shelter, a nesting site, and the probability of attracting a mate, it may settle. If not, it resumes its exploration of the surrounding area.

Some years ago, the German ornithologists R. Berndt and W. Winkel monitored the preferences of young pied flycatchers (*Ficedula hypoleuca*). Some of the birds were hatched and raised in deciduous woodland, others in coniferous woodland. Of those raised in deciduous woodland, 69 per cent settled in deciduous woodland, and of those raised in coniferous woodland, 79 per cent settled in deciduous woodland. Clearly, those pied flycatchers preferred deciduous woodland even though some of them had not lived in it when young and, presumably, they settled in coniferous woodland only when they found the deciduous woodland overcrowded (BREWER, 1988, p. 33). They chose it.

We can imagine birds or badgers visiting possible sites and settling when they find one to their liking, but much smaller animals than these choose their habitats. Crop pests, such as aphids, occur on some plants but not others, yet aphids are weak fliers, carried this way and that by winds and thermal currents. They disperse as winged forms, but within an hour or two they cease to fly actively and begin to fall, possibly steering themselves towards plants that exhibit some promising feature. If it turns out that the plant suits them they stay, if not they take off again, if necessary making several flights during the limited time remaining before their wing muscles degenerate (JOHNSON, 1963). Their habitat selection is more haphazard than that of a bird or mammal and many individuals perish without finding the food they seek but, as every farmer and gardener knows, aphids are highly successful at the species level.

Habitat selection is a matter of locating resources and these may be of quite different types. Biting midges (*Culicoides* species), for example, breed in water. Their thin, worm-like larvae swim about in ditches, ponds, and wet soil, but the female adults, no more than 1–4 mm long, feed on bird or mammalian blood, which contains proteins they need for egg production, and those notorious in parts of Scotland (and known in America as 'no-see-ums') favour human blood. They fly mainly in the evening and locate their victims by scent, taking full advantage of warm, sultry weather, when humans are perspiring (COLYER AND HAMMOND, 1968, pp. 77–78), but they cannot fly very far. Their habitat, therefore, must comprise suitable sites for egg-laying located close to places where humans congregate (and they are best controlled by identifying and destroying their breeding sites). Other species, of which there are about 800, specialize in the blood of different animals, and some tropical species transmit diseases (FREEMAN, 1973), the insects and the animals on which they feed comprising habitats for the parasites.

Habitats vary in size according to the species occupying them. For a species to survive, the habitat must support a population large enough to sustain itself by breeding. Should numbers fall below a certain threshold, some species lack the social stimulation to breed and their populations may collapse, even to extinction. This phenomenon was first described in 1949 by a team of ecologists led by W.C. Allee. Known as the Allee effect, it is believed to be what caused the extinction of the American passenger pigeon (*Ectopistes migratorius*). The minimum viable population (MVP) of any species is generally defined as that number which gives the population a 95 per cent chance of persisting for 100 years. This can be calculated by population viability (or vulnerability) analysis (PVA), and Figure 5.10 shows the result for mammals. Depending on the amount of variability over the area of the habitat, the graph shows, for example, that for a species with an average body weight of 1 kg (about the size of an adult hedgehog), the MVP is between about 8000 and 70 000. MVP can then be used to calculate the minimum area of habitat that population requires. Herbivorous mammals require a smaller habitat than carnivores, but for a 1 kg mammal in temperate latitudes, a viable herbivore population needs between about 5 and 80 km^2 and a viable carnivore

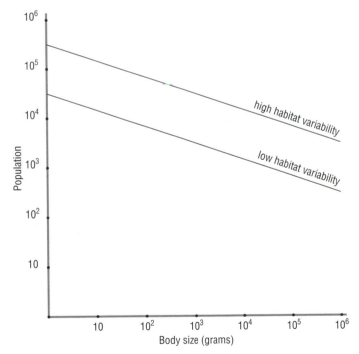

Figure 5.10 **Population size needed for a 95 per cent probability of persisting 100 years**
After Brewer, Richard. 1988. The Science of Ecology. Saunders College Publishing, Fort Worth, TX

population between about 50 and 1000 km^2, depending on the variability within the habitat area. The calculation is of obvious value in planning national parks and nature reserves.

Once established, a species can profoundly alter its habitat, with consequent effects on other species, but some species are more influential in this regard than others. Recognizing this gave rise in the 1960s to the concept of 'keystone' species, species that determine the character of the entire habitat by their presence or absence. It began when Robert Paine, an American marine ecologist, removed many individuals of a particular starfish species (*Pisaster ochraceus*) from the intertidal zone on a rocky shore. Mussels, the principal prey of the starfish, occupied the area and more than 300 other species, absent when the starfish was present, established themselves in the mussel beds. In the years that followed, more keystone species were identified. Keystone species are top predators and, although clearly important, they are not the only determinants of habitats. Among plants and microorganisms, species are now known to form 'functional groups', in which several species perform the same ecological function and one can be replaced by another. Ecologists are using this idea to develop 'patch models' of habitats in which varying ecological processes and relationships can be monitored (STONE, 1995).

Introduced species, often lacking the predators and parasites associated with them in their original habitats, can sometimes invade and disrupt habitats. Biologists distinguish between 'native' species that arrived by natural colonization in prehistoric times, 'naturalized' species that were introduced by humans but now maintain themselves without human intervention, and 'exotic' species that grow mainly or wholly because humans have planted and tended them. It is the naturalized species that can prove disruptive. Britain has only three naturalized and invasive tree or shrub species – sweet chestnut (*Castanea sativa*), introduced in Roman times; sycamore (*Acer pseudoplatanus*), introduced in medieval times; and the common rhododendron (*Rhododendron ponticum*), introduced more recently as an ornamental (RACKHAM, 1976, p. 19) – but there are several smaller plants, such as Japanese knotweed (*Polygonum cuspidatum*), that have become troublesome weeds and others that are in the process of becoming naturalized and invasive, sometimes with mixed results. *Buddleja*, for example, was introduced to Britain as an ornamental in the 1890s and by the 1930s was colonizing waste ground. It is now establishing itself widely. It has proved more attractive to butterflies than any native plant and by the 1980s a large variety of animal species were living in association with it.

Wildlife, the term embracing all undomesticated species (although some people confine it to animals), can be conserved in special locations, such as zoos and botanic gardens, but retaining it in the wild is preferable. This requires that habitats be protected, but if they are to survive, species must be provided with the full range of conditions they require. This means that before they can be protected habitats must be identified and analysed to determine their suitability. Once defined, they must then be managed to ensure they remain suitable, and one of the most important management tasks is the prevention of invasion by highly competitive, naturalized species. Rhododendrons, for example, are attractive, popular plants and people often object when conservationists cut them back quite savagely or even uproot them, but unless they were controlled, the invaders would quickly come to dominate the habitat and so destroy its value for many other species.

51 Biodiversity

At the United Nations Conference on Environment and Development (UNCED, also known as the 'Rio Summit' and 'Earth Summit'), held in Rio de Janeiro in the summer of 1992, governments agreed a Convention on Protecting Species and Habitats. Environmental groups and the press nicknamed it the 'Biodiversity Convention'. 'Biodiversity' is a contraction of 'biological diversity'. At the most general level it refers to the variety to be found among living organisms throughout the Earth, and the Convention reflects the widespread public sentiment that this variety enriches the planet and our own cultures by virtue of its existence and the desire that so far as possible it should be maintained. Accompanying these feelings is the fear that in fact species are going extinct at a rate unprecedented in recent history and, therefore, the variety of organisms is being eroded. These views, and the fear, are shared by most biologists.

Unfortunately, translating this rather vague desire into practical policies for maintaining variety raises formidable difficulties. In the first place, we must agree on a definition of 'biodiversity'. We might take it to mean genetic diversity, the entire global gene pool. In that case, it is not only species we must preserve, but the genetic variations within each population of every species, possibly amounting to some 10^9 different genes, of which we have limited knowledge of some of the 1 per cent that are expressed in the appearance of organisms (phenotypes) (PELLEW, 1995). Alternatively, we might seek to protect ecosystems by identifying and safeguarding each small area of habitat.

Most people are satisfied with an apparently less extreme requirement and define 'diversity' as 'species diversity', which is usually defined as the number of species in a particular area or community. In the case of biodiversity, it means the total number of species alive on Earth (ways of estimating diversity are outlined in section 42, headed 'Simplicity and diversity'). This requires us to prevent, so far as we can, the extinction of species. If we are to achieve this we must first determine the present rate at which species are going extinct and before we can do that we must know how many species there are.

There is an additional complication over the word 'species'. At one time this referred to a clearly defined concept. Literally, 'species', from the Latin *speculare*, to look, describes a group of organisms that resemble each other. Traditionally, taxonomists have applied the term to a group of organisms that are able to interbreed among themselves but are unable to breed with other groups. This is a useful definition, but it is not the only one. Nowadays a species defined in this way, by reproductive isolation, is usually called a 'biological species' or 'biospecies' for short.

The biospecies concept is essentially negative, in that it defines a species by the inability of its members to breed with non-members. A positive approach refers to the unique fertilization system that requires an organism to recognize as a mate only an individual sharing that fertilization system. In this definition, organisms mate specifically with members of their own species. This is known as the 'recognition species concept'.

Even so, the concept can be applied only to sexually-reproducing organisms. To expand it to asexual organisms there is a 'cohesion species concept'. It defines a species as a group of organisms that are genetically similar or comprise populations that are recognizable by their periodic increase and decrease in size.

Applying any definition based on reproduction requires fairly detailed and reliable knowledge of the behaviour of organisms and this may not exist. Increasingly, therefore, biologists are tending to use a 'phylogenetic species concept'. This defines a species as a group of organisms that share certain inherited physical characteristics not possessed in that combination by other organisms.

There are, however, groups of organisms that differ in form or structure – morphologically – from other groups, but without constituting a formal species. These are known as 'morphospecies'.

In discussions on biodiversity, 'species' usually means 'phylogenetic species'. This is seldom specified, however, and work is still continuing to develop a satisfactory definition of 'biodiversity'.

So far, about 1.6×10^6 species have been described and of those fewer than 10^5 are familiar, interesting, or pretty enough to have been studied in detail (PIMM ET AL., 1995). More are being discovered all the time. In 1988, two primates and one deer were discovered, another primate (a tamarin) was found in 1990, and since 1908 11 species of cetaceans (whales and porpoises) have been added to the global inventory, amounting to 13 per cent of all known members of that order (WILSON, 1992, p. 140). Between 1978 and 1987, the newly identified animal species were 5 birds, 26 mammals, 231 fish, and 7000 insects (PELLEW, 1995). Clearly, the species that have been found represent only a proportion of the total number, and estimating that number must necessarily involve an extrapolation from what is known to what is unknown. Many scientists have attempted the task, producing estimates of up to 100×10^6 species, but a conservative and probably more realistic estimate puts the total at $5–10 \times 10^6$. Robin Pellew, director of the UK branch of the World Wide Fund for Nature, has taken a median estimate of around 8×10^6, broken down as in Table 5.1 (PELLEW, 1995). The true total is likely to be higher, because the list omits a number of invertebrate groups, such as the annelids (worms), sponges (Porifera), and cnidarians (hydras and jellyfish). It also omits such plant groups as the ferns, mosses, and bryophytes. Further uncertainties arise from taxonomic disagreements between 'lumpers' and 'splitters' that can alter significantly the number of species recognized in certain groups.

Table 5.1 **Number of species described and the likely total number**

Group	No. described (thousands)	Estimate of total no. (thousands)
Viruses	5	500
Bacteria	5	400
Fungi	7	1000
Protozoa	40	200
Algae	40	200
Nematodes	15	500
Molluscs	70	150
Crustaceans	40	100
Arachnids	75	600
Insects	950	4000
Vertebrates	45	50
Higher plants	250	300
Total	1605	8000

Species do not persist indefinitely. Indeed, it was the realization that fossils are the remains or traces of organisms no longer to be found alive that led to the development of our present understanding of evolutionary processes. Estimating the number of species existing at present as a proportion of all the species that have ever lived is difficult, but the plants and animals alive today probably constitute around 2–4 per cent of all the species that have lived in the last 600 million years (MAY ET AL., 1995, p. 4).

Extinctions have not been spread evenly. There is a general background rate at which species go extinct, but there have also been at least five mass extinctions in which 65–85 per cent (and at the end of the Permian Period, around 225 million years ago, 95 per cent or more) of all species disappeared. Background extinctions have little or no effect on biodiversity, because their rate is matched or exceeded by that of speciation, in which one species divides into two and thus increases the number of species. Nor do mass extinctions produce more than a temporary reduction in the number of species, because historically they have been followed by the rapid adaptive radiation of surviving species into the vacated niches. The extinction of the dinosaurs, for example, was followed by rapid speciation among mammals. Today, the number of land-dwelling species alive is about twice the average of the last 450 million years.

The fact of extinctions makes it possible to estimate the average lifespan of species within different groups. Invertebrate species are estimated to survive for around 11 million years, marine animals 4–5 million years, and mammals around 1 million years (MAY ET AL., 1995, p. 3).

Such calculations provide a context, albeit a very approximate one, within which present concerns about biodiversity can be placed. If we have at least an idea of the number of species present, of that number as a proportion of all species that have ever lived, and of the rate at which species inevitably go extinct within different groups, we are in a much stronger position to address those concerns.

Species are not distributed evenly throughout the world, so the next step is to identify those areas with the greatest diversity. As Figure 5.11 shows, the number of species in a particular area may vary in two ways. The area possesses resources which organisms utilize, and although no two species (or functional groups) can have a full range of identical resource requirements, there is a small amount of overlap. We might imagine, therefore, a fully exploited habitat that supports seven species (the top drawing in Figure 5.11). If the resource base increases (perhaps the area is enlarged to include essentially similar habitat nearby), the number of species can increase (middle drawing). The number can also increase, however, if the species themselves are more higly specialized, so the resource requirement for each is narrower (bottom drawing). This is the case in many tropical forest areas, where very high species diversity results from local resource specialization among many closely related species (mainly of insects).

Even when areas of high diversity have been located, it does not necessarily follow that all are equally deserving of immediate protection. Common sense suggests we should first protect the area with the greatest number of species, then the area with the next largest number, and so forth, but in this case common sense is an untrustworthy guide. If the species in the area with the highest diversity evolved only recently, then they will still be closely similar genetically. Another area, with fewer species, might well contain a much more diverse assortment of genes (PELLEW, 1995). If the aim is to preserve genetic diversity, the second area would be a better place to begin, despite being poorer in species.

Sensible conservation requires detailed surveys not only of habitats and their inhabitants, but of the genetic constitutions of those inhabitants. This is a task for molecular biologists, who isolate, multiply, and compare sequences of DNA. Molecular ecology, a subdiscipline of molecular biology, is now well established in its own right.

Within habitats, many species tend to occur in 'clumps', rather than spreading themselves evenly. Animals which can choose their own habitats settle first in the area that suits them best and their

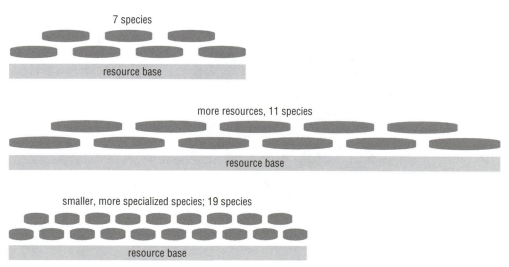

Figure 5.11 **Species richness**

numbers increase most rapidly inside that area. This is the 'core area' for that population and, as Figure 5.12 shows, it is surrounded by a wider area, more sparsely populated, from which individuals migrate into the core area as opportunity affords. This outer area is, therefore, a source area for individuals, but the rate of population increase is lower there than in the core area. Beyond that again there is an even more sparsely populated area, containing a 'sink' population of individuals that have moved away from the core area, perhaps because old age or illness renders them incapable of holding their own against more vigorous competitors. There is no population increase within this area. If part of the overall habitat is to be protected, but not all of it, so far as that species is concerned the core area is of greatest value and the outermost area can be sacrificed with little loss.

This much is obvious, but it can happen that owing to earlier disturbance of the habitat, the population is not where it would most like to be. The original core area may have been lost and the population is now elsewhere or distributed among several core areas each of which is inferior to the original one, and populations in some of these areas may survive only because they are periodically renewed by immigration from the other areas. Simply fencing off the entire area may be insufficient to halt a decline in the total population unless the original core area is first located and restored. Animals reintroduced into a habitat are more likely to survive and increase in number if they are released into the core area than the periphery (LAWTON, 1995).

So much is uncertain that all predictions of future extinction rates must be treated with great caution. Over the last few centuries, human activities, especially habitat destruction, have been the principal causes of present extinctions, but other factors are also involved. Madagascar, for example, has one of the most diverse populations on Earth, including 10 000 plant species, more than half of all chameleon species, and many primitive primates, yet a few thousand years ago it

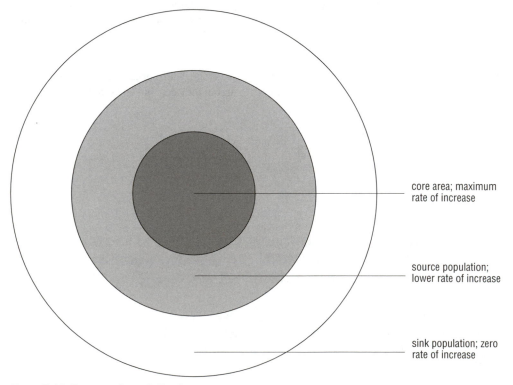

core area; maximum rate of increase

source population; lower rate of increase

sink population; zero rate of increase

Figure 5.12 **Range and population increase**

accommodated far more species than it does now. It was widely assumed that many of the lost animals were hunted to extinction by early human colonists. There is now reason to suppose that humans arrived during a time of increasing aridity, when habitats were already shrinking, so although the human effect was additive, it is likely that climate change alone would have caused the extinction of many species (CULOTTA, 1995).

Since 1600, 485 species of animals and 584 of plants have gone extinct. At present, the IUCN (International Union for Conservation, Nature and Natural Resources) lists as 'threatened' 3565 species of animals and 22 137 of plants (MAY ET AL., 1995, p. 11). Some of the threatened species will certainly disappear, but not all of them. Recent rates of extinction have been calculated as between 20 and 200 species per million years (E/MSY). Were all the threatened species to vanish within less than a century, this rate would increase to 200–1500 E/MSY (PIMM ET AL., 1995), which seems incredible.

Such a high rate of extinction would catastrophically reduce global biodiversity. It is unlikely to be so extreme, but nevertheless it is evident that much biodiversity will be lost unless positive steps are taken to prevent it. Before effective steps can be taken, however, a great deal of surveying remains to be done.

52 Fisheries

Fishing is the only form of hunting on which we continue to rely for a significant proportion of our food and, like hunters of old, we are beginning to quarrel on the hunting grounds as rival tribes compete for what appear to be dwindling stocks. In 1994, fishermen from France and Cornwall came to blows with Spanish fishermen in the Bay of Biscay where both groups were hunting tuna. The Spanish fishermen maintained that their traditional fishing method, using poles and lines, allowed them to catch bigger and better fish and that the French and Cornish boats, using drift nets they claimed exceeded the maximum length permitted in European Union waters, were catching smaller and therefore younger fish and so depleting the breeding population. The drift-netters replied to this that their nets were within the legal length, albeit only just, once allowance was made for the wide gaps between sections of net to allow marine mammals, such as dolphins, to pass. The real issue is more likely to have been the fact that by using drift nets the French and Cornish boats could catch three times more fish while employing half the number of fishermen.

Should this have been an issue? In a world claiming commitment to market economics are we entitled to criticize those who exploit technology and labour efficiently? Does it matter that so many Spanish coastal villages are economically dependent on fishing? The Cornish boats came mainly from Newlyn, which is no less dependent on fishing (and, ironically, most of the tuna landed in Britain is exported to Spain). It is hardly surprising that the quarrel became so bitter.

It erupted again a few months later, this time on the other side of the Atlantic, when the Canadians prevented Spanish boats from fishing for Greenland halibut (*Reinhardtius hippoglossoides*, in some reports incorrectly called 'turbot' by some British newspapers) on the edge of the Grand Banks. This is a large area, comprising a number of separate banks, south-east of Newfoundland, where the cold Labrador Current, flowing south, meets the warm Gulf Stream, flowing north. The resulting conditions favour plankton on which vast quantities of fish feed. The value of the Banks has been recognized ever since they were discovered by the Venetian navigator John Cabot (Giovanni Caboto) in 1497. Most of the Banks lie within Canadian territorial waters, but the Spanish boats were working on the small area that is in international waters. The Canadians argued that the Banks comprise a single, unified resource from a biological point of view and, since almost the whole resource was indisputably Canadian, they were entitled to protect it from over-exploitation even if this meant interfering with the right of fishermen to work in international waters. The Canadian action was illegal and placed friendly European governments in a difficult position, but in Cornwall Canadian flags were flown from so many fishing boats and public buildings that extra flags had to be imported from Canada and the Canadian High Commissioner visited the fishermen to thank them.

Less attention was paid to other disputes, but there were plenty in 1994. Icelandic gunboats were used to repel Norwegian trawlers. Fishermen were hurt in a disagreement between China and Taiwan over the waters around the island of Quemoy in the Taiwan Strait. Britain argued with Argentina over fishing rights around the Falklands (Malvinas). Namibia complained that EU vessels were taking lobsters in the Gulf of Aden.

When people warn of wars over resources, oilfields are what they usually have in mind, but serious disputes, if not literally warfare, have been occurring over fisheries for more than 20 years. It was in the 1970s that Iceland banned foreign vessels from fishing in waters over which it unilaterally claimed sovereignty. This led to three 'cod wars' between Britain and Iceland (ALLABY, 1977, p. 311).

Figure 5.13 illustrates the problem. Since 1972, the total world catch has increased by about 50 per cent, but it is based on a rather small number of species. As Table 5.2 shows 2, 20 species of marine and freshwater

Table 5.2 **The 20 most important species in the world's fish catch, 1995**

Species	% of total
Anchoveta	20
Chilean jack mackerel	12
Alaskan pollock	11
Silver carp	6
Atlantic herring	5
Grass carp	5
Common carp	4
Skipjack tuna	4
Chub mackerel	4
South American pilchard	3
Yesso scallop	3
Atlantic cod	3
Bighead carp	3
Largehead hairtail	3
European pilchard (sardine)	3
Yellowfin tuna	2
Pacific cupped oyster	2
Japanese anchovy	2
Atlantic mackerel	2
Caplin	2

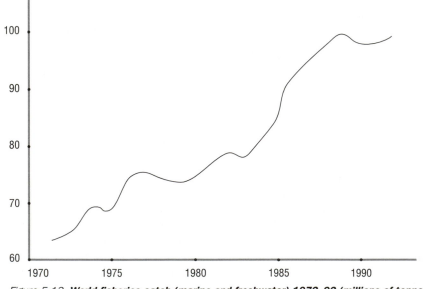

Figure 5.13 **World fisheries catch (marine and freshwater) 1972–92 (millions of tonnes)**

fish accounted for 38 per cent of the 1995 total world catch (of 112.9 million tonnes) and three species – anchoveta, Chilean jack mackerel, and Alaska pollack – accounted for 16 per cent of the world total, or 18.3 million tonnes.

Nor are the fish catches distributed evenly. China is the biggest fishing nation by far. Counting catches of marine and freshwater fish and farmed fish, in 1996 it produced almost 32 million tonnes, which is 26 per cent of the world total. Between them, and in descending order of production, China, Peru, Chile, and Japan produce 45 per cent of the world's fish (GILL, 1999).

Carrying capacity and yield

The number of individuals of a given species that can survive in good health in a specified environment without degrading that environment is the carrying capacity for that species in that environment. Obviously, it varies from species to species and from one environment to another.

Each breeding season (usually, but not necessarily, 1 year) new individuals enter a population. Some are born and some arrive as migrants. If the number of these individuals exceeds that of those which die during the same period, the recruits represent a net increase to the population. If they are harvested, the size of the population remains unchanged. This number is the maximum sustainable yield (MSY) for that population in that environment.

In practice, harvesting the MSY will lead to a population decline, because recruitment varies from season to season, but for harvesting purposes the MSY figure must be calculated in advance. Since it represents a maximum, lower recruitment in some seasons will cause it to over-estimate the size of the harvest that can be taken safely.

The optimum sustainable yield (OSY) allows a wide safety margin. It is usually calculated as half the carrying capacity.

Yield quotas for commercial fisheries are calculated as shares of the OSY allotted to individual fleets.

There is widespread concern that fish stocks may be unable to sustain so intensive a fishing effort, but estimates of the level of effort they might sustain are very uncertain. The maximum sustainable yield (MSY) for any harvested species is calculated as being equal to the number of individuals entering the population, by birth or migration, during the harvesting period, usually one year. Prudence requires an allowance to be made for unpredictable events that might deplete the population, so the optimum sustainable yield (OSY) is lower than the MSY based on the crude replacement rate. Recruitment into a population may vary considerably from year to year, so for many species an MSY is valid only for the year to which the data refer. This is especially so for fish, which experience good and bad breeding seasons, but it means that an OSY figure always refers to the past. It does not tell fishermen what they can catch next season, but what they should have caught some time ago. Consider the history of North Sea herring, for example. At one time this fishery underpinned many local economies but, as Figure 5.14 shows, it was never reliable. The stock (not the catch, note) was more than 2 million tonnes in the late 1960s, but within a decade had collapsed to little more than 100 000 tonnes. Then, during the 1980s, it recovered.

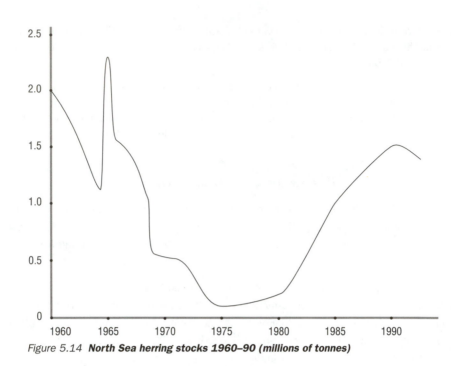

Figure 5.14 **North Sea herring stocks 1960–90 (millions of tonnes)**

While the stock was declining it was widely believed that over-fishing was the cause of the depletion, but it is more likely that the fluctuation was mainly natural, though possibly exacerbated by
fishing. Huge fluctuations in the South American anchoveta fishery are similarly due to natural
population changes, in this case caused by El Niño events.

Clearly, a retrospective OSY would be of limited use even if it could be calculated with anything
like precision, but it cannot. In order to calculate it, the size of the population and the rate of
recruitment into it must be known and in the case of fish this is not usually possible.

In practice, an MSY figure is usually based on the carrying capacity, which can be calculated
(BEGON ET AL., 1990, pp. 583–587). Even so, estimates vary. For many years the annual global
MSY was believed to be around 100 million tonnes (ALLABY, 1977, p. 315), but others calculated
it as 200 or even 400 million tonnes (KUPCHELLA AND HYLAND, 1986, pp. 267–268). For what it's
worth, the present annual catch, of around 113 million tonnes, is close to the earlier estimate of
100 million tonnes.

Fish continue to be caught by methods developed over many centuries. The hunt is, indeed, traditional. Hooked lines are still used, sometimes from the boat, as with Spanish tuna-fishing, mackerel
handlining in British waters, and trolling, in which the line is trailed astern of the moving boat,
and sometimes as long lines. Hundreds of metres long, these are baited and left in position for
several hours, or overnight, weighted at one end and buoyed at the other. The trawl net is a bag,
tapering from an end held open by 'otter boards' to a narrow 'cod end', that is towed through the
water. A ring or seine net is a curtain-like net, with weights along the bottom and floats along the
top, that is paid out to form a circle enclosing the fish. The seine net is converted to a purse seine
by drawing together the edges so the caught fish are entirely enclosed within it. A drift net is a
curtain, or series of curtains, that hang vertically and trap fish swimming into them. Figure 5.15
illustrates these main techniques.

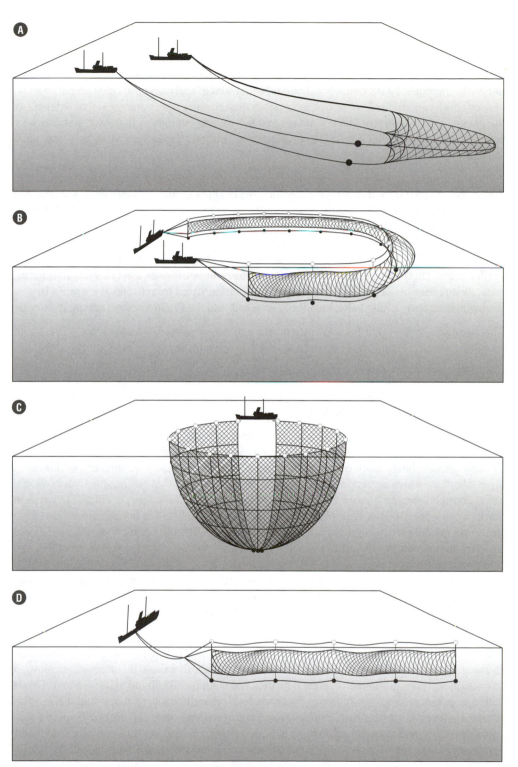

Figure 5.15 *Commercial fishing methods. A, Mid-water pair trawling. B, Ring (or seine) netting. C, Purse seining. D, Drift netting*

It is not the techniques that have changed, but the scale on which they operate and the ancillary equipment that supports them. Modern fishing boats are big and powered by large engines, which allow them to store and tow much larger nets than were once used. In some cases (see Figure 5.15A and B) two boats can collaborate, allowing even larger nets to be used. A large purse seine net could contain St Paul's Cathedral, London, with room to spare and will catch an entire shoal of fish. This hunting power is augmented by sophisticated navigation equipment that determines the position of a boat to within a few metres and fish-finding equipment that locates shoals very precisely. Commercial fishing still depends on luck, but to a far lesser extent than was once the case.

Such power and technical sophistication are expensive. Fully equipped fishing boats now cost in the region of £500 000 and consume a considerable quantity of fuel. So large an investment can be supported only by maximizing the catch. Modern fishermen, therefore, are compelled to travel far and land huge catches simply to meet the cost of their own operation. They are trapped financially as firmly as the fish are by the nets. The rise in the global catch is thus a consequence of increased fishing power driven economically and it is no coincidence that certain countries are predominant. Those are the countries that choose to encourage investment in their fishing industries. By no means are all of them industrialized countries. Apart from China, the world's largest fishing nation, Peru, Chile, India, Indonesia, Thailand, and the Philippines are among the top 12.

The situation, then, is that hunters armed with extremely powerful weapons are pursuing an unknown amount of game. It is not possible that they will hunt any species to extinction, because as its numbers decline it will become economically impossible to pursue it and populations will recover. What is possible, however, is that the more popular species may be hunted to commercial extinction. Herring, once a cheap food eaten mainly by poor people, almost disappeared from British shops during the 1970s and although it has now reappeared it is no longer cheap.

Regulation is necessary if stocks, and hence the industry itself, are to be protected. It is based partly on technical limitations placed on the equipment used and partly on quotas. Drift nets, for example, should be no more than 2.5 km long within EU waters, and minima are set for mesh sizes. Quotas are also agreed for all commercially important species, setting the total allowable catch and then sharing it among fishing nations. Despite the inevitable squabbles, the system works up to a point, but until now its success has extended only to the boundaries inside which governments can exercise jurisdiction. It was much more difficult, often impossible, to enforce in international waters. Attempts to reach a global agreement on the management of fish and other marine stocks continued, without much success, from 1974 to 1982 at the protracted UN Conference on the Law of the Sea and again from 1993 to 1995. Agreement was finally reached in August 1995, when a treaty committing nations to the regulation of fishing in international waters won unanimous approval and was forwarded to the UN General Assembly for adoption (HOLMES, 1995).

If the fishing effort is reduced there is good reason to suppose that depleted stocks will recover. In some species, when population size falls below a certain threshold it suffers an increase in parasitism and predation which deplete it further, reducing the likelihood of its recovery. This is called 'depensation' and it is known to affect whales, leading to concern that it might affect fish as well as these marine mammals. Studies of a variety of commercially important fish species indicates that they, and possibly all marine fish, do not experience such depensatory declines, suggesting that even though populations are severely depleted their numbers will increase if they are allowed to do so (MYERS *ET AL.*, 1995). If depletion were irreversible, it might be argued economically that fishermen should continue to hunt until it became commercially impractical to continue. If stocks can recover, on the other hand, fishermen can be told that if they reduce catch sizes for several seasons, perhaps drastically, the population will recover and catches can be allowed to increase again.

There is an alternative to hunting. Just as hunted mammals were domesticated and gave rise to animal husbandry, so fish can be raised in captivity. Most European salmonids (trout and salmon) are raised in this way, and the success of the fish-farming industry has lowered the price of their products so that what were once luxuries are now within the reach of many more people. Bass and most flat-fish can also be bred and raised in captivity. Fish farming is also traditional, in that it has been practised for many centuries in some parts of the world, and its popularity is increasing. Its disadvantage is that its products are more expensive than wild fish that can be caught in large amounts. Wild salmonids are caught in small numbers, or individually, which is why they can be farmed economically (see box).

Aquaculture

Fish can be farmed by a variety of methods. Freshwater ponds may be stocked with herbivorous species, such as carp, and fed with organic material either directly or indirectly by adding fertilizer (traditionally human sewage) to stimulate the growth of aquatic plants.

A coastal area, such as a natural bay, may be enclosed by permanent nets. Within the enclosed area, a stock of fish is protected from predators and its food supply augmented. This method is similar to ranching.

The most intensive methods involve controlled fish breeding and the raising of fish at high population densities in closed ponds, each pond holding fish of similar age. The fish are fed a highly nutritious diet and water temperature is controlled, to promote rapid growth to marketable size. Salmon, which spend part of their lives in fresh water and part at sea, are bred and raised in freshwater ponds, then moved to cages in sheltered salt water.

As with all animals stocked at high densities, disease can spread rapidly, with devastating consequences, and hygiene is extremely important. Water used for aquaculture must be kept clean and fish excrement removed promptly. Accidental leakages can cause pollution, and accidental escapes of exotic species, chosen on commercial grounds, can disrupt native populations in adjacent waters.

Should the costs of commercial fishing cause fish prices to rise beyond a certain threshold, farmed species will become cheaper and begin to appear in the shops. It is not inconceivable that one day hunting will cease to be viable and the fish we eat will be produced on farms.

53 Forests

Despite the stone, concrete, ceramics, metals, and almost endless list of other materials on which we depend, wood is still one of our most important resources. We build and furnish our homes with it and many of us heat them and cook with it as well. It is estimated that wood is the main or only fuel for 30–40 per cent of the people in the world (Tolba and El-Kholy, 1992, p. 165). Indeed, so high is dependence on fuelwood in the less industrialized countries that in the world as a whole slightly more fuelwood is produced than wood for industrial use (Tolba and El-Kholy, 1992, p. 166). Paper, cardboard, and much of our packaging is also made from wood. In 1997, the total world production of paper and board was more than 299 million tonnes (Lavallée, 1999).

Forests (*www.fao.org/forestry/forestry.htm*) are of such importance that on economic grounds alone it is in our interest to manage them with care. Unfortunately, however, much of the land they occupy is capable of producing food, and throughout history forests have been cleared mainly to provide agricultural land. Today, as Figure 5.16 shows, the proportion of the total land area that is forested varies widely from one country to another. This is not an entirely fair way to consider the situation, because countries also vary greatly in size and population density. Russia, for example, has 57 per cent of the world's boreal forest and Canadian forests cover an area greater than that of western Europe, but in neither country do forests occupy so large a proportion of the total land area as they do in Japan. At the other extreme, Ireland, only 6 per cent of which is forested, has large areas that are unsuitable for trees.

It is still true that agriculture, in one form or another, remains the immediate cause of most defor-estation. Shifting cultivation accounts for 45 per cent of all clearance of closed forest, commercial farming and grazing for a further 15 per cent each, dams and roads for 10–15 per cent, and forestry for 10 per cent. When allowance is made for planting, the net annual forest loss in the world as a whole is probably 11–12 million hectares. In the tropics, the drier and mountain forests are being lost much more rapidly than the rain forests (Persson, 1995). The losses are confined to the trop-ics, however. In temperate regions, other than the United States where the area is decreasing by about 300 000 hectares a year, the forested area is increasing, and the area is increasing overall. In Europe forests are expanding by about 200 000 ha a year and in Russia by 2 million ha a year. The expansion is due mainly to plantation forestry, but it involves no loss of natural forest. The decrease in the United States affects only plantation and areas of natural regrowth, not primary or old-growth forest (Allaby, 1999, pp. 162–163).

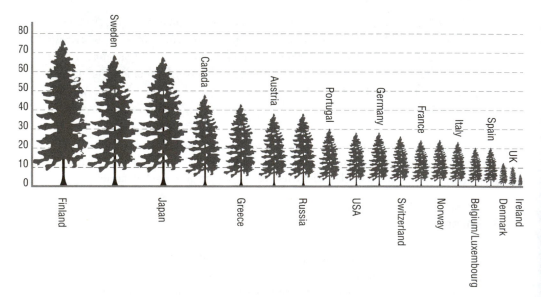

Figure 5.16 **Percentage of land area under forest in various countries**

Shifting cultivation was once widely practised in temperate regions, but now survives only in the tropics. It is a system of subsistence farming that begins with the clearance of trees and shrubs from an area. Such timber as may be of use is taken and the rest is commonly burned, clearing the ground and fertilizing it with wood ash. Crops are then sown and cultivation continues until, after a few seasons, tree seedlings and other 'weeds' become too troublesome to remove and yields start to fall. The operation then moves to another site, where it is repeated. Cultivation proceeds as a rotation, visiting sites in turn. Ideally, each site remains undisturbed for about 12 years between cultivations. This allows time for secondary forest and other vegetation to re-establish itself and the system can continue for many years with no serious depletion of soil fertility. Managed in this way, however, 1 km^2 of forest will support only about seven people (WHITMORE, 1975, p. 229). As demand for farmland intensifies, shifting cultivators have little choice but to use fewer sites and return to them more frequently. This leads to soil deterioration and the permanent disappearance of forest.

There is nothing new in the conflict between agriculture and forestry. At one time, most of the British Isles was forested. As the climate warmed during the early years of the present (Flandrian) interglacial, first conifers and eventually broad-leaved trees colonized the land. Figure 5.17 shows that from the time the climax forest was established until about three thousand years ago, oak (*Quercus* species) and hazel (*Corylus avellana*) were the dominant species over the largest area, with elm (*Ulmus* species) substituting for oak in the south-west of Wales and England and over most of Ireland, birch (*Betula* species) and pine (*Pinus* species) confined mainly to northern Scotland, and a smaller area dominated by small-leaved lime (*Tilia cordata*) in the Midlands and south of England. About 5000 years ago, there was a sudden and dramatic reduction in the amount of elm, called the 'elm decline'. Opinions vary as to its cause. Some authorities hold that the elm was destroyed by early farmers, who cut its branches to provide browse for their cattle, preventing the trees from flowering and producing the pollen from which past vegetation patterns are reconstructed. The decline is certainly associated with a proliferation of weeds typical of arable crops (RACKHAM, 1976, pp. 45–46). It may be, however, that the elms succumbed to disease. The 1970s outbreak of Dutch elm disease in Britain provides dramatic evidence of the speed with which disease can almost eliminate elms.

By AD 43, when the Romans invaded, the forest had been cleared from the lighter lands and farmers were starting to plough the heavier clay soils. During the centuries of their occupation, the Romans turned England into one of the most important agricultural countries in the empire, clearing vast areas of forest for the purpose. They also used timber for building and as fuel for the many industries they established. When the Anglo-Saxons arrived, after the Romans had departed, they inherited a land from which much of the primary forest had gone (RACKHAM, 1976, pp.50–52) and was being replaced by a secondary regrowth of forest on uncultivated land. The Domesday survey of 1086 portrays a countryside perhaps more wooded than it is today, but with many villages having no woodland nearby and many places, especially in the east, where small patches of woodland were miles apart. Large areas of the country consisted of farms with patches of woodland, much as it has remained (RACKHAM, 1976, pp.61–65). Clearing continued in the centuries that followed and by 1350 probably no more than 10 per cent of England was forested, the same as the present-day figure for the United Kingdom. The later construction of the navy, in Tudor times, and the rise of the iron and steel industries still later actually had little effect (ALLABY, 1986, pp. 95–97).

Where forests and farms compete for land, confrontation invariably presents victory to the farmers, for the simple reason that the economic return from farms always exceeds that from forests. If necessary, the farmers can buy out the foresters. Yet forests are a vital resource. In Britain, the state forests are owned and managed by the Forestry Commission, established in 1919 with the task of building up a strategic reserve of standing timber following the widespread felling that had occurred during the First World War, when timber imports were much reduced. The proportion of forests managed by the state varies, but national or federal governments manage at least some forests in most countries.

For many years, most Forestry Commission planting was of fast-growing conifers, sited on marginal land, usually in the uplands, that could be spared from agriculture. The relegation of

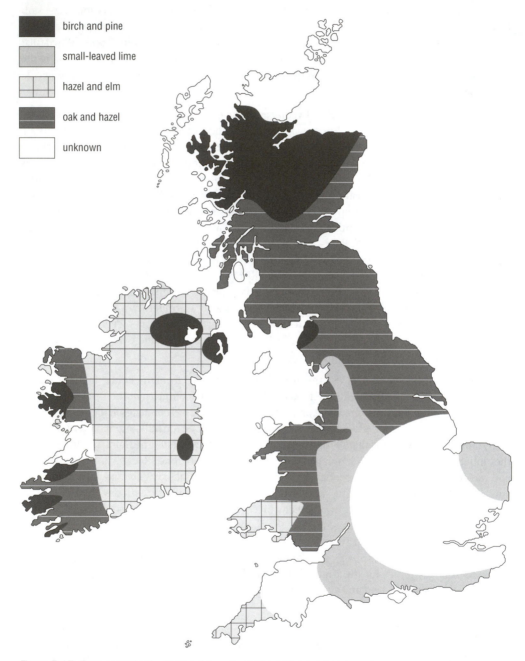

Figure 5.17 *Tree cover in the British Isles about three thousand years ago*
Source: Rackham, Oliver. 1976. Trees and Woodland in the British Landscape. J. M. Dent and Sons, London

forests to marginal land caused problems. Trees demand growing conditions not much different from those required by many farm crops, so species were chosen that could survive the relatively harsh environment allotted to them and, to protect them from severe weather, they were planted in solid blocks. These were visually intrusive in landscapes from which the primary forest disappeared centuries ago and was never replaced, partly because the plantations were very dark in colour against the paler background of upland pasture and partly because of the severely geometrical shape of the blocks. It was also the practice to plant an entire block at the same time, so its trees were all of the same age, and harvest the timber by clear-felling the whole area, leaving the unsightly debris to scar the landscape until the succeeding crop grew up to hide it.

Modern forestry practice avoids these causes of very understandable offence. Trees are planted at intervals, producing stands of mixed ages that are harvested by removing much smaller blocks from within the forest. The edges of blocks are less straight, giving the forests a more natural appearance, and conifer plantations are often surrounded by broad-leaved species that produce a lighter, gentler colour.

More recently, the policy has been to plant rather more broad-leaved trees, especially in the lowlands of England and Wales, and to take much greater account of the wildlife importance of forests and their value for public recreation. The national forest will remain predominantly coniferous, however. Large 'community forests' are planned for several areas, sited to allow access from major population centres and intended primarily for amenity and wildlife conservation purposes, as well as the commercial production of timber.

Conifers produce softwood. In addition to its wide range of uses as timber, it is also used to make chipboard, fibreboard, and paper of all kinds. All paper is made from wood grown in plantation forests. Traditionally, broad-leaved forests were managed to produce timber for construction and shipbuilding and also smaller wood to make furniture and implements. When charcoal was an important fuel it, too, was produced from small wood. Selected trees were allowed to grow to their full size, as 'standards', to supply timber. The others were coppiced. This involved cutting them almost to ground level, leaving a stump, called a 'stool'. The section of Figure 5.18 shows on

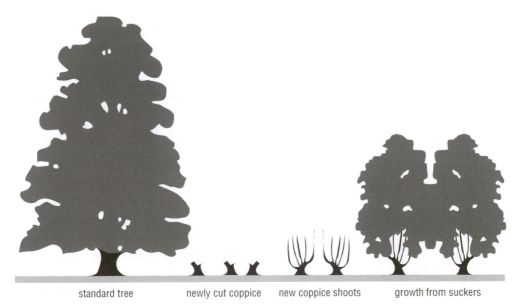

standard tree   newly cut coppice   new coppice shoots   growth from suckers

*Figure 5.18 **Traditional tree management***

the left the tree prior to cutting, in the middle the stool left after cutting, and on the right the appearance of a mass of thin shoots one year after cutting. With some species, the new shoots can grow more than 5 cm a day (RACKHAM, 1976, p. 22). Coppicing was practised rotationally, certain trees being cut each year, and about 12 years was allowed for the shoots to grow to useful size. Curiously, this apparently harsh management seems to prolong the life of many species. On more open sites, where trees were surrounded by pasture, they were more commonly pollarded, by cutting about 3 m above ground level to leave a 'bolling' from which the new shoots would grow. The effect was similar to coppicing, but pollarding protected the young shoots from damage by livestock. With some species, coppicing kills the stool but not the roots, from which new growth appears (the right-hand section of Figure 5.18). Coppicing has long been advocated by conserva-tionists, for two reasons. By preventing most of the trees from growing to their full size and producing an almost complete canopy of leaves that shade the ground, coppicing allows sunlight to penetrate to the woodland floor, which encourages the growth of woodland herbs and greatly increases the number of species the woodland supports. The other reason appears paradoxical, but is not. Coppicing is exploitive, in that it removes material (wood) without using fertilizers or even compost to restore the nutrients taken from the soil. In time, therefore, the soil is depleted. This inhibits the growth of aggressive plant species that are able to overwhelm others only if they receive ample nutrients, thus allowing a wide variety of less aggressive species to flourish. In the last few years, pro-coppicing conservationists have found new allies, because this management system seems ideally suited for the production of small wood for use as fuel for power stations.

Plantation forestry has proved the most effective way of ensuring a sustained supply of timber and wood products in temperate regions and it is now being introduced in the tropics. Plantation tree species are chosen for their rates of growth and the quality of their wood. In most cases this leads to the selection of species that are not native to the areas where they are grown. In temperate climates these exotic species are conifers and in the tropics they are often eucalypts, cypresses, and pines.

Many tropical plantations have succeeded, but there are difficulties. Not the least of these is the fact that rural communities in the tropics are more highly dependent on products they obtain from the natural forest than are people living in higher latitudes, and plantations may not supply them. Where they do supply them, commercial plantations must charge for materials, such as fuelwood, that formerly were free. Many fuelwood plantations have failed for this reason. Other plantations have failed because trees were planted without testing them in the environment first, and they did not grow well, because no markets could be found for their products, because corrup-tion depleted the resources needed for investment, or because the concept was imposed on people who rejected it rather than being developed with their full cooperation. Nor are plantations the only means available to produce wood on a sustainable basis. With proper management and protec-tion from fire, cleared areas of tropical forest will usually regenerate a natural secondary growth. This may be ecologically inferior to the original primary forest, but its timber is satisfactory (PERSSON, 1995a) for fuel use and many other purposes.

The pressures leading to the clearance of tropical forests are identical to those responsible for the removal of temperate forests. Some combination of natural regeneration and carefully planned plantation may be sufficient to secure a regular supply of timber, in the tropics as in higher lati-tudes, but we should not look to them to solve wider problems.

In particular, plantation and regeneration forests have little effect on the rate of forest clearance. Forests are cleared, as they have always been, to provide land for agriculture, not because of the demand for forest products. Agricultural land is needed by impoverished people who hope to farm it at a little above subsistence level. That pressure would be reduced by increasing employ-ment and so providing people with an alternative to farming. Their wages would then purchase food grown more efficiently on a smaller area of land by adequately capitalized methods. Like large-scale agriculture, however, most commercial forestry is highly mechanized and provides little employment.

Sound forest management provides us with forest products on a sustainable basis. The forested area is increasing in many high-latitude countries, including Britain, and may well be increasing overall. If we wish to reduce the rate at which primary tropical forests are being cleared, better forest management will not suffice. Far more radical economic, social, and political reforms will be needed.

54 Farming for food and fibre

Farming began, with the deliberate cultivation of cereals, rather more than 10 000 years ago in the Middle East (HARRIS, 1996), possibly at about the same time in Taiwan (BENDER, 1975, p. 223), and perhaps 5000 years ago in Central America (BENDER, 1975, p. 10). The cultivation of non-cereal crops, such as pulses, began at about the same time or soon afterwards. Why farming began is uncertain, but once established it spread through lands between the Tigris and Euphrates Rivers and by about 6000 years ago food productivity had increased sufficiently to support urban populations (BRAIDWOOD, 1960). In the space of 4000 years, the new technology spread through-out western Europe (BENDER, 1975, p. 13). The dog was the first animal to be domesticated, in a process that began at least 12 000 years ago, followed by sheep, goats, and pigs rather more than 7000 years ago, and western cattle (descended from *Bos primigenius*, the aurochs) about 1000 years later (CLUTTON-BROCK, 1981, pp. 34, 56, 60, 66, 72).

It was not a sudden transition, with people abruptly changing their way of life as soon as trav-ellers or immigrants explained the new technology to them. Hunters developed a close relationship with and understanding of the animals they hunted. Perhaps they held them in enclosures until, gradually, they were tamed. People who gathered food might observe that spilled seed sprouted and start to scatter it deliberately. As they did so, they would have inadvertently commenced the process of selective breeding. Wild cereals, for example, shed their seeds as soon as they ripen. This makes them difficult to gather and people would probably have selected those individual plants that retained their seeds a little longer, and these would have supplied a higher proportion of the seeds that were sown than they represented in the wild population. Eventually, this led to the immediate ancestors of our modern cereals, from which the seeds must be removed by a separate operation (threshing) after harvest. Other food plants developed from individuals chosen from the wild population because they possessed certain desirable properties, such as size or flavour. In those days neighbouring communities would have met in the course of their hunting and gathering, and interesting information would have travelled quickly.

Crop cultivation encouraged the establishment of settled communities on the better land and this led in turn to emigration and the colonization of adjacent areas. Shifting cultivation will support only a limited number of people, so where it was practised an increase in the population meant some had to leave and start a new community elsewhere. For communities using more efficient farming methods it was probably impracticable to cultivate fields more than about one hour's walk from the village. Again, an increase in population would necessitate emigration (BENDER, 1975, pp. 13–14). Early farmers continued to rely on wild foods for a surprisingly long time. Although they were probably fed ceremonial meals prior to their ritual killing, people whose bodies have been recovered from Danish bogs had eaten a variety of wild grasses and herbs as well as culti-vated grains, and those people lived in the Iron Age, Tollund man around the time of Christ and Grauballe man around AD 310 ± 100 years (GLOB, 1969, pp. 33, 57). Since we still rely on hunting as the principal means of obtaining fish, wild birds are still shot and eaten, and in rural areas people continue to collect edible wild fungi and gather blackberries, sloes, and other berries from hedgerows, it appears we have not entirely abandoned the old ways of obtaining food even now.

Those who till the soil require tools. At first these were simple sticks for digging and hoes for removing weeds, but the technology advanced dramatically with the introduction of the plough. Probably invented by the Babylonians, this was used in ancient Egypt (see Figure 5.19A) and the modern plough (Figure 5.19C) is its direct descendant. Drawn by oxen, in some parts of the world to this day, horses or, on modern farms, by a tractor, the plough cuts into the soil, aerating it and

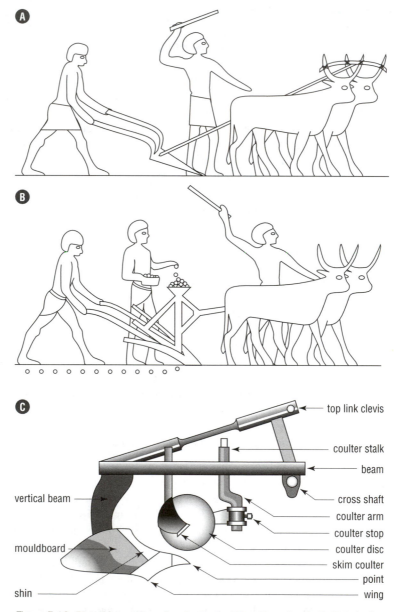

Figure 5.19 **Ploughing and sowing. A, Ancient Egyptian plough. B, Babylonian plough and seed drill (c. 1316 BC). C, Modern single-furrow plough**

improving its drainage. The principal innovations in modern ploughs are the mouldboard, which turns over the ploughed soil, breaking weeds and burying their foliage, and the coulter. This is a disc (or on some ploughs a knife) that cuts the soil to a predetermined depth ahead of the point and wing of the share. The plough shown cuts a single furrow, but most tractor-drawn ploughs cut several at the same time. As long ago as Babylonian times, ploughing was being combined with sowing by attaching a seed drill, comprising a container into which seed is poured and from which it falls through a hole and down a tube into the ground (see Figure 5.19B). The seed drill did not reach England until Jethro Tull introduced it, amid considerable controversy, in about 1730 (PARTRIDGE, 1973).

Ploughing aerates the soil, stimulating the activities of soil organisms, but by exposing it to wind and rain ploughing also accelerates weathering. This increases the rate of soil formation, but it also leads to erosion. Where rainfall is insufficient for the needs of crop plants, irrigation may be necessary. Supplied carelessly, or in ignorance of the risks, this can lead to salinization, especially in the seasonally dry climates of the Mediterranean, where irrigation is most needed. Environmental problems caused by farming emerged in ancient times with salinization in the Tigris and Euphrates valleys, and the severity of soil erosion in Greece was described by Thucydides (c. 465–400 BC) (HYAMS, 1976, pp. 92–94). Today, it is estimated that in the world as a whole about 80 per cent of agricultural soils are to some extent eroded (THOMPSON, 1995). Even in Britain, where the mild, maritime climate favours agriculture and limits soil erosion, most cultivated land near the top of slopes is to some extent eroded.

When soils show signs of failure farmers have two choices. They may extend the area they cultivate into previously uncultivated, 'virgin' land, or they may intensify production on existing land. In practice, they do both.

Intensification is achieved by using fertilizers to augment soil nutrients and pesticides to reduce crop losses, and by breeding new crop varieties and breeds of livestock that respond better to the changed circumstances and consumer demand than the varieties and breeds they replace. This type of intensification began in Europe in the eighteenth century and accelerated greatly in Europe and North America from the 1940s, when governments provided the economic stability that allowed farmers to invest in their farms and systems of grants and subsidies that encouraged them to do so.

A version of this type of intensification was introduced much more rapidly in Latin America and Asia. New varieties of wheat were developed at the Centro Internacional de Mejoramiento de Maíz y Trigo, in Mexico, and at Washington State University in the United States, by a research programme that began in 1943. By 1977 these wheats were growing on 30 million hectares, nearly half the wheat-growing area in developing countries, and yields increased from about 8 tonnes per hectare for traditional varieties to 15–20 tonnes per hectare. Indian wheat production tripled between 1969 and 1979. Research to develop new rice varieties began in the early 1960s at the International Rice Research Institute in the Philippines, and the first of the new rices, IR–8, entered commercial use in 1966. The new rice varieties, of which there are many, are now grown widely in Asia. They grow and ripen more quickly than traditional varieties, so two or even three crops a year can be grown on the same land. Yields have risen from less than 2 to up to 16 tonnes per hectare. This transformation of low-latitude farming was described in the *Indicative World Plan for Agricultural Development*, a massive document prepared by the Food and Agriculture Organization of the United Nations (FAO). Journalists nicknamed it the 'Green Revolution' (ALLABY, 1977, pp. 101–133). Improvement has occurred more slowly in sub-Saharan Africa, partly because of a lack of new varieties of millets and sorghum, but also because of war and drought.

Dramatic though the results of intensification have been, they have generated problems of their own. Some have been social and economic, where the sudden improvement in the profitability of farms has led landowners to seize all the gains for themselves, leaving tenants no better off than before, or to evict tenants altogether. The provision of credit facilities and efficient infrastructure has often delayed improvements in the more remote areas. The increased use of fertilizers and pesticides has also caused pollution.

Several researchers also pointed out that the operation of farm machines and the production and application of agricultural chemicals consume energy. Since food can also be described in terms of energy they constructed energy budgets suggesting that intensification exacted a price. Average-sized British dairy farms consumed 26.5 GJ ha^{-1} and yielded 14.6 GJ ha^{-1}, pig and poultry farms consumed 44.8 and yielded 14.1 GJ ha^{-1}, and broiler chicken farms consumed 58.9 and yielded 5.87 GJ ha^{-1}, in each case the energy coming from fossil fuels (LEACH, 1975, pp. 110–111, 125). Parallel studies by William Lockeretz at the Center for the Biology of Natural Systems, Washington University, St Louis, published in 1975, reached similar conclusions in respect of US agriculture. Much was made of these findings at the time, the 1970s being a period of anxiety regarding the security of oil supplies to the industrialized countries, but it can be argued that it is sensible to use inedible fuels to produce food, provided the agricultural system is not so extreme as to become economically nonsensical. This does not resolve the issue, of course, because even if this is a wise use of fuel, fossil fuels are not inexhaustible and so energy-intensive agricultural systems may prove unsustainable in the long term. Fuel is used in agriculture in the manufacture of tools, machines, and other equipment, to power machines, and in the manufacture of fertilizers and pesticides. It is not inconceivable that alternatives could be substituted for all these uses. Electrical power from nuclear generation, for example, might power factories and drive machines. This consumes no fossil fuel, and the uranium on which it depends, though also non-renewable, is at present abundantly available, and should it become scarce in the future, by that time fission generation may have given way to another means of power production. Hydrogen, ethanol, or methanol might power farm machines. The feedstock chemicals used in fertilizer and pesticide production are widely available naturally; fossil fuels are used to supply them only because at present these provide the cheapest and most convenient source.

Energy budgeting in agriculture

In the 1970s, several academic studies were made comparing the amount of energy used in the production of food for human consumption and the energy value of that food. In Britain, Gerald Leach prepared a report, *Energy and Food Production*, that was published in 1975 by the International Institute for Environment and Development. It was this report which stimulated vigorous debate in Britain, but similar debates took place in the United States and other countries.

Leach collected data for the output of food from British farms. He then made allowances for food lost by wastage after harvest to produce a 'net edible output'. This was converted into energy values using standard conversion figures used in dietetics. He had now produced figures representing the energy content of home-grown food entering the public supply, the output energy.

Input energy took account of all fuel and electricity purchased by food producers and 40 per cent deducted because it was for domestic use. Agricultural chemicals were assessed in terms of the energy per tonne required to produce them. Machinery was assessed by the energy consumed during all the processes leading to their manufacture. In use, it was assessed by the fuel consumed for standard operations, such as ploughing, harrowing, and combining. Foodstuffs for livestock were assessed on the assumption that they had been grown on farms in Britain, or farms indistinguishable from British ones, with similar energy requirements. Transport, services, and the construction of farm buildings were also included, and so was human labour, measured as the food required by the workers.

Finally, the energy required to produce unit amounts of food (the input) was compared with the food energy in the edible product (the output). Leach calculated that the average ratio of energy output : energy input in specialist dairying was 0.38; in cattle and sheep rearing 0.59; in sheep farming 0.25; in pig and poultry farming 0.32; and in cereal-growing 1.9. After processing, the figures were even more startling. White, sliced bread had an energy output : input ratio of 0.525, broiler chicken 0.10, battery eggs 0.14, and winter lettuce grown under glass 0.0017.

Expanding the area of cultivated land is more difficult. Almost all the land suitable for agriculture is being farmed, and although there could be some extension into marginal lands, yields from them are unlikely to be great and might do little more than compensate for yield reductions on degraded land.

As Figure 5.20 shows, the amount of food produced per head of population has remained fairly constant through the 1990s, but its balance is changing. In the developed countries over-production and the desire for wildlife and landscape conservation in rural areas have led to a reduction in food output, but in the less developed countries output continues to increase steadily. Figure 5.21 shows that overall output of cereals has remained generally steady, although there was a drop in output of coarse grains, used to feed livestock, in 1993–94. In *Agriculture: Towards 2010*, a review of world agriculture it published in 1992, the FAO (*www.fao.org/*) showed

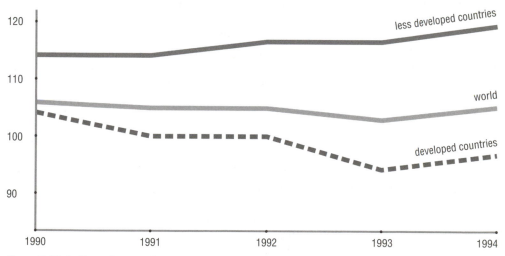

Figure 5.20 **Indices of per capita food production 1990–94 (1979–81 = 100)**

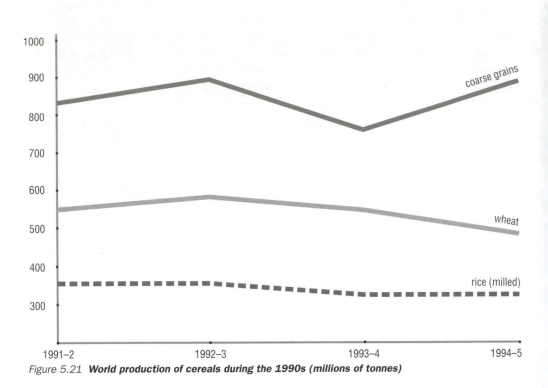

Figure 5.21 **World production of cereals during the 1990s (millions of tonnes)**

that over 30 years world food output per person had risen 18 per cent and the number of malnour-ished people had decreased from nearly 950 million to 800 million in 20 years. It predicted that by 2010 this figure will fall to 650 million. Poverty was the principal reason for hunger, as it has always been. Between 1990–92 and 1995–97 the extent of malnourishment decreased in 37 coun-tries and the number of people going hungry decreased by 40 million. The number of malnourished people remains unacceptably high, estimated at 790 million in 1999, but it is falling in line with the FAO projection.

It is not only food that farmers produce. Fibres, such as cotton, wool, silk, and flax, are also import-ant. Wool is closely linked with meat production, of course, and silk is produced very intensively in indoor units, although land is needed to grow the mulberry and other trees that supply the leaves on which the larvae feed. In 1997, the world produced 52 700 tonnes of silk. Cotton and flax, on the other hand, compete directly for land with food crops. Flax is little used nowadays, although interest in it has revived somewhat, but in 1998 world production of cotton amounted to 18.6 million tonnes. Annual world cotton output is slightly lower than that of artificial fibres, which is predicted to continue increasing.

In years to come it is desirable that we improve the nutritional status of the very poor. This implies a continuing increase in food production, although the primary need is for economic advance-ment in particular countries. The need to grow more food on essentially the same area of land might suggest changes in our land-use priorities, but this is not so simple as it seems. We might argue, for example, that fibre crops, such as cotton, are no longer needed. Synthetic fibres provide adequate substitutes, and abandoning cotton-growing would free land for food production. The same case is sometimes made for cash crops, such as tea, coffee, and tobacco, which are grown for export. Perhaps tobacco-growing will be abandoned one day, but for reasons of public health, not economics. Cash crops, and cotton, allow countries to earn currency with which to build their manufacturing industries. It is difficult to see how they could advance from their present level of economic development without exporting agricultural produce. Were the tea, coffee, and cotton plantations to close, the land would not necessarily produce food, because unemployed people could not afford to buy it. Hunger would persist.

It is also maintained by some that world food production could increase dramatically were we to abandon, or at any rate drastically curtail, livestock production. This is because pigs and poultry, and to a lesser extent cattle, are fed cereals that could be eaten directly by humans, the animals achieving a maximum of about 10 per cent conversion efficiency. In other words, of the grain they eat, 90 per cent or more is lost in respiration. There is some truth in this, but it is only part of the story. Not all livestock consume grains and large areas of the world are suitable only for the production of pasture. Grass is the only practicable crop over much of Britain, for example, and it is fed to cattle and sheep. Reducing the number of cattle and sheep would not increase the amount of food produced unless people were prepared to accept food made directly from grass. It is technologically possible to convert grass into food suitable for human consumption, but it probably costs more than leaving animals to do it by themselves and consumers might not be prepared to pay more for what appeared to be an inferior substitute. It is not even true that all pigs and poultry eat only grain suitable for human consumption. Most do, in intensive units to produce cheap meat and eggs, but they can also eat spoiled grain and other foods that are not suitable for human consumption and would otherwise be wasted. Indeed, this was their traditional agricultural role.

Many people go hungry in the world, but their number is decreasing. Gains are being made and the agricultural improvements achieved in the last thirty years are truly impressive. With food production, as with so many of the dilemmas facing us, however, the most obvious solutions are often wrong in a situation that is much more complex than it sometimes seems.

55 Human populations and demographic change

When Charles Darwin read a long essay by Thomas Robert Malthus (1766–1834), the first edition of which was published in 1798, he found it provided him with a mechanism to drive natural selection. The struggle for existence, Darwin wrote, was 'the doctrine of Malthus applied with manifold force to the whole animal and vegetable kingdoms; for in this case there can be no artificial increase of food, and no prudential restraint from marriage' (DARWIN, 1859, Ch. 3).

Today, there is widespread anxiety about the rate at which the human population of the world is increasing. The fear is not new. The Zoological Society of London was founded, in 1825, partly to identify animals suitable for domestication and acclimatization to new parts of the world, in order to increase the food supply for growing populations (BURGESS, 1967, pp. 87–88). From figures in parish registers, the population of England and Wales is estimated to have been about 5.2 million in 1695 and 9.2 million by 1801. After 1801 the figures are based on census returns. They show that by 1851 the population had risen to 17.9 million and by 1901 to 32.5 million (TRANTER, 1973, pp. 41–42).

It is little wonder that the increase caused alarm and that the increase in world population continues to do so. As the familiar graph in Figure 5.22 illustrates, the rate of increase appears to be exponential. Exponential or geometric growth occurs when the increment accrued during each period is added to the total before the next increment is calculated. It is contrasted with arithmetic growth, in which increments are calculated against the original quantity, accumulated separately, and added to the original quantity at the end. In financial terms, these are known as compound and simple interest respectively.

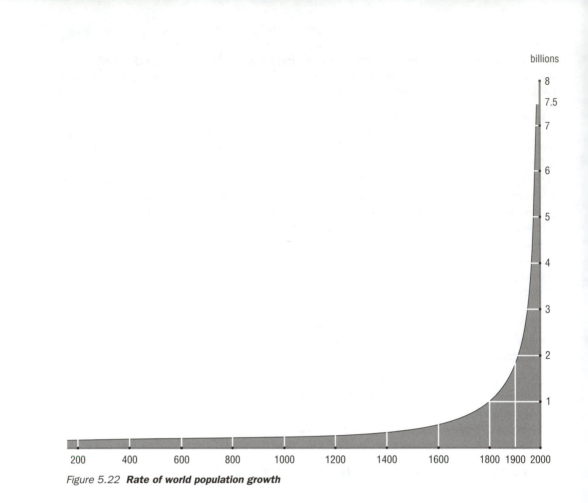

billions

8
7.5
7
6
5
4
3
2
1

200   400   600   800   1000   1200   1400   1600   1800 1900 2000

Figure 5.22 **Rate of world population growth**

This was the distinction on which Malthus based his argument (MALTHUS, 1798). He pointed out that parents are capable of producing more than the 2 children required to replace them in the population. If, for example, each pair of parents produces 3 children in each generation, and we start with two pairs (to avoid inbreeding!), the number in each generation proceeds as: 4; 6; 9; 13.5; 20.25; 30.375. . . . Using information from populations in the then sparsely populated United States, he went on to assert that human populations double in size every 25 years (a 2.8 per cent annual rate of increase). If a quantity is increasing exponentially, the time taken for it to double can be calculated approximately by dividing 70 by the percentage rate of increase; if the annual rate of increase is, say, 2.0 per cent, the doubling time will be $70 \div 2 = 35$ years.

People require resources to sustain them, the most obvious being food, but Malthus asserted that food output can increase only arithmetically.

> The population of this island is computed to be about seven millions, and we will suppose the present produce equal to the support of such a number. In the first twenty-five years the population would be fourteen millions, and the food being also doubled, the means of subsistence would be equal to this increase. In the next twenty-five years the population would be twenty-eight millions, and the means of subsistence only equal to the support of twenty-one millions. In the next period, the population would be fifty-six millions, and the means of subsistence just sufficient for half that number (MALTHUS, 1798, pp. 74–75).

In other words, the geometrical rate of population increase inevitably outruns the arithmetic rate at which resource availability can be made to increase. Population is then held in check, through hunger and disease or voluntarily by producing fewer children. This is the 'Malthusian limit' toward which our numbers are said to be heading. All present fears over global population increase derive from this line of argument.

The argument sounds persuasive, but it was not really fear of overpopulation that led Malthus to advance it. He was seeking to refute ideas current among certain intellectuals in the years following the French Revolution, and held by his own father. They believed that new scientific discoveries and radical political and economic change offered the promise of improvements in the living standards of ordinary people that could barely be imagined. Malthus claimed to prove mathematically that 'population must always be kept down to the level of the means of subsistence' (MALTHUS, 1798, p. 61). From this it followed that attempts to improve living standards, by increasing the availability of resources, must lead to a population increase sufficient to consume the additional resources. This would leave people worse off than before, because a greater number of them would be living in want. The kindest strategy was to treat the poor harshly and on no account seek to alleviate their lot. The Malthusian argument was used to defeat a campaign to institute a minimum wage and led many politicians to believe that relief for the poor simply produced more wretchedness. In particular, Malthus urged in the succeeding years, no allowances should be paid for children (ALLABY, 1955, pp. 163–166).

By 'proving' that poverty can never be eliminated, the Malthusian characterization of human demography places responsibility for their plight firmly in the hands of the poor. If they wish to prosper, they must reduce their number. Should they fail to do so they will have only themselves to blame when their children starve, as assuredly they will. Malthus regarded the poor as feckless. Had he not done so, he might have observed that according to his argument the rich, with ample access to resources, should multiply until they exhausted those resources and reduced themselves to poverty. On the contrary, however, they tended to have fewer children than poor people. Malthusianism regards poverty as the fault of the poor, whose ignorance of their own best interest will defeat any attempt by the rich to help them.

Contemporary 'neo-Malthusians' accept the Malthusian premise, but moderate it by advocating programmes for population control, mainly through the provision of contraceptive devices and information. They also point to the disproportionate allocation of material resources to the wealthy: the European and North American standards of living are sustained by vastly greater consumption of resources than are available to the citizens of poor countries, and Western economies appear to grow mainly by artificially stimulating consumption. A more equitable distribution of access to resources would, in their view, increase the likelihood that the great majority of the world's people might live in decency, their essential needs met. Anti-Malthusians reject this line of reasoning. They maintain that rather than there being a kind of levelling down, which in any case would be politically and probably economically impossible, poor countries should be helped to expand their economies and, by implication, their levels of resource use. This, they believe, would improve living standards directly and reduce rates of population growth indirectly, and the availability of resources would increase as demand required. Resources, according to this line of reasoning, should not be regarded as 'renewable' or 'non-renewable', so much as flexible, with virtually unlimited capacity for the substitution of one that is abundant for another that is less so.

In the event, Malthus was wrong. Advances in agricultural methods increased output to a far greater extent than he imagined possible and living standards improved greatly. Today, with a population of 58 million, the British people enjoy much better nutrition than they did when there were only the 7 million of them he described. In 1791, the population of England and Wales was increasing annually at about 1.0 per cent (TRANTER, 1973, p. 41), not the 2.8 per cent on which he based his calculations, and as prosperity increased during the nineteenth century the rate of increase never exceeded 1.8 per cent (between 1811 and 1821). For most of this century it has remained below 1.0 per cent and at present it is below 0.7 per cent. At the time Malthus wrote his essay, no country had experienced the 'demographic transition'. This is the process by which death rates fall in a previously stable population, but birth rates remain unchanged. This generates an increase in the population. Eventually, birth rates also fall, until birth and death rates balance once more and numbers stabilize.

In the world as a whole, however, population is certainly increasing rapidly. As Figure 5.23 shows, it has risen from around 1 billion in 1850 to 5.6 billion in 1994, and the United Nations officially designated October 16, 1999, as the date of birth of the world's six-billionth citizen. According to the United Nations median estimate, it is likely to reach 8.25 billion by 2025. Steep though the increase is, the curve in Figure 5.23 is less dramatic than that in Figure 5.22. It is also more credible, because only the crudest guess can be made of the size of the world population many centuries ago.

Exponential growth appears on a graph as a curve that rises at a very shallow angle for a long time, then suddenly turns upward until it rises almost vertically. The curve in Figure 5.22 is exponential and that in Figure 5.23 might be, but in this case it is less certainly so. Biological populations sometimes increase exponentially, in a J-shaped curve, then collapse. That is what some people fear may befall the human population. More commonly, though, their growth follows an S-shaped

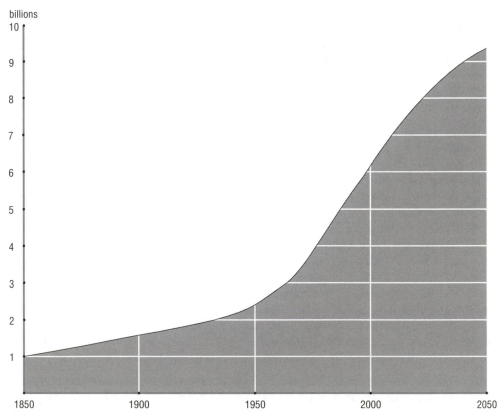

Figure 5.23 **World population (billions) 1850–2025 (median estimate)**

curve, rising rapidly for a time, but then stabilizing owing to density-dependent factors. The curve in Figure 5.23 could develop into an S-shaped curve.

There is good reason to suppose this is what is happening. When fears of a 'population explosion' were popularized in the late 1960s, the rate of increase was about 2.0 per cent, doubling the world population every 35 years. More recently, as Figure 5.24 illustrates, it has been falling. The fall has been erratic, but the trend is clearly downward. Between 1997 and 1998 it fell from 1.47 per cent to 1.42 per cent (HAUB, 1999). Some forecasts predict a decline in the world population, starting in about 2050.

Population growth is confined, almost entirely, to the less industrialized countries – in other words, to the poor. This may seem to support the Malthusian model, but it is equally characteristic of a demographic transition. Improved health services and nutrition led to a reduction in perinatal, infant, and childhood mortality. Death rates fell, but the same number of babies continued to be born. That is why the population increased so sharply. If health improvements are sustained, parents come to accept that fewer babies need be born to provide the number of offspring needed to help with the work and care for them in their old age. If education is then made compulsory for all children, child labour is reduced. Children cease to be productive workers and become economic dependants, reducing further the incentive to have large families. The most important reform then is to improve educational opportunities for girls and employment opportunities for women. This allows women the choice either to stay at home bringing up a large family, or to work outside the home and contribute directly to the household budget while enjoying the social life and prestige paid employment brings. Hardly surprisingly, this is what many women choose. Provided adequate contraceptive advice and materials are easily obtainable, birth rates then start to fall.

This is the demographic transition. Many countries have implemented the necessary reforms, and the total fertility rate (the number of children a woman will bear during the course of her reproductive life) is falling. In the 1950s, the world average fertility was 5 children. By the late 1990s it had fallen to 2.7, and the annual increase in population had fallen to 78 million, from 90 million in the 1980s. In 1999, the fertility rate in 61 countries was below the 2.1 needed to maintain

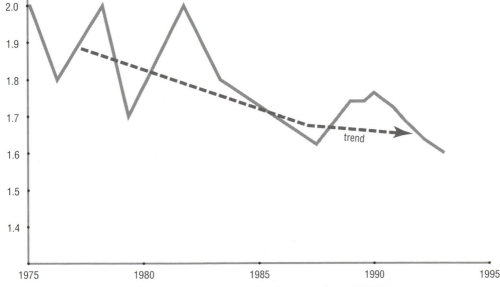

Figure 5.24 **Estimates of the rate of global population increase since 1975 (%)**

a constant population size. In most of Europe, the Caribbean countries, and eastern Asia including China, populations are set to decrease and in others it will increase, but only slowly. In Thailand, for example, the fertility rate fell to 2.3 from 6.4 in 1960. Since the 1980s it has fallen in Ghana from 6.0 to 5.4 and in Kenya from 8.0 to 5.4. (HAUB, 1995). This change is happening everywhere, even in the poorest countries. The fertility rate in Bangladesh has fallen from 6.2 to 3.4 children in ten years. There are 18 countries in which the population may decrease by 15 per cent by 2050. In Bulgaria, for example, the 1999 population is 8.3 million and its predicted 2050 population has been revised from 7.8 million to 6.7 million (PEARCE, 1999).

It appears that the rate of population increase peaked in the 1960s. The 1999 UN medium population projection for 2050 has been reduced from 9.4 billion to 8.9 billion, but some demographers believe this is too high. They calculate that the world population will peak at about 7.7 billion in 2040 and then enter a long decline leading to a population of less than 6 billion by 2100 and about 3.6 billion by 2150 (PEARCE, 1999).

There is inevitably a delay in any such transition. While the birth rate exceeds the death rate, causing a population to grow, it is children who are being added and consequently the average age of the population decreases. As the children enter the reproductive stage of their lives their children will cause the population to continue growing. More parents imply more children, even if the new parents have fewer children than did their own parents. Signs that this phase of the change is ending appear as a reduction in the number of children. In 1990 there were 623 million children in the world below the age of five. In 1995 this figure was 614 million. The change had begun, but the world population will continue to increase for some time – at least until about the middle of the next century. For the same reason, when the increase ends the resultant population will be much larger than it was before the increase began. The British population is no longer growing, but it has increased from 7 to 58 million. As populations decrease, nationally and globally, the average age of the population increases. This can impose burdens on societies caring for people who are too old to be economically active.

A rapid increase in population also places a heavy burden on financial resources, because the economically active members of the population must support a larger number of dependants. This leaves less money for investment in productive industries and thus inhibits the economic development needed to sustain the reforms that will reduce the rate of growth..

Increasing prosperity will lead to a much increased demand for material resources. The challenge of the coming decades will be to supply them without imposing too heavy a burden on the natural environment. Environmentally, we have no alternative. Land is degraded primarily by the very poor seeking to scratch a living from it in ways that deplete it. Industrial pollution is much greater from factories equipped with obsolete plant whose managers must cut corners if they are to compete with more modern plants overseas. Even pollution by transport is reduced if the vehicles are new and incorporate modern devices to improve fuel efficiency and minimize emissions. The industrialization of the presently less industrialized countries may cause environmental problems, but a lack of industrialization and consequent poverty will not lead to a more wholesome environment.

56 Genetic engineering

Organic farmers are familiar with *Bacillus thuringiensis*. They have been using it for many years as a kind of living insecticide. It was first identified in 1911; cultures were being sold commercially in France in 1938 and it was used to control Japanese beetle in the United States in 1939 (GOLDSTEIN, 1978, p. 173). *Bacillus thuringiensis* kills leaf-eating caterpillars. It does so by producing a protein that turns into a lethal poison when the insect ingests it. Different strains of *B. thuringiensis* produce toxins effective against different insects, so farmers often spray a mixture of strains.

Even with a substance as environmentally safe as *B. thuringiensis*, however, spraying is a crude and wasteful way to deal with pests. It would be much better if the plants produced the active ingredient themselves. No pest could then escape the poison, because every leaf would be poisonous, and the grower would not have to invest in spraying equipment and wait for suitable weather conditions before using it. In the case of *B. thuringiensis* this is possible, because the poison is a naturally produced protein, encoded genetically. Identify the relevant bacterial gene, remove it, insert it into the plant cell nucleus, and all being well the plant will now produce the protein. It is nowhere near so simple as this makes it sound, but scientists have mastered the necessary techniques and the *B. thuringiensis* gene has been or is in the process of being introduced into varieties of a number of crop plants. Tobacco was the first to receive it, in 1987 (VAECK *ET AL.*, 1987), followed by rice in 1993, potatoes in 1995, and then tomatoes, cotton, maize, and sunflower. This is genetic manipulation, popularly known as 'genetic engineering', the deliberate alteration of organisms by the introduction of genetic material previously alien to them. The technology promises to make agriculture much more sustainable (PLUCKNETT AND WINKELMANN, 1995).

DNA (deoxyribonucleic acid), the compound comprising the material of heredity, consists of four bases, or nucleotides: the purines adenine (A) and guanine (G), and the pyrimidines thymine (T) and cytosine (C). These are linked to form two chains wound helically; A can pair only with T, and C with G. The bases work in threes, called codons or triplets, each codon specifying a particular amino acid. A gene consists of a sequence of codons, with particular codons marking the start and end of the sequence. It is rather like a sentence written using only these four letters, but beginning with a capital letter and ending with a full stop. A functional gene (many genes have no apparent function) encodes all the amino acids required to construct a particular protein. In addition to the proteins from which muscle and tendon are made, enzymes and antibodies are also proteins. Since protein molecules vary greatly in the number of amino acid molecules they contain, genes also vary in size. In plants and animals, the genetic code is contained in the cell nuclei (although certain cell organelles have DNA of their own). In prokaryotes, such as bacteria, there is no nucleus and the DNA is carried 'free' in a single chromosome. Bacteria also contain plasmids, loops of DNA that pass readily from one cell to another.

Once the gene encoding a desired protein has been identified it can be excised from its long DNA strand by means of restriction enzymes. These evolved in bacteria as a defence against viral attack, which they restrict (hence their name) by severing DNA at certain nucleotide sequences, each restriction enzyme attacking particular sequences. When some restriction enzymes cut DNA they leave two nucleotides from one DNA strand protruding beyond the end of the other strand. This is known as a 'sticky end', because these protruding nucleotides will anneal to a complementary pair protruding from another DNA strand. If AC protrudes, for example, a strand with protruding TG will bond to it and this will happen even if the DNA comes from quite unrelated organisms. This allows excised segments of DNA to be recombined in new ways, as recombinant DNA, with the help of a ligase, another enzyme that strengthens the bonds by which the segments are annealed.

Use a restriction enzyme to break open a plasmid, leaving appropriate sticky ends, add a gene, with complementary sticky ends, excised from another organism, and the excised gene will be inserted into the plasmid, the ends of which will rejoin to form a closed loop. The plasmid then contains all its original genes, plus the introduced gene. Because plasmids readily cross bacterial cell walls, the altered plasmid can be introduced into bacteria. These can then be cultured, the plasmids replicating as the cells themselves replicate, and the bacteria containing the modified DNA injected into another organism. Figure 5.25 illustrates the process. The introduced gene may be a replacement for a defective gene in the host. This is the basis for gene therapy. If it is a gene alien to the host, the host is called 'transgenic', because it carries one or more genes from a different species (TUDGE, 1993, pp. 206–214).

Plasmids are commonly used as vectors, the means by which genes are transported into a host, but there are others. Deactivated viruses and 'naked' DNA are sometimes used. They are also cloning vectors, producing an identical copy, or clone, of the desired gene each time they replicate, but genes can now be replicated by themselves before being introduced into a vector.

This is made possible by the polymerase chain reaction (PCR), the invention for which Kary B. Mullis won the 1993 Nobel Prize for Chemistry. DNA is added to a solution containing 'primers', one of several possible DNA polymerase enzymes, and rich in purines and pyrimidines. The solution is warmed to 95 °C, at which temperature the two DNA strands separate. It is then cooled to about 50 °C and the primers anneal one to each end of the separated strands. It is warmed to about 70 °C, and the polymerase builds a complementary DNA strand to each separate strand, forming double strands. At each repetition of the process the amount of DNA is doubled.

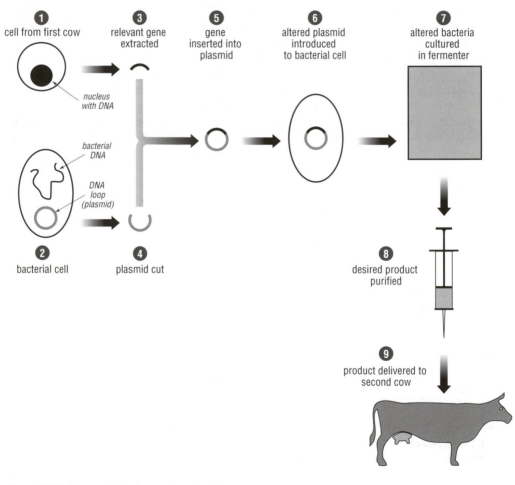

Figure 5.25 *One method of genetic engineering*

Despite all that has been written in the press about gene therapy for humans, installing functioning genes in human bodies has proved difficult and so far the successes are modest (MARSHALL, 1995), although there is good reason to hope that eventually gene therapy will bring significant benefits (CRYSTAL, 1995). In plants, on the other hand, there has been much greater success, especially in conferring resistance to pests and diseases.

This is not really new. Plant breeders observed in the first years of this century that disease resistance is heritable in a Mendelian fashion and can be bred into commercial varieties. Today the genetic details of such resistance have been discovered and appropriate genes can be introduced into plants that can benefit from them (STASKAWICZ *ET AL.*, 1995).

GM Foods

On August 10, 1998, Arpad Pusztai, a specialist in lectins working at the Rowett Research Institute, near Aberdeen, said in a TV interview in *World in Action* that his research had found that potatoes, genetically modified to contain a gene for the production of a lectin found in snowdrops, had stunted the growth of rats fed on them. This claim led to his suspension, and later to his retirement (he was well past retirement age), because his work had not been published and there was doubt over his interpretation of its results. On February 12, 1999, British newspapers reported that scientists from 13 countries had backed Pusztai's findings and were calling for his reinstatement. The scientists were revealed as friends and colleagues of Pusztai, some with known anti-GM views, but it was too late and the issue developed into a major food scare. Greenpeace and the Soil Association called for an outright ban on genetically modified (GM) foods and as sensational, antiscientific press stories, based on misinformation from environmentalist groups, increased public alarm, the environmentalists were able to claim to represent public opinion. Environmental activists destroyed GM crops being grown under test conditions to determine their effect on the surrounding environment.

A team of scientists from the Royal Society examined the Pusztai experiments and concluded his study was flawed in its design, execution, and analysis. Eventually, the Pusztai paper was published in *The Lancet*, despite strong scientific opposition (LODER, 1999). The effect of the anti-GM campaign was to remove all GM foods and foods containing GM ingredients from British shops. Millions of people, in several countries, had been consuming GM foods for a number of years and no evidence whatever had emerged that GM foods were harmful to health or to the environment.

In 1999 it was estimated that in the world as a whole GM crops were being grown on almost 40 million hectares. This was an increase of 44 per cent over the area growing GM crops in 1998 and occurred mainly in China, Argentina, Canada, and South Africa.

Is it safe to grow such genetically engineered plants? Scrupulous tests are conducted before a licence is granted to release any genetically manipulated organism into the environment. In the case of disease resistance, there is no risk to other species. The resistance genes that allow plants to destroy invading pathogens (disease-causing organisms) have no effect other than this and are highly specific. In the case of transgenic plants carrying *B. thuringiensis* genes there is the possibility that pests might acquire resistance to the toxin. They have not done so to toxin delivered by conventional application, probably because the toxin breaks down so rapidly that insects are not exposed to it for long enough for resistance to emerge. If entire plants produce the toxin insects may be exposed to it for much longer. So far, however, there has been no sign of pest resistance.

The most widespread application of genetic manipulation has been to introduce resistance to herbicides, and especially to glyphosate, a compound marketed as Roundup. Non-GM crops of maize and soybeans must be sown into ground that has been treated with herbicide to kill emerging weeds. Herbicide treatments must then be repeated several times during the growing season, using sprays that are relatively inefficient because they must not harm the crop. GM crops can be sown without prior spraying and herbicide can be used, without risk to the crop, when the weeds have emerged and are at their most vulnerable. The result is that much less herbicide is needed. Weed control improves, less tillage is required, and soil erosion is reduced. In the United States in 1999 herbicide-resistant soybeans were being grown on about 14 million hectares (ABELSON AND HINES, 1999).

Many aspects of plant growth, including flowering and fruit ripening, are regulated by ethylene (C_2H_4), a plant growth substance ('hormone') they synthesize and secrete. Certain of the genes involved in ethylene synthesis and recognition have been identified and they may be manipulated to modify ethylene-induced effects (ECKER, 1995). There is also research directed at introducing into crop plants the genes carried by *Rhizobium* bacteria that allow them to fix atmospheric nitrogen, with huge potential for reducing the demand for inorganic nitrogen fertilizer.

The first phase of GM crop plants have been designed to cope better with pests, weeds, and diseases, so farmers are the principal beneficiaries of their introduction. The benefits are not confined to farmers in the industrialized countries. The introduction of disease resistance in bananas, potatoes, and cassava (MOFFAT, 1999) will improve yields in tropical countries. Already, though, crops are being modified to improve their nutritional qualities (DELLAPENNA, 1999), in a second phase that will bring immediate benefits to consumers.

Bacterial cultures, animals, and plants, are also used to synthesize useful substances. Researchers are seeking to develop edible plants that have been genetically manipulated to produce vaccines, for example (MOFFAT, 1995). People eating the appropriate plants would be vaccinated automatically against a whole range of diseases. At first the plants would not be among those forming part of the ordinary diet, so they would have to be eaten specially, like the plants used in herbal medicine, but eventually some dietary plants might also deliver vaccines. Plants are also being modified to produce useful industrial materials, such as cotton plants that yield a polyester-like fibre and rapeseed plants that produce several industrial chemicals. Both plants and bacterial cultures are being used to remove chemical pollutants and others are being genetically manipulated to make them do so. Plants are being modified to tolerate soils contaminated with heavy metals. This will allow crops to be grown on land that until now has been agriculturally useless. The metals would accumulate in the plants, from which those of commercial value could be extracted. Plants storing unwanted metals would be incinerated. In time the land would be cleansed of the contaminants much more cheaply than the usual alternative, which is to remove the soil to a landfill site that can accept hazardous waste (MOFFAT, 1999a).

The environmental implications of these developments are generally benign. Clearly, reducing the susceptibility of crop plants to pests and fungal diseases suggests they can be grown with pesticide applications much reduced or, perhaps, eliminated entirely. This would be environmentally beneficial, because accidents can happen even when safe pesticides are used. If crop plants are rendered resistant to herbicides, it will be possible to use herbicides in circumstances where they cannot be used at present. Used, as the herbicides would be, among growing crop plants, this is unlikely to cause environmental harm and if it increased crop yields, in principle land might be released for other uses, such as conservation. Wildlife is already returning to American farms growing GM crops. Field trials of herbicide-tolerant sugar beet conducted in Britain in 1998 found herbicide applications were reduced to two sprayings with glyphosate (Roundup), from the customary 5 to 7, using up to nine different compounds. The GM crop attracted more insects than adjacent plots growing non-GM sugar beet, conserved water better, and reduced soil erosion (WILSON *ET AL.*, 1999).

There is no reason to suppose genetically manipulated plants will be unwholesome. They are developed to be eaten, after all and, despite its 'high-tech' image, genetic manipulation is as old as agriculture itself. Few, if any, of the plant and animal products we eat are genetically identical to their wild ancestors. They were selectively bred, and selective breeding is a form of genetic manipulation. Indeed, the concern of some conservationists to maintain populations of animal breeds and crop plant varieties that have fallen from popularity centres on the genetic difference between them and their more commercially profitable rivals.

For some people, ethical issues emerge with transgenic breeds and varieties. Vegetarians and vegans may object to plants carrying genes derived from animals, and the presence of genes derived from pigs or cattle may offend others on religious grounds (ALLABY, 1995, pp. 224–227).

Ethylene and genetic engineering

Ethylene (C_2H_4) is chemically the simplest of all plant growth regulators. It is involved in many processes linked to growth, ripening, and response to stress. Studies of *Arabidopsis* plants have identified the genes producing the proteins that sense and respond to ethylene. This makes it potentially possible to genetically modify a wide range of plants in order to control such matters as growth, root production and pattern, and the ripening of fruit.

Within the next few years it will also become possible to genetically modify plants grown for their oils. By introducing genes identified in non-cultivated species and inserting them into cultivated plants, it will become possible to regulate the types of oil each crop variety produces. The proportion of saturated fatty acids can be reduced, for example, in oils intended for human consumption, and oils with valuable industrial uses can be produced. These developments far exceed what would be possible by conventional plant breeding.

Ornamental plants will also change. Flower colour is genetically controlled. Entirely new colours can be generated as well as plants that grow flowers of different colours side by side.

There is a risk, but it is economic rather than environmental. This is that particular transgenics, especially plant varieties, might be grown on so large a scale as to constitute genetic monocultures over wide areas. Imagine, for example, that a bread wheat was developed that could fix its own nitrogen and thus required no nitrogen fertilizer. Growing costs would be substantially reduced, as a consequence of which wheat prices might fall, and older varieties would become uneconomic and disappear. An attack by a pathogen to which they were not resistant might then prove catastrophic. This is not fanciful. In 1970, southern corn leaf blight, caused by a newly emerged strain of the fungus *Helminthosporium maydis*, attacked the American maize crop, a substantial part of which consisted of varieties sharing identical cell cytoplasm, which is where the fungus struck. The epidemic was contained, with difficulty and helped by a change in the weather, but not before around 15 per cent of the crop was lost. That episode led scientists to examine the genetic vulnerability of a range of important crop plants (NATIONAL ACADEMY OF SCIENCES, 1972). Increasing that vulnerability would clearly be unwise.

The use of organisms, genetically manipulated or not, to remove pollutants is of obvious environmental benefit, provided its success is not construed as a licence to release pollutants where this could be avoided. (Examples of the use of bacteria in cleaning up oil spills are given in section 61, headed 'Pollution control'.) Similarly, the use of genetically manipulated organisms as sources of industrial chemicals and other products may reduce environmental damage compared with that associated with more traditional production methods.

A risk remains, however, that essentially novel organisms resulting from genetic manipulation may establish themselves in the wild with consequences that were not predicted during their development. The case is cited of calicivirus, imported by Australia from Eastern Europe to discover how effective it might be in killing rabbits. It was planned to investigate this experimentally on Wardang Island, off South Australia, and if it proved successful to release it on the mainland in the late summer or early winter of 1997 or 1998. In October 1995, however, the virus escaped, possibly in insects carried by a mass of cold air, and killed large numbers of wild rabbits on the mainland. If this can happen, might it not also happen with a genetically novel organism?

So far at least, it has not happened, and present indications are that such organisms, modified for human purposes rather than for their own evolutionary advantage, do not compete well with others. Nevertheless, were such an organism to become well established it might persist for a very long time, and the ecological havoc that has been wrought by some introduced species does not fill ecologists with great confidence. Genetic modification is permanent, because it is transmitted from one generation to the next. It is for this reason that all such releases must comply with strict safety rules.

Genetic 'engineering' will bring many changes to our lives in the next few years. Many of these changes will be environmentally beneficial.

End of chapter summary

Like all other animals we use the environment, adapting it to our own purposes, and that environment includes other organisms. If we are to exploit our environment successfully we need to know how species come into being. The principal mechanism for this is natural selection and it is one we have utilized since our ancestors first began farming. Farmers have always selected plants and animals for the traits that make them desirable and this has altered the species genetically. Evolutionary processes also determine the effects of selection pressures introduced by human activities, sometimes inadvertently. The resistance of insects and plants to pesticides and of bacteria to antibiotics is an entirely natural response to selection pressure.

Genetic engineering, or the use of modern, science-based techniques to introduce desired characteristics into species has outraged some environmental groups and they have used their essentially ideological opposition to generate concern among an otherwise uninformed public. Since this has been inflated into an issue of international proportions it is important to understand what the techniques involve and the purposes for which they are used.

Applying the concepts derived from ecology, it is possible to estimate the way our behaviour affects natural ecosystems such as forests and natural populations such as fish stocks. The same concepts also allow us to understand the way our artificial agricultural and urban ecosystems function.

Much concern has been expressed over the last thirty years about the rate of human population growth. This, too, can be understood, at least up to a point, in ecological terms. These propose reasons for human population growth and suggest that it should stabilize naturally. This is what is now happening.

End of chapter points for discussion

How should sea fisheries be regulated?
What are the likely consequences of the campaign against genetically modified foods?
What is 'biodiversity'?
How can we best increase food availability in the less developed countries of Africa, Asia, and Latin America?

See also

Ocean circulation (section 16)
Irrigation, waterlogging, salinization (section 25)
Erosion (section 29)
Wildlife species and habitats (section 49)
Biodiversity (section 50)
Fisheries (section 51)
Forests (section 52)
Restoration ecology (section 59)

Further reading

The Engineer in the Garden. Colin Tudge. 1993. Jonathan Cape, London. A simple and entertaining account of genetic engineering and the possibilities it offers.

An Essay on the Principle of Population as it Affects the Future Improvement of Society. Thomas Robert Malthus. First published in 1798, the book is still in print, published by Penguin. There is so much talk about Malthus that it is useful to find out what he actually wrote.

Extinction Rates. Edited by John H. Lawton and Robert E. May. 1995. Oxford University Press, Oxford. An authoritative but somewhat technical account of the most recent thinking on biodiversity and the loss of species.

An Introduction to Evolutionary Ecology. Andrew Cockburn. 1991. Blackwell Scientific Publications, Oxford. A clear and straightforward explanation of its subject and not so technical as to be inaccessible to those with only a rudimentary knowledge of genetics.

On the Origin of Species by Means of Natural Selection. Charles Darwin. First published in 1859, the book is still in print. Reading what Darwin actually wrote is by far the best way to approach his work, and his writing is very easy to follow.

The Problems of Evolution. Mark Ridley. 1985. Oxford University Press, Oxford (paperback). A clear explanation of evolutionary theory and some difficulties associated with it.

Soil and Civilization. Edward Hyams. 1976. John Murray, London. A classic on soil, its treatment, and the problems arising from its erosion.

Notes

1 The scaling effect occurs because area is calculated by squaring a value and volume by cubing it. A sphere with a diameter of 4 has a surface area $(4\pi r^2)$ 50.26 and a volume $(4/3\pi r^3)$ of 33.51, a volume: area ratio

of 1:1.50. If the diameter is 6, the surface area is 113.10 and the volume is also113.10, so the ratio is 1:1. The larger the sphere, the smaller is its surface area in relation to its volume.

2 Figures are from *Britannica Book of the Year 1998*, published by Encyclopedia Britannica, Chicago, p. 129. Although they refer to 1995, the list of the top 20 species changes little from year to year. In 1995, the total world catch for all 20 species was 42 682 002 tonnes.

References

Abelson, Philip H. and Hines, Pamela J. 1999. 'The plant revolution', *Science*, 285, 367–368.

Allaby, Michael. 1977. *World Food Resources, Actual and Potential*. Applied Science Publishers, London.
—. 1986. *The Woodland Trust Book of British Woodlands*. David and Charles, Newton Abbot.
—. 1995. *Facing the Future*. Bloomsbury, London.
—. 1999. *Ecosystem: Temperate Forests*. Facts on File, New York.

Begon, Michael, Harper, John L., and Townsend, Colin R. 1990. *Ecology*, 2nd edn. Blackwell Science, Oxford.

Bender, Barbara. 1975. *Farming in Prehistory*. John Baker, London.

Berry, R.J. 1977. *Inheritance and Natural History*. Collins New Naturalist, London.

Braidwood, Robert J. 1960. 'The Agricultural Revolution', *Scientific American*, May 1960, W.H. Freeman and Co., San Francisco.

Brewer, Richard. 1988. *The Science of Ecology*, 2nd edn. Saunders College Publishing and Harcourt Brace, Orlando, FL.

Burgess, G.H.O. 1967. *The Curious World of Frank Buckland*. John Baker, London.

Burton, Maurice and Burton, Robert. 1977. *Inside the Animal World*. Macmillan, London.

Clutton-Brock, Juliet. 1981. *Domesticated Animals From Early Times*. Heinemann and British Museum (Natural History), London.

Cockburn, Andrew. 1991. *An Introduction to Evolutionary Ecology*. Blackwell Scientific Publications, Oxford.

Colyer, Charles N. and Hammond, Cyril O. 1968. *Flies of the British Isles,* 2nd edn. Frederick Warne and Co., London.

Corbet, G.B. and Southern, H.N. (eds) 1977. *The Handbook of British Mammals*, 2nd edn. Blackwell Scientific Publications, Oxford.

Crystal, Ronald G. 1995. 'Transfer of genes to humans: early lessons and obstacles to success', *Science*, 270, 404–410.

Culotta, Elizabeth. 1995. 'Many suspects to blame in Madagascar extinctions', *Science*, 268, 1568–1569.
—. 1995a. 'Shades of rain-forest green', *Science*, 269, 31.

Darwin, Charles. 1859. *On the Origin of Species by Means of Natural Selection*. Opening of the final paragraph of the book.

Dawkins, Richard. 1978. *The Selfish Gene*. Granada Books, London.

DellaPenna, Dean. 1999. 'Nutritional genomics: manipulating plant micronutrients to improve human health', *Science*, 285, 375–379.

Ecker, Joseph R. 1995. 'The ethylene signal transduction pathway in plants', *Science*, 268, 667–675.

Ford, E.B. 1981. *Taking Genetics into the Countryside*. Weidenfeld and Nicolson, London.

Freeman, Paul. 1973. 'Ceratopogonidae (Biting midges, "sand-flies", "punkies")', in Smith, Kenneth G.V. (ed.) *Insects and Other Arthropods of Medical Importance*. British Museum (Natural History), London.

Gibbons, Ann. 1995. 'When it comes to evolution, humans are in the slow class', *Science*, 267, 1907–1908.

Gill, Martin. 1999. 'Fisheries', in *Britannica Book of the Year 1999*. Encyclopedia Britannica, Chicago, p. 130.

Glob, P.V. 1969. *The Bog People*. Faber and Faber, London.

Goldstein, Jerome (ed.). 1978. *The Least Is Best Pesticide Strategy*. The JG Press, Emmaus, PA.

Hammerstein, Peter and Hoekstra, Rolf E. 1995. 'Mutualism on the move', *Nature*, 376, 121–122.

Harris, David R. 1996. 'The origins and spread of agriculture and pastoralism in Eurasia: an overview', in Harris, David R. (ed.) *The Origins and Spread of Agriculture and Pastoralism in Eurasia*. University College London Press, pp. 552–573.

Haub, Carl V. 1995. 'Demography', in *Britannica Book of the Year 1995*. Encyclopedia Britannica, Chicago, pp. 255–256.
—. 1999. 'Demography', in *Britannica Book of the Year 1999*. Encyclopedia Britannica, Chicago, p. 300.

Holmes, Bob. 1995. 'Fishing treaty wins unanimous approval', *New Scientist*, 12 August 1995, p. 7.

Hyams, Edward. 1976. *Soil and Civilization*. John Murray, London.

Johnson, C.G. 1963. 'The aerial migration of insects', in Eisner, Thomas and Wilson, Edward O. (eds) *The Insects*. W.H. Freeman and Co., San Francisco.

Kendeigh, S. Charles. 1974. *Ecology with Special Reference to Animals and Man*. Prentice-Hall, Englewood Cliffs, New Jersey.

Kupchella, Charles E. and Hyland. 1986. *Environmental Science*. 2nd edn. Allyn and Bacon, Boston.

Lavallée, H.-Claude. 1999. 'Wood products', in *Britannica Book of the Year 1999*. Encyclopedia Britannica, Chicago, p. 182.

Lawton, John H. 1995. 'Population dynamic principles', in Lawton, John H. and May, Robert E. (eds) *Extinction Rates*. Oxford University Press, Oxford. pp. 147–163.

Leach, Gerald. 1975. *Energy and Food Production*. International Institute for Environment and Development, London.

Loder, Natasha. 1999. 'Journal under attack over controversial paper on GM food', *Nature*, 401, 731.

Malthus, Thomas. 1798. *An Essay on the Principle of Population As It Affects the Future Improvement of Society*. Penguin edition 1985.

Marshall, Eliot. 1995. 'Gene therapy's growing pains', *Science*, 269, 1050–1055.

May, Robert M., Lawton, John H., and Stork, Nigel E. 1995. 'Assessing extinction rates', in Lawton, John H. and May, Robert E. (eds) *Extinction Rates*. Oxford University Press, Oxford.

Mlot, Christine. 1995. 'In Hawaii, taking inventory of a biological hot spot', *Science*, 269, 322–323.

Moffat, Anne Simon. 1995. 'Exploring transgenic plants as a new vaccine source', *Science*, 268, 658–660.
—. 1999. 'Crop engineering goes south', *Science*, 285, 370–371.
—. 1999a. 'Engineering plants to cope with metals', *Science*, 285, 369–370.

Morell, Virginia. 1995. 'Cowardly lions confound cooperation theory', *Science*, 269, 1216–1217.

Myers, R.A., Barrowman, N.J., Hutchings, J.A., and Rosenberg, A.A. 1995. 'Population dynamics of exploited fish stocks at low population levels', *Science*, 269, 1106–1108.

National Academy of Sciences. 1972. *Genetic Vulnerability of Major Crops*. NAS, Washington.

Nowak, Martin A., May, Robert M., and Sigmund, Karl. 1995. 'The arithmetics of mutual help', *Scientific American*, June 1995, pp. 50–55.

Nowak, Martin A. and Sigmund, Karl. 1992. 'Tit for tat in heterogeneous populations', *Nature*, 355, 250–252.
—.1993. 'A strategy of win-stay, lose-shift that outperforms tit-for-tat in the Prisoner's Dilemma game', *Nature*, 364, 56–58.

Oldroyd, D.R. 1980. *Darwinian Impacts*. Open University Press, Milton Keynes.

Parkinson, Claire L. 1997. *Earth From Above: Using Color-Coded Satellite Images to Examine the Global Environment*. University Science Books, Sausalito, CA.

Partridge, Michael. 1973. *Farm Tools Through the Ages*. Osprey Publishing, Reading.

Patterson, Colin. 1978. *Evolution*. Routledge and Kegan Paul in association with the British Museum (Natural History), London.

Pearce, Fred. 1999. 'Counting down', *New Scientist*, 2 October 1999, pp. 20–21.

Pellew, Robin. 1995. 'Biodiversity conservation – why all the fuss?', *RSA Journal*, CXLIII, Jan–Feb, p. 54.

Persson, Reidar. 1995. 'Myths and truths on global deforestation', *IRD Currents*, May 1995, Swedish University of Agricultural Sciences, Uppsala. pp. 4–6.
——. 1995a. 'Tropical plantations – success or failure?', *IRD Currents*, May 1995. Swedish University of Agricultural Sciences, Uppsala. pp. 7–9.

Pimm, Stuart L., Russell, Gareth J., Gittleman, John L., and Brooks, Thomas M. 1995. 'The future of biodiversity', *Science*, 269, 347–350.

Plucknett, Donald L. and Winkelmann, Donald L. 1995. 'Technology for sustainable agriculture', *Scientific American*, Sept. 1995, p. 150.

Preston-Mafham, Rod and Preston-Mafham, Ken. 1984. *Spiders of the World*. Blandford Press, Poole, Dorset.

Rackham, Oliver. 1976. *Trees and Woodland in the British Landscape*. J.M. Dent and Sons, London.

Ridley, Mark. 1985. *The Problems of Evolution*. Oxford University Press, Oxford.

Rose, Steven. 1995. 'The rise of neurogenetic determinism', *Nature*, 373, 380–382.

Simmons, I.G., Dimbleby, G.W., and Grigson, Caroline. 'The Mesolithic', in Simmons, Ian and Tooley, Michael (eds) 1981. *The Environment in British Prehistory*. Duckworth, London.

Staskawicz, Brian J., Ausubel, Frederick M., Baker, Barbara J., Ellis, Jeffrey G., and Jones, Jonathan D.G. 1995. 'Molecular genetics of plant disease resistance', *Science*, 268, 661–667.

Stock, Christina. 1995. 'Flying in the fact of tradition', *Scientific American*, June 1995, pp. 11–12.

Stone, Richard. 1995. 'Taking a new look at life through a functional lens', *Science*, 269, 316–317.

Thompson, Dick. 1995. 'Soil — financial asset or environmental resource?', *RSA Journal*, CXLIII, 5461, Royal Society of Arts, London. pp. 56–67.

Thompson, Harry V. and Worden, Alastair N. 1956. *The Rabbit*. Collins New Naturalist, London.

Tolba, Mostafa K. and El-Kholy, Osama A. 1992. *The World Environment 1972–1992*. Chapman and Hall, London, on behalf of UNEP.

Townshend, J.R.G., Tucker, C.J., and Goward, S.N. 1993. 'Global vegetation mapping', in Gurney, R.J., Foster, J.L., and Parkinson, C.L. (eds) *Atlas of Satellite Observations Related to Global Change*. Cambridge University Press, Cambridge.

Tranter, Neil. 1973. *Population Since the Industrial Revolution: The Case of England and Wales*. Croom Helm, London.

Tudge, Colin. 1993. *The Engineer in the Garden*. Jonathan Cape, London.

Tuomisto, Hanna, Ruokolainen, Kalle, Kalliola, Risto, Linna, Ari, Danjoy, Walter, and Rodriguez, Zoila. 1995. 'Dissecting Amazonian biodiversity', *Science*, 269, 63–66.

Vaeck, Mark, Reynaerts, Arlette, Hofte, Herman, Jansens, Stefan, De Beuckeleer, Marc, Dean, Caroline, Zabeau, Marc, Van Montagu, Marc, and Leemans, Jan. 1987. 'Transgenic plants protected from insect attack', *Nature*, 328, 33–37.

Webb, S. David. 1995. 'Biological implications of the middle Miocene Amazon seaway', *Science*, 269, 361–362.

Whitmore, T.C. 1975. *Tropical Rain Forests of the Far East*. Oxford University Press, Oxford.

Wilson, Edward O. 1992. *The Diversity of Life*. Penguin Books, London.

Wilson, Michael A., Hillman, John R., and Robinson, David J. 1999. 'Genetic modification in context and perspective', in Morris, Julian and Bate, Roger (eds) *Fearing Food*. Butterworth-Heinemann, p. 67.

Environmental Management

When you have read this chapter you will have been introduced to:

- wildlife conservation
- the history of zoos, nature reserves, and the idea of wilderness, and controversies surrounding them
- pest control
- restoration ecology
- world conservation strategies
- pollution control
- transnational pollution

57 Wildlife conservation

Consider a population of a certain species that occupies a particular range. The population is distributed fairly evenly throughout the range and utilizes the whole of it. Then something happens to fragment the range. Perhaps a network of roads is made through it, or parts of it are ploughed for agriculture or afforested, or rivers intersecting the range become so polluted that individuals drinking from them or trying to swim across them are killed. Whatever the cause, and human activities of one kind or another are nowadays the most frequent, the effect is to divide the population into several groups. These are isolated from one another by barriers they cannot cross.

They cannot cross them, but other things can. Suppose, after a year or two, there is a drought or an unusually severe winter, or perhaps a disease transmitted by insects, or some other chance occurrence that affects all the separate groups and kills many individuals. The population is now much more severely fragmented, its groups very isolated, and each of them may comprise too few individuals to constitute a viable breeding population. Such a sequence of events, illustrated schematically in Figure 6.1, is quite common and leads to the extinction of that species within

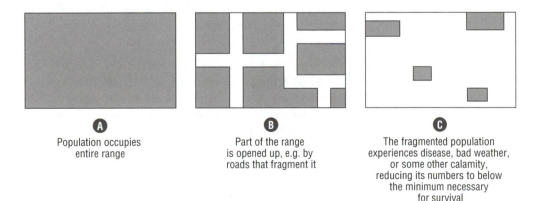

A	**B**	**C**
Population occupies entire range	Part of the range is opened up, e.g. by roads that fragment it	The fragmented population experiences disease, bad weather, or some other calamity, reducing its numbers to below the minimum necessary for survival

Figure 6.1 **Effects on a population of fragmentation of habitat**

that range. It explains why conservationists place so much emphasis on the need to preserve habitats as the best means to ensure the survival of species.

Loss or fragmentation of habitat is a common reason for extinction, but traditionally conservation efforts have been directed toward species. It is species that are considered to be endangered, rather than habitats. This is reflected in the Red Data Books, started in the 1960s by the International Union for Conservation, Nature and Natural Resources (IUCN also known as the World Conservation Union) (*www.iucn.org/icon_index.en.html*), and the World Conservation Monitoring Centre (WCMC), which introduced what is still one of the principal schemes for classifying 'rarity', the other being the Endangered Species Act 1973, in the United States. The IUCN classification, which is currently being revised, categorizes species by the degree of threat facing them. Categories include 'possibly extinct', 'endangered' for those likely to become extinct if present threats continue, 'vulnerable' for those likely to become endangered if present threats continue, 'rare' for those that are uncommon but not necessarily at risk, 'no longer threatened' for those from which threats have receded, and 'status unknown'.

The scheme has succeeded admirably in drawing attention to the species it lists, but on other grounds it is hardly satisfactory. It is biased heavily toward the better-known species, and new species are added as field biologists report them, rather than on the basis of comprehensive reviews. Only birds have been studied fully. For the remainder, the status of about half of all mammal species has been considered, probably less than 20 per cent of reptiles, 10 per cent of amphibians, 5 per cent of fish, and still fewer of invertebrates (MACE, 1995). As an alternative, it has been suggested that all species be regarded as endangered in the absence of clear evidence to the contrary, but such a scheme would not avoid the need for much more detailed information regarding the less familiar species that limits the value of the Red Data Book (*www.wcmc.org.uk/data/database/rl_anml_combo.html*) approach. Nor do the Red Data Book or Endangered Species Act propose any time-scale for the threats they list, a vagueness that leaves them open to varying interpretations.

Perhaps it is a mistake to concentrate on species, a concept that may be at once too precise and too imprecise to be helpful. Its excessive precision makes it unworkable, because biologists know far too little about most species to be able to apply it in sensible conservation programmes. They opt instead for the conservation of the habitats in which particular species occur. This is a more practicable approach, although one not immune from controversy.

The imprecision of the species concept is revealed at the genetic level. Advances in genetics have led to the concept of the gene pool, which is defined as the complete assemblage of genetic information possessed by all the reproducing members of a population of sexually reproducing organisms. Many conservation biologists now maintain that it is gene pools which should be conserved, rather than species.

In most cases it is not too difficult to decide what constitutes a species (but see section 50, on Biodiversity). Humans are sufficiently different from all other animals to be classified as a species, for example, as are house mice, blackbirds, red admiral butterflies, seven-spot ladybirds, and countless more. Genetically, it is more complicated, and a species is defined by a supposedly typical representative. We are told, for example, that the genetic difference between an average human and an average chimpanzee is smaller than the difference between two humans at the extreme limits of human variability. Humans and chimpanzees differ in less than 1 per cent of their genetic material (in fact about 0.6 per cent), a genetic distance that places them well within the range of sibling species. Taxonomically, there is a strong argument for placing both species within the same genus (PATTERSON, 1978, p. 173). Were humans in need of conservation, we would need to decide whether the preservation of, say, the population of Cumbria, England, would meet the case. Yet Cumbrians are not genetically identical to Devonians, let alone to the inhabitants of more distant parts of the world, although humans comprise only one species. The species, then, is a convenient but rather crude categorization.

Figure 6.1 shows how the fragmentation of a range may leave a population as small, isolated groups that are no longer reproductively viable. Figure 6.2 shows the possible consequences of such fragmentation on the gene pool. The diagram shows a habitat shared among three species. Members of these species intermingle to a limited extent by moving from one part of the habitat to another. Species 1 and 2 each consist of three populations, and species 3 of four populations. Populations of a species can interbreed, but they are not genetically identical, so there is much more movement among populations (shown only for species 1). The populations of species 1 occupy separate areas, but those of species 2 and 3 occupy areas that meet (b and c of species 2), overlap, or are contained one within another (a and c of species 3). Situations like this are not unusual, especially among marine species, and raise the question of just what it is that species conservation aims to conserve. It is an acute problem with whale conservation (DIZON ET AL., 1992).

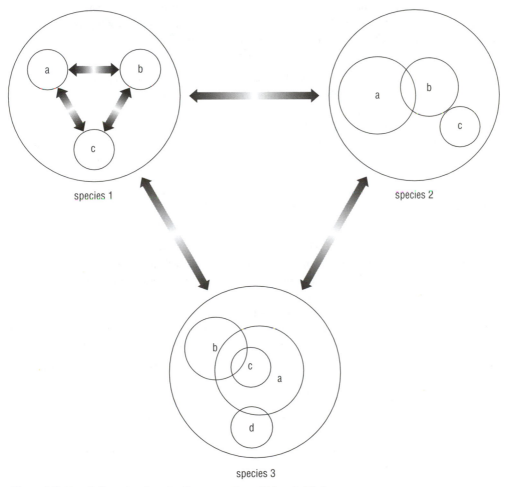

Figure 6.2 **Population structure for three species within a habitat**

Suppose habitat fragmentation destroyed part of the area occupied by one of these species in a way that isolates one or two of the populations. This will produce several gene pools that are impoverished in respect of the total gene pool for all populations. Within each of these gene pools there will be recessive alleles, some of them deleterious. While individuals could mate with members of other populations, most offspring were heterozygous for those genes, so the advantageous dominant allele was the one expressed. In the depleted gene pool, however, recessive alleles have a greater chance of meeting and, of course, they will be expressed in offspring homozygous for those genes. This is the most likely cause of inbreeding depression, and over several generations it reduces population size through early death and infertility. It is usually difficult to calculate how large a population must be to avoid inbreeding depression, but there can be no doubt that provided threats are removed, if the population is genetically healthy its numbers will regulate themselves and it will be safe.

Faced with the risk of inbreeding depression, it is tempting to introduce individuals of the same species from another region, perhaps from another part of the world entirely. This raises a new risk, albeit a less common one, of excessive outbreeding. Some years ago, individuals belonging to two Middle Eastern subspecies of ibexes (*Capra ibex aegagarus* and *C. i. nubiana*) were released in what is now Slovakia in the hope of invigorating the Tatra mountain ibex (*C. ibex*), which had been hunted to extinction but reintroduced from Austria. The subspecies interbred successfully enough, producing fertile hybrids, but whereas the native ibex mated in winter, giving birth to young when food was abundant, the hybrids mated in autumn. The young were born in winter, died, and the population became extinct (COCKBURN, 1991, p. 297).

Despite the risks, species can sometimes be rescued from the very brink of extinction, provided the causes of their decline are clearly identified. The North American bison, or buffalo (*Bison bison*), is a well-known example. With a range that once extended from northern Canada to Mexico and an estimated population of 75 million, by 1883 commercial hunting for meat and hides had reduced the species to about 10 000 individuals. From this low level the breeding of captive animals increased numbers. Some are in private herds and others have been reintroduced, as half-wild animals, to the National Bison Range in Montana and elsewhere (BREWER, 1988, pp. 605–606). Similar programmes have also saved the European bison, or wisent (*Bison bonasus*), herds of which now live in various parks and wild in the Białowieska Forest, Poland (NOWICKI, 1992, pp. 10–11).

Most of the arguments in favour of wildlife conservation are economic, as they have always been. It may be that among the species of which at present we know little there are some that may one day be domesticated for food or other commodities, or yield pharmaceutical or other valuable products. We should not deny our descendants the right to choose whether such species should be exploited. This is an apparently objective argument, but one that is likely to carry little weight with economists, who generally disapprove of investments based on nothing more substantial than the hope that benefits may accrue at some time in the future to people who are not yet even born.

Others offer an aesthetic argument. The world would be a poorer place without the pleasures of watching birds and butterflies, the sight of a meadow ablaze with flowers, the sound of birdsong. Arguments along these lines sound weak, but in fact are strong, because most of us sympathize with them. Unfortunately, however, they begin to weaken as the defence moves away from the most popular species. It can be argued that the world would be a poorer place without slugs, malarial mosquitoes, and the HIV virus. Indeed, the argument is the same, but people may take a little more persuading of its validity.

Still other people maintain that all species have a right to live. It is an opinion which is held strongly, but it raises considerable philosophical difficulties. Do species 'live' at all, or is it individuals that live? If it is individuals, what precisely do we mean by a right to live, since all individuals must die? Is it possible to confer rights without also imposing obligations which, in this case, conflict with them? If all animals have a right to live, should not the lioness respect the rights of the gazelle?

Some environmentalists propose a contextual reason. They maintain that complex networks of ecological relationships may be disrupted by the extinction of component species and that such disruptions may have widespread and unpredictable repercussions. Those repercussions may be economic or aesthetic, but they may also be biological, possibly to the extent of reducing the capacity of the global environment to sustain humans. Is this feasible? No one can say.

Whatever the reason, most people accept that the conservation of wildlife is desirable. Achieving this objective is difficult, requiring a much deeper understanding of the natural world than we possess at present. Nevertheless, we must do what we can with such knowledge as we have, and there have been successes to encourage us.

58 Zoos, nature reserves, wilderness

Zoos have had a curious history. They began as menageries, collections of living wild animals made for various reasons. In the twelth century BC, the Chou dynasty emperor Wen maintained a collection of animals from all parts of the Chinese Empire, presumably to reflect his authority over far-flung regions with exotic fauna. Ancient Mesopotamian (FOSTER, 1999) and Egyptian rulers were especially keen on menageries and the Romans maintained huge collections, many for use in gladiatorial combat. A few ancient menageries were used to study animals, but the great majority served only as entertainment or as a source of impressive animals, often large cats, to emphasize the political power of their owner. The menagerie established by the English king Henry I (1100–35) at Woodstock, in Oxfordshire, was later moved to the Tower of London and taken from there, in 1829, to form the nucleus of the collection at Regent's Park Zoo.

Wherever zoos were opened to the public they became highly popular but, despite assertions of their educational value, they remained entertainments. The zoo was a place where parents could spend a fine afternoon with their children. To make them clearly visible at close quarters, the animals were often housed in cramped and quite unsuitable accommodation. In modern times this has led many people to denounce zoos as 'prisons' in which wild animals are cruelly confined for no valid reason.

Unfortunately there remain some disreputable zoos that justify such criticism, but the reputable zoos exist today primarily for conservation purposes. Zoos remain open to the public, partly because nowadays they really do offer educational facilities but more importantly because they depend on entrance charges to help with their operational costs. Keeping wild animals, adapted to markedly different climates and diets, is an extremely expensive business.

Botanic gardens have a parallel history. They too have developed from collections of exotica gathered by plant collectors. After appropriate acclimatization and development, many became popular garden cultivars. Nowadays, botanic gardens are also concerned primarily with conservation.

Plants and animals are protected while they remain within botanic gardens, zoos, and aquaria. If they can be bred in captivity, then it may be possible to reintroduce species to places where they have become extinct or where surviving populations are declining. There have been successful reintroductions, but there have also been failures. In the 1970s, for example, the Hawaiian goose, or ne-ne (*Branta sandvicensis*), was apparently rescued from extinction by a captive breeding programme from a small stock held by the Wildfowl Trust at Slimbridge, England, and funded by the Worldwide Fund for Nature (WWF). More than 1600 birds were released on the islands of Hawaii and Maui and the WWF claimed the release as a success (STONEHOUSE, 1981, p. 96). By the early 1990s, however, only four birds survived from the 1600 released (RAVEN *ET AL.*, 1993, p. 360). The failure was probably due to the restricted gene pool represented by the small breeding stock. The geese succumbed to inbreeding depression. Reintroductions are also likely to fail if the pressures leading to the decline of the wild population continue to operate or if, in the absence of the wild population, the habitat has been altered in ways that render it no longer hospitable. Even where these criteria are satisfied, there is a danger that in the course of its captive breeding a species will have been modified in ways that reduce its ability to survive in the wild. Animals are usually prepared for release, essentially by teaching them how to find food, shelter, and mates. Care must also be taken to ensure that captive-bred individuals do not carry diseases, acquired in captivity, to which they but not the wild populations are immune.

Questions also arise over precisely what is being captively bred for reintroduction. In the light of modern genetic understanding, the species concept is inadequate if the aim is to maintain as high a level of genetic diversity as possible. Breeding programmes for both plants and animals now involve karyotyping, the comparison of chromosomes. This can reveal differences between populations of the same species. It has led to the recognition, for example, of two genetically distinct populations of orang-utan separated by a geographical barrier, although both belong to the same species, *Pongo pygmaeus*. The distinction will be lost if the two interbreed, so it is important to reintroduce pure-bred individuals to their native populations. It has been discovered that more than 20 per cent of orang-utans in zoos are hybrids of the two populations and so, despite the rarity of this species, they are not permitted to breed (TUDGE, 1993, pp. 267–268). 'Genetic fingerprinting' is also used to categorize organisms in fine detail.

Zoos and botanic gardens do not have unlimited space to keep whole plants and animals, but there are other ways in which species can be conserved. Suitable restriction enzymes make it possible to cut DNA into small fragments which can be recombined with plasmids and inserted into bacteria that are then cultured. This technique can be used to store, as fragments, the entire genome of selected individuals as a genomic library (TUDGE, 1993, p. 212). At present it is not possible to reconstruct individuals from such a store, but one day it may become so and meanwhile their genetic material is secure.

Many rare or endangered plants are preserved in seed banks, where seeds are desiccated to a water content of about 4 per cent and stored at 0 °C, the quality of the seeds being checked from time to time by germinating them. Stored seeds usually remain viable for 10–20 years. Of course, the security of the plants depends on that of the store and there are fears that lack of funds threatens to make some seed banks into 'seed morgues' because of staff shortages and, in some cases, too small a quantity of seeds to warrant the risk of thawing and attempting to germinate them (FINCHAM, 1995). 'Recalcitrant' seeds cannot be treated in this way, because desiccation destroys them and they can be stored for only a few days. Where possible they are preserved as growing plants, but in some cases they can be held more economically as tissue cultures.

Nature reserves offer a different approach to conservation, protecting habitats directly and the species occupying them by implication. There has been much debate among ecologists over the relative merits of the wide variety of features that may qualify an area for protection as a reserve. One widely accepted aim is to establish a set of reserves representative of every type of habitat within a country or region, sites being selected on the basis of their flora, fauna, or geological features. Reserves may be publicly or privately owned and managed by agencies of national or local government or by voluntary bodies. In Britain, the Royal Society for the Protection of Birds, the Royal Society for Nature Conservation and its affiliated county naturalists' trust in England

and Wales, and the Scottish Wildlife Trust manage many hundreds of nature reserves. Because they exist solely to conserve valued areas, public access to reserves may be controlled or denied, although open public access is allowed wherever possible.

Reserves vary greatly in size, mainly because sites are acquired as opportunity arises in the form of patches of land for which landowners have no commercial use or which they are prepared to relinquish out of sympathy for the aims of conservationists. Although this is clearly the best that can be achieved, and implies no criticism of them, the somewhat haphazard patchwork of small, isolated reserves that results might be thought unsatisfactory. The link between habitat fragmentation and species extinction is well established and suggests that in the case of nature reserves, the bigger the better.

It is not necessarily so, and ecologists have not yet resolved what has been nicknamed the 'SLOSS' debate, 'SLOSS' being an acronym for 'single large or several small'. There is no general answer. Some species, such as grizzly bears and tigers, require large areas, and a large reserve is likely to support a greater number of species than a small one.

The choice, though, is not between large or small areas, but between one large reserve or several small ones with the same combined area. If small reserves are preferred, a further choice must be made, illustrated in Figure 6.3. Should the reserves remain isolated, like islands, or should they be linked by corridors? Ecological studies of actual islands and of 'islands' produced when habitats are fragmented have provided information that will provide guidance in particular situations. In the Brazilian Amazon, the fragmentation of forest into isolated patches was followed by a doubling in the number of frog species, and after seven years in their patches they seemed to be thriving. Bird and insect numbers declined, however (CULOTTA, 1995a). It has also been found that compared with a single large reserve of the same area, several small reserves between them support more species of mammals and birds in East Africa, mammals and lizards in Australia, and large mammals in the United States (BEGON ET AL., 1990, pp. 790–791).

Should 'island' reserves be isolated or linked by corridors? Since small, isolated populations may be prone to inbreeding, corridors that are ecologically similar to the islands may provide opportunities for migration, thus increasing outbreeding. In Britain, hedgerows have often been described as corridors, ecologically resembling woodland edge, linking isolated patches of woodland, and have been valued for that reason, but there is little reason to suppose they are used for migration. Corridors are narrow and an animal might be wary of moving along one for fear of predators in the hostile environment to either side. The exception to this might be large

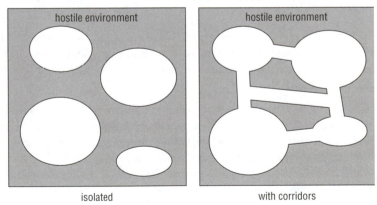

Figure 6.3 **Island wildlife refuges**

predators themselves. They routinely travel considerable distances and corridors would conveniently guide them to prey more or less trapped in the islands. Diseases and parasites might also move along corridors (BREWER, 1988, p. 636). These considerations do not detract from the value of corridors used to link otherwise separated parts of the same range. Conduits built beneath roads for the use of migrating toads and of mammals patrolling their ranges have proved very effective at reducing road fatalities.

Nature reserves protect relatively small areas of habitat. National parks protect very large areas. National parks were defined by the IUCN (International Union for Conservation, Nature and Natural Resources) in 1975 as large areas of land that have not been altered materially by human activities and are of scientific, aesthetic, educational, or recreational importance. They are managed by the state, and the public are welcome to visit them provided their activities do not conflict with conservation policies. British national parks, which were designated before the IUCN definition was written, are rather different in that most of their area is privately owned and farmed. National parks are large enough to meet the needs of many species, but even they are not big enough for some. The Yellowstone National Park, occupying almost 9000 km^2, may not provide sufficient space for its grizzly bear population and for this reason it has been proposed to link the park to several national forests and the Red Rock Lakes National Wildlife Refuge to produce a 'greater ecosystem' (BREWER, 1988, p. 637).

Finally, entire areas of wilderness may be afforded protection. A wilderness is an extensive tract that has never been occupied permanently by humans or used by them intensively and so exists in something close to a natural state. Such areas are rare in Europe, but less so in North America and other continents. Their protection includes a prohibition on the construction of roads into or through them and controls on the number of people visiting them at any one time.

Natural communities or living organisms are not static. Left to itself, a nature reserve, national park, or in some cases even a wilderness area will gradually change. Species will disappear and others replace them, possibly altering radically the character of the entire system. When grassland, including prairie, is protected from grazing and fire, for example, it tends to develop into scrub and eventually forest. This raises yet another controversy among conservationists, some holding that protected areas should be allowed to develop naturally, others that they should be managed so that they continue to support the species by which their value was defined in the first place. People who believe areas should remain unchanged from the condition they were in when their importance was first recognized are sometimes described as 'preservationists' and contrasted with 'conservationists', who seek to prevent industrial and urban development that would destroy or degrade habitat, but not to interfere unduly with other ecological changes which occur naturally.

In practice, most reserves and parks are managed. Management may involve such tasks as culling species that become too numerous, clearing waterways of plants that might choke them and deplete the amount of oxygen dissolved in their water, and allowing natural fires to take their course or even firing areas deliberately.

Different as they are, all these approaches to species conservation share the same objective and complement one another. Seed banks, gene banks, and genomic libraries store the genetic diversity of living organisms under strict control and without occupying land that might be converted to other uses regardless of the protests of conservationists. Zoos, aquaria, and botanic gardens store living plants and animals for purposes of study and, albeit controversially, as a source of individuals for reintroduction. Operating at different scales, nature reserves, national parks, and wilderness areas conserve entire biological communities.

On 1 March 1872, Yellowstone became the first national park in the world. Since then much has been learned about the need for conservation and the most appropriate means for achieving it. Scientists and managers are still learning, now more rapidly than ever before, and we may anticipate that in years to come conservation methods will continue to advance.

59 Pest control

Farmers have always had to contend with pests which feed on their crops in the field or after harvest, and for many years they have relied mainly on toxic chemicals to achieve a satisfactory level of control. In the 1930s the principal substances used were based on nicotine, arsenic, and cyanide. They were highly dangerous to humans and to wildlife, but evoked no public alarm, although crime writers were fond of using 'weedkiller' as a fictional murder weapon. A new generation of organic compounds began to replace them in the 1940s. These were much less toxic to mammals. DDT is about as poisonous to humans as aspirin, but it is a great deal more difficult to ingest a lethal dose of it.

Problems with the new insecticides soon started to emerge. As early as 1945, scientists suspected that DDT might have an adverse effect on wildlife and in 1947 seven British workers died of poisoning after working with DNOC (dinitro-*ortho*-cresol). This led to legislation controlling pesticide use, in the Agriculture (Poisonous Substances) Act 1952. During the 1950s the effects on wildlife increased and in 1961 certain substances used as seed dressings, to prevent fungal infestation of seeds prior to germination, were withdrawn (CONWAY ET AL., 1988). The publication of Rachel Carson's *Silent Spring* (CARSON, 1963), in 1962 in the United States and 1963 in Britain, aroused public awareness of the hazards associated with insecticide use, but it told scientists nothing of which they were not already aware and irritated many of them by exaggerating the seriousness of the problem.

That problem arose primarily from the biomagnification, or bioaccumulation, of chemically stable compounds as they passed along food chains, but also from their lack of specificity. Organochlorines, the first generation of organic insecticides, of which DDT is the best-known member, succeeded partly because of their persistence. Once applied, the insecticide remained on and around crop plants, to poison any insects that came into contact with it. Predators eating prey exposed to a sublethal dose accumulated the insecticide until it reached harmful concentrations. At the same time, organochlorine compounds were toxic to a wide variety of arthropods. As well as killing members of pest species they also killed arthropod predators of those species.

As Figure 6.4 shows, however, the agricultural effect of the new pesticides was dramatic. Yields rose sharply and post-harvest losses fell. In the tropics, where the climate makes food storage much more difficult than in temperate climates, rodents, insects, and fungi can destroy 8 per cent of stored potatoes, 25 per cent of cereal grains, 44 per cent of carrots, and 95 per cent of sweet potatoes before they reach the market (GREEN, 1976, p. 98).

DDT was first used not in food production, however, but to control such insect vectors of disease as the human body louse (*Pediculus humanus corporis*), which transmits typhus, and the *Anopheles* mosquitoes that transmit malaria. In 1946 there were 144 000 cases of malaria in Bulgaria and in 1969 there were 10, in Romania the number of cases fell from 338 000 in 1948 to 4 in 1969, and in Taiwan from 1 million in 1945 to 9 in 1969 (GREEN, 1976, p. 100). DDT is still used in some countries against malaria mosquitoes, but its effectiveness is restricted by the number of species that have become resistant to it. As early as 1946, houseflies in northern Sweden were immune to DDT and by the 1950s mosquitoes and lice were becoming immune in southern Europe and

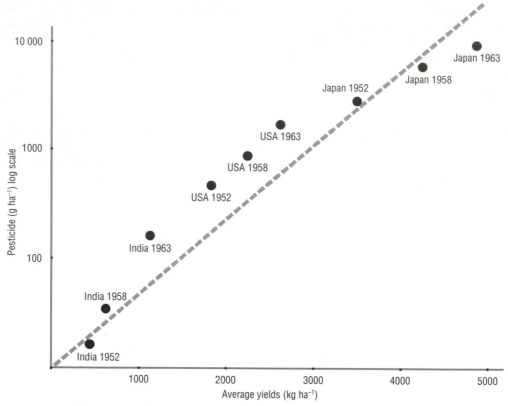

Figure 6.4 *Pesticide use and crop yield*
Source: Green, M.B. 1976. Pesticides: Boon or bane? *Elek Books, London*

Korea (MELLANBY, 1992, pp. 53–60). It was estimated that in 1980 the world was spending almost US$640 million a year to control insect disease vectors, yet 100 million new cases of malaria occur every year and almost 1 million people die (ABRAMOV, 1990).

Taken together, the adverse effects on wildlife and rapid acquisition of resistance by pest species have led many people to speculate about the possibility of abandoning entirely the use of chemical pesticides. This has not happened, of course. In 1986–87 British cereal farmers spent £110 million on herbicides, £85 million on fungicides, £4 million on insecticides, and £15 million on the treatment of seeds (TYSON, 1988). There are now literally hundreds of pesticide compounds on the market.

Alternatives to the chemical control of pests have been developed, but in parallel with developments in the formulation and application of pesticides themselves. Predictably, the highest environmental impact resulted when broad-spectrum poisons were pumped from sprayers not very different from lawn sprinklers. Crops were drenched with huge quantities of pesticide. The upper surfaces of leaves were thoroughly coated, but the undersides were largely missed and most of the pesticide fell to the ground where it poisoned harmless or beneficial organisms and could drain into waterways — and in mosquito control programmes insecticide is sprayed directly on to the surface of stagnant water to kill larvae. Pesticides also travel by air, forming microscopic aerosols that can be carried long distances.

Over the past twenty years all the industrialized countries have banned or severely restricted the agricultural and horticultural uses of organochlorine compounds. Traces of them still remain in the environment, because of their great stability, but concentrations are very low. They have been detected in ground water at more than 0.1 parts per billion in some parts of Britain (Tyson, 1988) and minute traces, at the very limit of measurement, have been found in rivers in Northern Ireland, but there they are believed to have come from the domestic use of wood preservatives, not from farms or factories (Mason, 1991, p. 179). They have been replaced by progressively more specific compounds, designed to poison only target species. At the same time, new pesticides are required to undergo very rigorous environmental testing before they are licensed for use. Testing can take five to ten years from the time a potentially useful compound is identified, during which time its fate must be traced in the soil, air, and water of every environment in the world in which it is likely to be used. Once it is marketed, its environmental effects continue to be monitored (Allaby, 1990, pp. 36–37).

More efficient application methods were also sought. The most promising of these were based on ultra-low-volume (ULV) sprayers. Some worked electrostatically, but in the simplest the pesticide is pumped from a reservoir on to the centre of a toothed disc, resembling a cog-wheel. The disc spins, spreading the pesticide to the edge where it flows along the teeth, leaving the disc as a fine stream that quickly breaks into minute droplets all of much the same size (see Figure 6.5).

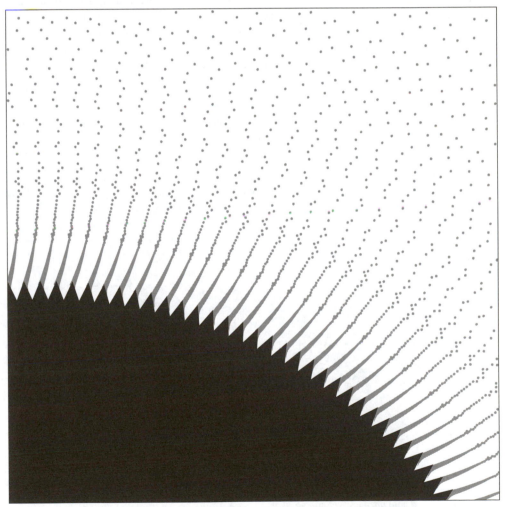

Figure 6.5 *Even-sized droplets from the teeth of an ultra-low-volume pesticide sprayer*

The droplets form a mist that drifts into the crop, coating all surfaces of the plants but without contaminating the ground. The sprayers themselves are made from plastic, the pump driven by a torch battery, and the discs can be changed to alter the size of the teeth and thus the size of the droplets, as appropriate to the pesticide and crop (see Figure 6.6). ULV sprayers achieve better pest control than conventional sprayers and use 1–10 per cent of the amount of pesticide. They must be used with care, because they require a more highly concentrated pesticide solution and so expose the operator to greater risk, but compared with other sprayers their environmental impact is greatly reduced.

Biological control offers an entirely different way of dealing with pests. It has been applied most widely to glasshouse crops, because it is in glasshouses that pests cause the most serious damage and where they most rapidly acquire resistance to insecticides. In the best-known method, the pest is attacked by its own natural parasites and predators, bred for the purpose and introduced. First, the pest is introduced to the crop and allowed to become established. This provides a food supply for the predator or parasite, which is introduced next. The pest is not eliminated, but once its population is reduced to an economically tolerable level the pest's own enemies prevent it from increasing. A range of agents for biological control are now produced on an industrial scale in many countries to deal with a number of species of mites, aphids, thrips, caterpillars, mealybugs, and others. Other pests are controlled by bacteria, fungi, nematodes, protozoa, and viruses. By 1986, 63 per cent of British glasshouse-grown cucumbers were being protected biologically from two-spotted mites and 55 per cent from whitefly, and 14 per cent and 43 per cent of tomatoes from those two pests respectively. Biological control is also used, but to a smaller extent, in fruit orchards (PAYNE, 1988).

The sterile male technique has been used against the screw worm, a fly that attacks cattle, various fruit flies, tsetse fly, cockchafer, codling moth, onion maggot, and others (LACHANCE, 1974). It involves breeding the pest species, separating the males, and sterilizing them, usually by irradiation. Then they are released to mate unproductively with females, which lay unfertilized eggs.

Pheromones are used to trap certain pests. These are chemical attractants by which males and females locate one another for mating. Synthetic pheromones released in the right place at precisely the right time draw large numbers of insects into traps, where they can be killed.

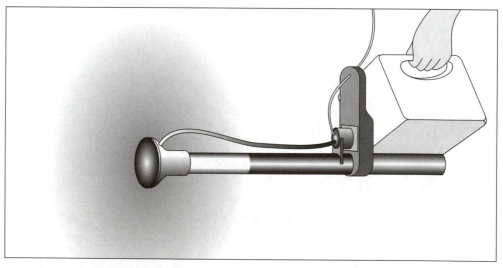

Figure 6.6 *A hand-held ultra-low-volume sprayer*

Integrated pest management

Modern pest control uses all appropriate methods, including pesticides, but is based primarily on detailed knowledge of the life cycle and behaviour of the pest. Its aim is not to eradicate the pest, but to control its population at a level below that at which economic damage becomes intolerable. It is often called integrated pest management.

The pea moth (*Cydia nigricana*) mates in summer. Its eggs hatch after 9–16 days and the larvae quickly bore into pea pods. They spend three weeks there, feeding, then fall to the ground, make cocoons, and remain in the soil until they emerge as adults the following year. They are especially vulnerable to insecticide during the 24 hours between hatching and entering pods.

Prior to mating, females attract males by releasing a scent (a pheromone). This substance has been analysed and synthesized. The synthetic pheromone is placed in traps with sticky floors, located among the rows of peas, and the grower checks the traps every day while the pea plants are in flower. On the day the traps are full of male moths, the grower knows eggs are about to be laid and, therefore, larvae will start emerging 9–16 days later. The crop is sprayed twice, one week after the males are found in the traps and again three weeks after that.

The lygus bug (*Lygus hesperus*), a serious pest of cotton, is dealt with in a similar way, by using nets to sweep the crop in search of the insects. When the ratio of bugs to cotton buds exceeds a certain threshold, the crop is sprayed.

These are examples of integrated pest management (IPM). Its success requires detailed knowledge of the pest and its ecology, and workers trained to monitor populations reliably. These difficulties are not insuperable, but they have delayed the widespread adoption of IPM.

Synthesized compounds that mimic juvenile hormones have also been tried. Juvenile hormones are produced by insects while they are immature. When they cease to produce them they mature. If they are exposed to compounds with a similar effect, they fail to mature and so do not mate.

Pests, weeds, and plant and livestock parasites must be controlled. In the world as a whole, the need to increase the amount of food available means that losses from all causes must be minimized. Even in the industrialized countries, where farmers are capable of producing more food than their own markets demand, a relaxation of control would not be acceptable. As yields fell, more land would be needed for cultivation, and it is the practice of agriculture itself that has the most serious effect on wildlife habitats and the environment generally. Reduced yields would also mean that food prices would rise. Abandoning control is not an option and would not necessarily bring any environmental improvement.

This means that pesticides will remain in use for many years to come, but in reducing amounts of less environmentally hazardous compounds. Better application methods will allow adequate control to be achieved with less pesticide, and alternatives to chemical control, including those made possible by genetic manipulation, will become available for an increasing range of targets. Meanwhile, pesticides themselves are far more specific than they were and great care is taken to ensure they cause no harm to non-target species.

In the past, pesticides have caused environmental damage. This is already much reduced and in the future we may expect it to fall still further. These necessary improvements have resulted from detailed studies of the biology and ecology of pest species that allow infestations to be identified early and the pests to be attacked with considerable precision. Increasingly, the development of crop varieties that are genetically modified to render them tolerant of herbicides and resistant to insect pests and to viral and fungal diseases will allow the use of pesticides to decline in years to come.

60 Restoration ecology

Environmentalists sometimes complain that once an area becomes degraded and its environmental quality reduced, it is lost for ever. This makes a good campaigning argument, but it is untrue. Many environments will recover naturally in time and others will develop into new environments no less interesting and valuable than those they have replaced. Long-abandoned quarries are often of considerable ecological and geological interest. Even land poisoned by industry may eventually acquire new ecological value. Until 1919, for example, household washing soda was manufactured by the Leblanc process. The British soda industry was concentrated in Lancashire, where it caused appalling air pollution and generated large amounts of toxic and very alkaline wastes, with a pH of nearly 14. These were dumped, in some places forming a layer several metres deep. Then the Leblanc process was replaced by the Solvay process. This also produces alkaline wastes, but they have been disposed of more carefully. After seventy years, the old Leblanc waste sites have weathered until now they support a wide variety of lime-loving plants, including many orchids, and are of considerable botanical importance (MELLANBY, 1992a, pp. 64–66).

Human intervention can restore other sites to their original state, or to something closely approaching it, or rehabilitate them to a state different from the original, but much better than their degraded state. Spoil heaps from mining can be transformed into areas supporting a rich flora appropriate to the surrounding environment, but not necessarily identical to the communities the land supported originally.

Restoration and rehabilitation call for a detailed understanding of community ecology. In most cases, ecologists concentrate on plant ecology, because if viable plant communities can be established animals typical of such communities will enter them of their own accord. The branch of ecology specializing in this work is called restoration ecology (*darwin.bio.uci.edu/~sustain/ EcologicalRestoration/index.html*) and it has been described as the 'acid test' for ecology, because it calls on ecologists not simply to take ecosystems to pieces to see how they work, but to assemble them and make them work (BEGON *ET AL.*, 1990, p. 607).

Restoration ecologists all over the world are watching the progress of the largest restoration project ever attempted. If it succeeds there are many places where it may be repeated. It began in the mid-1990s in the Everglades.

At one time, most of this large area in southern Florida (see Figure 6.7) was flooded for at least eight months of the year and much of it for more than that. Every year, during the wet season Lake Okeechobee overflowed and water flowed slowly south in what was effectively a river 80 km wide and less than 1 m deep, covered in algae and passing through swathes of saw grass. It was called a 'river of grass'. The swamp environment was rich in wildlife but inhospitable to humans, and about half the area was drained early in this century. Then, in the 1960s, the US Corps of Engineers began building 1600 km of channels with levees to carry the water away more quickly, some of it to be stored in 'water conservation areas'.

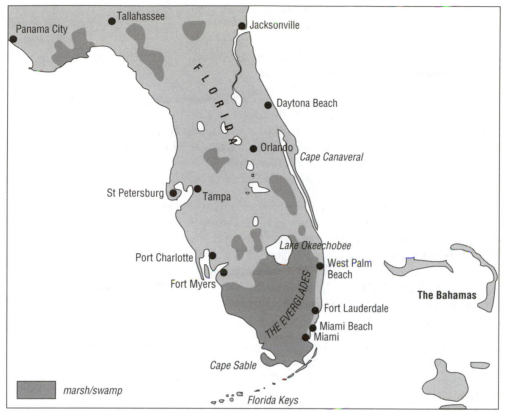

Figure 6.7 *Florida, showing the location of the Everglades*

No one predicted the extent of the consequences. The marshes dried, and became more saline, and populations of wading birds and other vertebrates fell by up to 90 per cent. Water in Florida Bay became anoxic, threatening fish stocks, and it was feared that if the depletion of aquifers continued it might lead to water shortages in the cities they supply.

Restoration involves lowering the levees, changing the straight channels back into river meanders, and eventually, over the next 15–20 years at a cost of about $2 billion, returning the area to natural wetland that floods and drains according to the rainfall. The work is being carried out by the Corps of Engineers in collaboration with a number of federal and state agencies and the Everglades National Park, and it proceeds slowly and cautiously. It is especially important to monitor closely the quality of the water that is released to flow into the area. If the water is polluted, the contamination could affect a wide area and wildlife would not return. Basing the operation on 'adaptive management', the planners deal with one small area at a time and make a minor change then wait to see what happens before proceeding further (CULOTTA, 1995).

Not all restoration involves major environmental engineering. It can be subtle and invisible to the naked eye, especially where the purpose is to remove pollutants by 'bioremediation'. The industrial detergents used to emulsify the crude oil after the first major oil-spill incident, when the *Torrey Canyon* went aground off the Isles of Scilly in 1967, may have done more harm to marine organisms than the oil itself. Since then much has been learned about the fate of oil in the sea. In particular, it has been found that over 30 genera of bacteria and fungi feed on hydrocarbons, converting them to carbon dioxide, water, and their own cell matter. These microorganisms are common in most environments and as long ago as 1973 some biologists were suggesting their help might be enlisted in dealing with oil spills. There were several successful trials, but it was not until March 1989 that this idea could be really tested. That was when the *Exxon Valdez* spilled 41 million litres of crude oil into Prince William Sound, Alaska, contaminating approximately 2000 km of the intertidal zone along the rocky coast.

No microorganisms were introduced, but the growth of those already present was stimulated by adding fertilizer to provide the nutrients the oil could not supply. The fertilizer, amounting to about 50 tonnes of nitrogen and 5 tonnes of phosphorus, was applied in the summers between 1989 and 1992. Careful monitoring and comparisons between treated and untreated sites indicated that the treatment was effective, produced no adverse side-effects, and that it might have been even more successful with higher fertilizer applications (BRAGG *ET AL.*, 1994).

Plants can also be used to remove pollutants. Reed beds are being established in some places to purify water before its discharge into rivers or lakes. The reeds (*Phragmites communis*) are planted into gravel or soil in a pit sealed with an impervious liner. The plants transport oxygen to the root area, where water flowing through the pit is purified by aerobic bacteria, solid wastes are composted in the layer of dead leaves and stems from the reeds, and water is treated by anaerobic bacteria in the surrounding soil (MASON, 1991, p. 69). Reeds are used in this way to treat sewage, nutrient-enriched water leaching from farmland, and water contaminated with metals that drains from abandoned mine workings.

This use of plants is called 'phytoremediation' and it can also be applied in terrestrial habitats. *Brassica juncea*, a variety of mustard, accumulates such metals as selenium, cadmium, nickel, zinc, chromium, and lead. Under field conditions, after several years it had reduced selenium levels by up to 50 per cent in the uppermost metre of soil.

In trials at Rothamsted Experimental Station, England, alpine pennycress (*Thlaspi caerulescens*) was found to accumulate zinc and cadmium until these metals accounted for several percent of the weight of the plants. *Thlaspi caerulescens* can tolerate the poisons because it and other plants produce phytochelatins, small peptide molecules that bind metals in less toxic forms and, in some plants, transport them into cell vacuoles where they are stored safely. Mercuric reductase, an enzyme that detoxifies mercury, is produced by certain bacteria, and the gene encoding it has been transferred to *Arabidopsis thaliana* (thale cress) plants. Thus transformed, the plants grew in a solution of mercuric chloride that killed other plants. They convert the mercuric chloride into elemental mercury, which they release slowly into the air as mercury vapour, at what biologists hope are harmless concentrations. Selenium is also released into the air by cabbage, broccoli, and some other plants, as dimethyl selenide, which is harmless (MOFFAT, 1995).

Plants with phytoremediation potential are especially common in the tropics and subtropics, possibly because the toxic metals protect them against herbivorous insects and microbial parasites. There is a risk that the plants might also poison small mammals, but on contaminated land, where they would be grown, those mammals are already in danger. The plants are harvested and are then either dried and buried, or burned. Energy from the burning of the biomass fuel can be sold and the ash from them contains the metals they accumulated. Some of these are of commercial value and can be sold. Sales of energy and metals offset much of the cost of this treatment and may even make it profitable. This is the technique currently used to treat contaminated soil. Dried plants and ash have much less mass than the soil from which the metals were removed, so if they cannot be sold burying them costs much less than excavating the soil and disposing of it.

Plants can be used to obtain metals, including thallium and gold, from low-grade deposits. This is called 'phytomining' and it causes far less environmental disturbance than conventional mining (BROOKS ET AL., 1999).

Excavation and burial is the alternative to phytoremediation for restoring mine spoil and tailings heaps and many abandoned industrial sites. Antinuclear campaigners often criticize the cost and technical difficulties inherent in the decommissioning of nuclear plants that have reached the end of their useful lives, but these are well known and modest compared with the those of decommissioning other industrial installations.

At Oakville, Ontario, in 1983, falling demand for petroleum led to the closure of a Shell refinery that had been processing 44 000 barrels (about 8 million litres) of oil a day. It took Shell six years and cost an estimated Can$4 million to restore the 222-ha site to residential and commercial use. All the remaining oil was removed and the plant and buildings dismantled. Wells were dug to monitor ground water, the soil was analysed and either treated to clean it or excavated and removed, and the soil population studied to determine whether the soil could support plants. This was the first refinery site to be restored, but it will not be the last (ALLABY, 1990, p. 102).

At one time, factories were not decommissioned when they were no longer needed; they simply closed, often because they had failed and their owners were bankrupt. Anything that could be sold was removed, but the rest was left to decay. Even if the buildings found new uses, no thought was given to the ground around them, where for many years metals may have been stored and chemicals spilled. In the early 1990s, the British government proposed the compilation of a register of such industrially contaminated land at 100 000 sites. When it was realized that this would seriously inhibit attempts to regenerate inner cities by developing those sites, the scope of the proposed register was first reduced and finally, on 17 February 1993, the plan was abandoned altogether. Much of the environmental degradation we are now trying to remedy has been inherited for this reason. The problem will diminish over the years, as old sites are restored, and under present planning laws permission for the industrial use of land is not granted unless a detailed, funded scheme for site restoration is included in the application, with firm assurances that it will be implemented.

Restoration ecology, and the bioremediation and phytoremediation techniques it employs, make this a practicable requirement. As restoration ecologists learn more about the way viable biological communities can be established on previously degraded land, even the most recalcitrant sites may recover. At the same time, restoration ecology provides the understanding that allows restoration plans to be structured into the life history of present and future industrial operations.

61 World conservation strategies

By the late 1960s it was clear to those engaged in the emerging environmental movement that the world faced problems which could be resolved only at a global level, an idea that quickly resonated with public opinion. The issues arising from the combined effects of population growth, resource depletion, and environmental degradation, were explored in countless books and articles and summarized, perhaps most lucidly, by Paul and Anne Ehrlich in *Population, Resources, Environment*, a book they published in 1970 with a revised edition in 1972 (EHRLICH AND EHRLICH, 1972). 'A Blueprint for Survival', published as the entire January 1972 issue of *The Ecologist* magazine (GOLDSMITH ET AL., 1972), attracted much attention, as did *The Limits to Growth*, the popular report of a computer model of the world, sponsored by the Club of Rome and also published in 1972 (MEADOWS ET AL., 1972).

Such publications reflected the intense intellectual fervour that formed the background to the first major conference on a single topic to be held by the United Nations, in Stockholm in June 1972. The UN Conference on the Human Environment was attended by delegations from almost all member states, with the exception of the USSR and its East European allies, as well as non-governmental groups, who held meetings and events of their own. A team from *The Ecologist* and Friends of the Earth published a daily newspaper, *The Stockholm Conference Eco*, which was distributed by bicycle to the hotels where delegates were staying; after its first few days permission was granted for it to be handed out in the conference centres.

A book setting out the issues to be debated was commissioned by the Secretary-General of the conference, Maurice Strong (WARD AND DUBOS, 1972). Somewhat less apocalyptic in tone than other environmentalist literature, it ended with a chapter on 'strategies for survival'. This emphasized the need for sovereign nations to collaborate in research and programmes of action. 'If this vision of unity — which is not a vision only but a hard and inescapable scientific fact — can become part of the common insight of all the inhabitants of Planet Earth, then we may find that, beyond all our inevitable pluralisms, we can achieve just enough unity of purpose to build a human world' (WARD AND DUBOS, 1972, p. 297).

The Stockholm Conference produced a Declaration on the Human Environment, adopted by the General Assembly in 1973, and led to the establishment, also in 1973, of the UN Environment Programme (UNEP), based in Nairobi. This was an entirely new UN agency, charged with coordinating global monitoring of the environment and international programmes for environmental improvement. Over the more than twenty years of its existence, UNEP has brokered treaties and conventions on a wide range of topics, from pollution reduction in partially landlocked seas (the Regional Seas Programme) to the 1987 Montreal Protocol on Substances that Deplete the Ozone Layer, and the 1992 United Nations Framework Convention on Climate Change (*www.unfcc.de/*) with the Kyoto Protocol that was added to it in 1997.

Progress was clearly being made, but there was still no broad framework of defined objectives against which individual schemes could be evaluated. The task of developing one was assumed by the International Union for Conservation of Nature and Natural Resources (IUCN), based in Switzerland, with Robert Allen, a former editor of *The Ecologist*, as its compiler and editor. A large team of ecologists, conservationists, and writers contributed ideas and outlines. UNEP and the World Wildlife Fund (WWF, now the Worldwide Fund for Nature) cooperated and gave financial assistance, and the document was prepared in collaboration with the Food and Agriculture

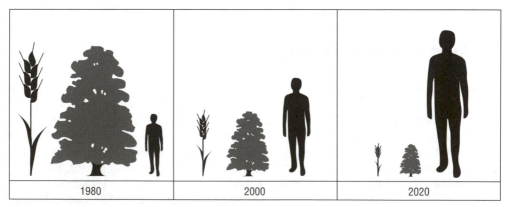

| 1980 | 2000 | 2020 |

Figure 6.8 **Living resources and population**
Source: World Conservation Strategy, Introduction: Living resource conservation for sustainable development. IUCN, UNEP, and WWF, 1980

Organization of the UN (FAO) and the UN Educational, Scientific and Cultural Organization (UNESCO). The result was the *World Conservation Strategy: Living Resource Conservation for Sustainable Development* (IUCN, 1980), published in March 1980 as an Executive Summary, Preamble and Guide, and Map Section, as separate booklets accompanying the *World Conservation Strategy* itself, all contained in a folder. It was directed to government policy-makers and their advisers, conservationists, and all those involved professionally in economic development.

The *World Conservation Strategy* took as axiomatic what by then had become the orthodox environmentalist diagnosis, that as human numbers continue to increase, each person will be entitled to a dwindling share of the resources upon which human life depends. This prognosis was presented graphically as a man growing bigger between 1980 and 2020, while a tree and wheat plant beside him grew smaller (see Figure 6.8). It also pointed out that access to resources is not shared equally, again illustrated graphically as 1 Swiss person being equivalent to 40 Somalis in terms of resource consumption (see Figure 6.9).

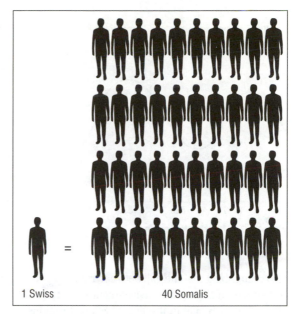

Figure 6.9 **Resource consumption by rich and poor**
Source: World Conservation Strategy, Introduction: Living resource conservation for sustainable development. IUCN, UNEP, and WWF, 1980

Having outlined problems arising from the loss or degradation of agricultural soils, forests, coastal wetlands and freshwater systems, and genetic diversity, the *Strategy* set out a list of objectives and action that might be taken at the national and international level. This included a recommendation that each country produce a national or several subnational strategies of its own, modelled on the *World Strategy*.

Britain was one country which did. Following the pattern of the *World Conservation Strategy*, it was published in 1983 as three documents: a brief summary, an overview, and the the full report of nearly 500 A4 pages (WWF *ET AL.*, 1983). The British response dwelt on the 'post-industrial society' that appeared to be emerging and urged the rebuilding of those industries which meet 'real' domestic and export needs (WWF *ET AL.*, 1983, para. 7). It based this call on its judgement of the position of the economy in the fourth Kondratieff cycle. This was the approximately 50-year alternation of periods of prosperity and decline identified by the Russian economist Nikolai Kondratieff (1892–*c*. 1931), illustrated in Figure 6.10. At the time the report was prepared, the British economy was clearly in decline and it was only by assuming the validity of the Kondratieff model that the point could be estimated at which full-scale depression would be reached and followed by recovery. What the model did show, however, was a slow but steady economic advance which meant people were a little more prosperous at each peak and a little less poor at each trough than they had been at the peak and trough of the preceding cycle.

Both reports sought ways to achieve steady economic growth without causing injury to the natural environment, especially the countryside and wildlife, and without so depleting the resources on which industry and human welfare depend as to block growth at some time in the future. They

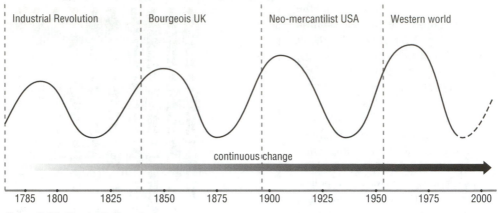

Figure 6.10 **Kondratieff cycles**
After Introduction to **The Conservation and Development Programme for the UK. 1983. Kogan Page, London**

also recognized that the gap between rich and poor countries, and groups within countries, must be reduced if the global environment is to be protected adequately. The most acute problems were seen to reside in the less developed regions of the world. Without development poverty would continue to exacerbate them.

This view provided the starting point for another influential report that approached the world situation from a different point of view. The Brandt report (INDEPENDENT COMMISSION ON INTERNATIONAL DEVELOPMENT ISSUES, 1980), produced by a commission of senior politicians and economists led by Willy Brandt, a former chancellor of West Germany, drew attention to the pressure on resources that could result from population increase (INDEPENDENT COMMISSION ON INTERNATIONAL DEVELOPMENT ISSUES, 1980, pp. 105–116), although it did not predict a depletion of mineral resources generally, only of oil supplies. It argued that economic development could trigger a demographic transition, and was concerned about the large-scale migrations it believed population growth would cause and the fate of migrants. It diagnosed the economic disparity between rich and poor as the gravest problem facing the world and its elimination through development as the solution. Its recommendations were directed to this end.

The *World Conservation Strategy* and its British sequel placed great emphasis on 'sustainability'. This was not a novel concept, but they drew it to the attention of the politicians to whom their reports were addressed, and it was the report of yet another international commission (WORLD COMMISSION ON ENVIRONMENT AND DEVELOPMENT, 1987) that finally brought the word into common use. The World Commission on Environment and Development, or Brundtland Commission after Gro Harlem Brundtland, its leader, drew together and aimed to reconcile the two strands of environmental conservation and economic development, and supplied what came to be the generally accepted definition of 'sustainability': 'Sustainable development is development that meets the needs of the present without compromising the ability of future generations to meet their own needs' (WORLD COMMISSION ON ENVIRONMENT AND DEVELOPMENT, 1987, p. 43). 'Sustainable development', the Report said a few pages later, 'requires the conservation of plant and animal species. ... [and] requires that the adverse impacts on the quality of air, water, and other natural elements are minimized so as to sustain the ecosystem's overall integrity.' This led to an expansion of the definition:

> In essence, sustainable development is a process of change in which the exploitation of resources, the direction of investments, the orientation of technological development, and institutional change are all in harmony and enhance both current and future potential to meet human needs and aspirations (WORLD COMMISSION ON ENVIRONMENT AND DEVELOPMENT, 1987, p. 46).

Although difficult to define in more precise economic terms, the concept of sustainable develop-ment quickly became a necessary ingredient of all papers and reports dealing with the environmental implications of economic development and industrialization. It was central to the preparations made for the sequel to the Stockholm Conference, held in 1992. In the twenty years separating the two, the United Nations had held other large conferences: on human settlements in 1976; on desertifi-cation in 1977; and on new and renewable sources of energy in 1981. The major conference, covering these topics and more, was the UN Conference on Environment and Development (UNCED) held in Rio de Janeiro in June 1992 and nicknamed the 'Rio Summit' or the 'Earth Summit'.

Attended by leaders from 178 countries, UNCED was the largest summit meeting ever held. It concluded with a number of agreements. The Convention on Protecting Species and Habitats (the so-called Biodiversity Convention) was signed on behalf of 167 countries and the Framework Convention on Climate Change was also accepted by more than 150 countries. It was also agreed that the Rio Declaration on Sustainable Development, signed at the Conference, would be passed together with Agenda 21, the outline of a programme for action, to the UN Sustainable Development Commission, a body the General Assembly was asked to authorize.

Not everything was agreed. Decisions on forestry, desertification, and fish stock management were postponed for a later conference. Nor did all subsequent discussions run smoothly. Nevertheless, UNCED was regarded as a considerable success and some governments produced documents relating what had been agreed in Rio to their own countries and policies. The British government published its *Sustainable Development Strategy*.

It is never likely to be easy to persuade national governments to cooperate in matters that affect their perception of sovereignty, which is generally taken to mean their inalienable right to assert the interests of their own peoples. Yet much was achieved between about 1970, when the exis-tence of a complex of problems that could be addressed effectively only at a global level was first widely recognized, and 1992. Virtually all governments had come to accept the need for inter-national collaboration and environmental legislation. Problems identified at the global level were being discussed and made the subjects of intensive scientific research. Sustainable development, whatever it might mean in practice, was held to be the necessary route to future environmental stability. It would be no exaggeration to say that during this period, from 1970 to 1992, the atti-tude of world leaders changed radically.

62 Pollution control

All the strategies for conserving the environment called for pollution to be reduced, but achiev-ing any significant improvement meant that politicians and the public had to address the economic issues raised by pollution control. In a market economy, goods and services are produced in response to consumer demand, which in most cases is sensitive to price. If prices rise, demand falls, and where alternative products or services are available at different prices, consumers will tend to prefer the cheaper. This strongly encourages producers to minimize their costs in order to keep prices as low as possible and, therefore, only those costs actually incurred in the course of production and distribution, such as materials, fuel, wages, administration, transport, and marketing, were counted in the retail price. These are the internal costs.

Every manufacturing process generates waste and by-products with no commercial value, and every product eventually wears out and is thrown away. Some products, such as detergents and the propellants used in aerosol cans, are thrown away immediately, in the course of their ordi-nary use. Others, such as coal, generate and release by-products in the course of their ordinary use. Wastes and by-products were traditionally released into the environment at every stage in the production and use of goods and services. If not always free, disposal was cheap.

What the environmental debate revealed, however, was that such disposal does incur costs. Most obviously, water polluted by industrial or domestic discharges may have to be purified for return to the public supply, and this increases the cost of that supply. Less obviously, in the sense that it is more difficult to quantify, polluted air may harm the health of some people, requiring them to seek medical treatment that must be paid for, by the community at large or by themselves depending on the system of health care, quite apart from the cost they pay in terms of suffering. Such costs as these, and there are many, are borne either by the public as a whole or by individuals. They are not borne directly by the suppliers or consumers of the goods and services that gave rise to them. They are external costs.

In the jargon of economists, pollution control seeks to internalize these external costs. Where the costs arise in the course of production, they should be charged to the producer. This is the basis of the 'polluter pays' principle. Of course, this may increase production costs and, therefore, the price paid by the consumer, but this is considered fair. People who do not consume that particular product or service no longer contribute to the costs incurred through the disposal of its wastes and by-products.

Where the environmental costs arise from consumption, matters are rather more complicated, because in practice it is usually impossible to charge for the disposal of individual items. One alternative is to encourage, or in some cases compel, the consumer to dispose of wastes in ways that minimize the final disposal cost or maximize opportunities for recycling. This is the reasoning behind bottle, paper, aluminium can, and clothing banks, and the system in some countries, such as Germany, that requires householders to sort their domestic refuse into separate containers. A more radical approach is to require the producer to assume responsibility for, and bear the cost of, final disposal of the product. At the end of its life, for example, a car might be returned to the factory that made it.

Reducing sulphur emissions

Coal-burning is a major source of sulphur dioxide emissions. These can be reduced by flue-gas desulphurization, or by burning the coal in a fluidized bed.

Flue-gas desulphurization works by reacting gaseous sulphur dioxide (SO_2) with lime (calcium hydroxide, $Ca(OH)_2$) to produce calcium sulphate ($CaSO_4$). The flue gas is passed through a lime bath and the insoluble $CaSO_4$ is precipitated. The process is efficient, but generates large amounts of waste $CaSO_4$ and requires a large supply of lime. This is obtained by quarrying and then kilning (heating) limestone, a process that drives off carbon dioxide ($CaCO_3$ + heat $\rightarrow$ CaO + $CO_2 \uparrow$).

In a fluidized bed, powdered coal is mixed with powdered limestone and the mixture burned in a chamber through which air is forced under pressure from below, making the mixture behave like a fluid (hence the name). The forced circulation of air and separation of particles ensure more efficient combustion than in a conventional furnace, and at a lower temperature. Efficient combustion reduces emissions of unburned hydrocarbons, the lower temperature reduces the oxidation of nitrogen to nitrogen oxides, and sulphur dioxide reacts with the limestone. SO_2 emissions are reduced by about 90 per cent.

Before any system of pollution control can be implemented, its costs must be quantified, and this is not straightforward. The cost of pollution abatement increases sharply as emissions fall, imposing an upper limit on the improvement that can be achieved at a price the public is willing to pay (RAVEN ET AL., 1993, pp. 116–121). Just how much people will pay depends on a comparison between the cost of the pollution and the cost of reducing it. Pollution costs can be calculated, for example as the cost of health care and lost working time attributable to pollution, although this is difficult and usually controversial, because it relies on epidemiological studies that yield probabilities, not certainties, and are open to varying statistical interpretations (TAUBES, 1995). Nor does pollution exact the same price in all places. Smoke from a particular factory causes much less harm in the countryside, far from any other factory, than it would in a city where it mingled with smoke from many other factories. Is it just, therefore, to require all factories to observe the same emission limits regardless of the actual harm they cause? It can be argued that similar costs must be imposed on all factories to prevent some enjoying a commercial advantage over their rivals. It can also be argued that lower costs in certain places would encourage the more even distribution of industry, favouring regions that are otherwise economically disadvantaged.

In the real world, pollution abatement proceeds as a series of compromises between the clean environment the public demands, the degree of improvement industries are technologically capable of achieving, and the overall effect on prices and national economies. It is supported by national and international legislation. This explains in detail what is expected and protects responsible producers from those prepared to undercut prices by ignoring environmental considerations. There is now a vast amount of environmental legislation, and exporting companies must observe the laws obtaining in all the countries to which their products are sent.

Industry has learned to accept environmental constraints and it would be wrong to suppose it necessarily hostile. After all, factory owners and managers breathe the same air, drink the same water, and visit the same countryside as everyone else. They share the general desire for environmental improvement, and many members of the public urging that improvement are also their employees.

Complacency is the vice guaranteed sooner or later to lead an industrialist into bankruptcy. Industrialists are opportunists and soon began to realize that constraints could be turned to advantage and costs into profits. Markets were found for some of the substances recovered from waste streams and in future we may expect some economic surprises. Agricultural crop plants require sulphur as an essential nutrient, for example. Until now they have received an adequate supply in the form of sulphur dioxide dissolved in rain. The sulphur dioxide is an industrial emission that contributes to acid rain. As it is recovered from exhaust gases to reduce acid rain damage, crop plants will be deprived, so perhaps the recovered sulphur can be sold to farmers as fertilizer. Acid rain would be reduced to some extent and farmers would have to pay for what they were used to receiving free.

More immediately, pollution abatement has become an industry in its own right. The manufacture, installation, and maintenance of the necessary equipment is a specialized and profitable enterprise. Large companies must provide themselves with laboratories to determine the environmental effects of their products, and those laboratories must be equipped and staffed. Many are much better equipped than university laboratories. This need has made work for the manufacturers of laboratory equipment and consumable supplies, and provided employment for scientists and laboratory technicians.

Removing pollutants once they have been generated is, at best, an interim measure and only some of the recovered substances have any commercial value. The search, therefore, is for technologies that generate fewer pollutants in the first place. Such technologies would be more easily sustainable. They would recover, recycle, or reuse materials as part of their primary process, substitute process materials to take account of their environmental effect (such as using water-based rather than solvent-based paints), and in some cases modify the product itself (HOOPER AND GIBBS, 1995). The obvious sense of this is recognized by governments and intergovernmental bodies such as the Organization for Economic Cooperation and Development (OECD). Since the goal of environmental improvement is socially popular and promises reductions in public expenditure, some governments are now providing practical support for environmental technologies (CLEMENT, 1995). As Figure 6.11 shows for the European Union, there is considerable variation in expenditure from one country to another, although the figure makes no allowance for the relative sizes of national economies.

The concept of 'cleaner technology' emerged in the late 1970s and led in some countries to a reduction in pollution and consumption of raw materials that was clearly discernible a few years later. Although supported by governments, industries paid for much of the investment themselves, as Figure 6.12 shows. Amounts vary, but in countries with the most stringent environmental regulations annual expenditure on pollution control is about 1.5 per cent of GNP, of which industry pays around 25 per cent, or 0.4 per cent of GNP (TOLBA AND EL-KHOLY, 1992, p. 358).

Pollution control may be profitable for those selling it and cleaner technologies may improve industrial profitability once they are installed, but those benefits cannot be obtained unless there is capital available to invest in them and a highly trained workforce to install and operate them. For this reason, the environmental gains have been most marked in the wealthy, industrialized

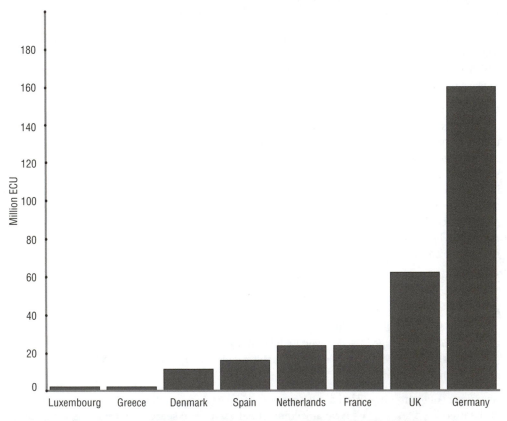

Figure 6.11 *Government assistance for environmental technologies in the EU 1988–90*
After Clement, Keith. 1995. 'Investing in Europe: Government support for environmental technology',
Greener Management International, *January, p.45*

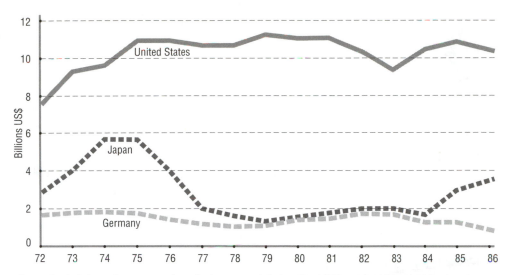

Figure 6.12 **Private investment in pollution control during the 1970s and 1980s (1980 prices)**
Source: Tolba, Mostafa K. and El-Kholy, Osama A. 1992. The World Environment 1972–1992. *Chapman and Hall,*
London, on behalf of UNEP

countries. Poor countries simply cannot afford to abandon existing industrial plant while it remains capable of producing goods, albeit inefficiently, nor to install equipment to recover pollutants.

Carbon dioxide emissions provide one way to measure differences between countries. Older industrial and power generation plants burn fossil fuels as their primary source of energy and in poorer countries fossil fuels also provide most domestic heating and cooking. Modern plant, both industrial and domestic, uses energy more efficiently, so it consumes less fuel for each unit of energy delivered, and it is more likely to rely on alternatives to fossil fuels, most notably nuclear power for electricity generation. The more carbon dioxide that is released for each unit of national income, the poorer and more technologically backward the country. As Figure 6.13 shows clearly, emissions and prosperity follow one another closely. China produces seven times more carbon dioxide than the United States for each US dollar of its income.

This situation is unsatisfactory and China began to do something about as long ago as 1979. By 1987 a system of pollution charges had been extended to the entire country. Since 1988 efforts have been directed at improving the policing of the system. This begins by setting standards for a range of pollutants. There are more than a hundred standards for discharges into water, atmospheric emissions, waste disposal, and noise. Those who exceed them must pay a charge based on category of pollution. After three years, continued failure to comply results in a 5 per cent annual increase in the charge, and a double charge for any new enterprise exceeding the standards that was built after the legislation was passed. A delay of more than twenty days in paying a charge incurs a penalty of 0.1 per cent per day. Money raised by the charges is invested in pollution control and, although the charges are lower than the cost of installing control equipment, so that many managers are content to pay them and carry on polluting, they have led to environmental improvements in the most heavily polluted cities and regions and provided employment for more than 40 000 people (POTIER, 1995).

Nevertheless, pollution remains severe. The area affected by acid rain increased from 18 per cent of the total land area in 1985 to 40 per cent in 1998, due to dependence on coal during a period of rapid economic growth. Scientists calculated that unless emissions from Chinese coal-fired plants

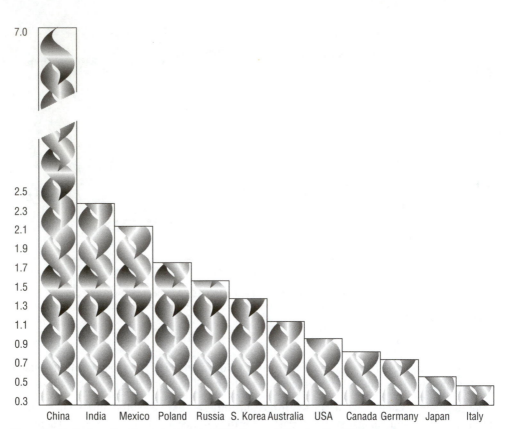

Figure 6.13 **Carbon dioxide emissions in 1988 (kg per US$ of national income)**
Source: Nowicki, Maciej. 1992. Environment in Poland. *Ministry of Environmental Protection, Natural Resources and Forestry, Warsaw*

were reduced, by 2020 deposited acid would overwhelm the ability of soils to cope over a large area, on a scale equivalent to or exceeding that in the 'black triangle' of central Europe. It would cost more than $23 billion per year for 20 years, about 2 per cent of China's gross domestic product, to remedy the situation. Scrubbers to remove sulphur dioxide would have to be installed on all new and all major existing smokestacks and power plants, and greater energy efficiency and a switch to less polluting fuels would have to be encouraged (HOLMES, 1999). Coal accounts for 75 per cent of total Chinese energy consumption and about 22 per cent of Chinese homes use uncleaned coal for cooking and heating. As well as sulphur dioxide, domestic cooking fires emit arsenic, fluorine, lead, and mercury.

There has been great concern over levels of pollution in the industrialized regions of Eastern Europe. These, too, result from old, inefficient industrial plant and over-reliance on low-quality coal as a primary energy source. The governments of those countries are no less determined than others to remedy the problem. Poland, where carbon dioxide emissions are relatively greater than those in Russia, has implemented a programme to reduce its economic reliance on heavy industry, move from coal to domestic and imported natural gas as a principal fuel, and modernize its old power stations. At the same time, it is encouraging energy conservation in industry, transport, and domestic heating and cooking (NOWICKI, 1992, p. 6).

The impetus arising from the environmental concerns of the 1960s and 1970s led to the conservation strategies of the 1980s and, from them, to the concept of 'sustainability'. This is now being translated into everyday industrial practice first in the industrialized countries but increasingly in the less industrialized and newly industrialized countries. Improvements are already becoming evident and, provided the momentum is maintained, more will appear in years to come.

63 Hazardous waste

Wastes are ordinarily disposed of in landfill sites or by incineration. These methods are not satisfactory for certain types of waste, however. Under British law, the Control of Pollution Act 1974 defines wastes and regulates their disposal according to the risk this presents.

Controlled wastes, categorized as household, commercial, or industrial, must be collected and disposed of in such a way that they cannot cause pollution. Agricultural wastes, explosives, and wastes from mines and quarries are not defined as controlled wastes.

Special wastes include prescription drugs, substances with a flash point below 21 °C, and other materials that present disposal difficulties. Wastes that are poisonous, but do not fall within the definitions of special waste are classed, rather loosely, as toxic wastes.

Hazardous wastes are also defined rather loosely. They include substances similar to special wastes, but listed under the Transfrontier Shipment of Hazardous Wastes Regulations 1988. Difficult wastes include all special wastes together with certain metallic wastes and wastes that are physically difficult to handle. The final category covers clinical wastes (BMA, 1991, pp. 22–24).

All the definitions are vague, but they do impose on all manufacturers a legally binding obligation to state the precise chemical composition of wastes they discharge and to list any toxic substances they contain. The best disposal method then depends on the nature of the waste. Some toxic substances can be made safe by diluting them, for example, but others must remain completely isolated from the environment (MELLANBY, 1992, pp. 55–67 and Appendix 3).

The imposition of more stringent controls on the disposal of wastes has led to the emergence of companies specializing in waste disposal. They have the expertise to categorize waste accurately and the facilities to render it safe. Because one disposal facility accepts wastes from many sources, this greatly simplifies the policing of the regulations. From time to time 'cowboy' operators seek to undercut the prices of legitimate companies, but any pollution they cause can be traced back to them fairly easily.

Traffic in waste

As industrialized countries tightened the regulations controlling the disposal of wastes that could harm human health or pollute the environment, specialist companies emerged to handle them. Specialist disposal is much more expensive than dumping wastes on land or discharging them into water, however, and an alternative method also emerged. Companies collected wastes and shipped them overseas. They went either to countries with adequate disposal facilities and underused capacity, or to developing countries that lacked proper means of disposal, but welcomed the foreign-currency payments they received for accepting them. Sometimes the wastes arrived unlabelled or incorrectly labelled and caused serious harm locally.

The international trade in toxic wastes became a matter of great concern and on March 22, 1989, the Convention on Transboundary Movements of Hazardous Waste was adopted at a conference held in Basel, Switzerland, under the auspices of the UN Environment Programme. Known as the Basel Convention, it was signed by 34 countries, with general endorsement from an additional 105. It established the right of every country to refuse to accept cargoes of hazardous wastes, established rules for the notification of planned shipments, and obliged the governments of originating countries to ensure that the recipient countries have adequate storage and treatment facilities. Wastes that are dumped illegally can be returned to their country of origin.

On March 25, 1994, signatories to the Basel Convention agreed that from the end of 1997 it would be illegal to export wastes to less developed countries for recycling, although the EU planned to continue to export wastes that presented no risk. At a further meeting, held in Kuching, Malaysia, in February 1998, more than 300 officials from 117 countries agreed on a list of materials defined as 'hazardous', and on a list of countries that were permitted to trade toxic waste among themselves.

64   Transnational pollution

Drifting air masses, continental rivers, and ocean currents are no respecters of national boundaries contrived by politicians and generals. They move where physical forces take them, and if they carry a load of pollutants those travel with them. Consequently, there is a limit to the pollution abatement that individual nation-states can achieve in isolation.

Free trade imposes a further constraint. Consider the case of two widely separated countries; call them A and B. Country A, heavily industrialized, suffers severe pollution of its air and fresh water, and its government, responding to pressure from a concerned citizenry, enacts legislation to protect the environment within its borders. This increases manufacturing costs. Country B, on the other hand, is either just industrializing and experiences little pollution or is prepared to tolerate such pollution as it has. Following the legislation in country A, its manufacturing costs are lower and country A is therefore presented as a tempting market it can penetrate with little difficulty. Country A is now in a difficult position. If it does nothing, its trading account will deteriorate and so will its overall economy. In the end, its citizens will suffer. If it seeks to protect its domestic manufactures by imposing tariffs and duties it risks distorting trade patterns and triggering reprisals.

The conclusion in 1994 of the Uruguay Round of negotiations on the General Agreement on Tariffs and Trade (GATT) represented a major liberalization of world trade and the replacement of GATT itself by a new body, the World Trade Organization (WTO). Some environmentalists feared it might lead to the relocation of industries from regions with stringent environmental legislation to those with lax controls, or that controls in the present industrialized countries might be relaxed. Others feared that economic development would be inhibited in the less industrialized countries through an insistence by countries with strong environmental regulations that similar regulations be imposed on the manufacture of goods they imported. These dangers have been recognized and ways are being sought to address them (VON FELBERT, 1995).

It is in the interest of all countries that, so far as possible, environmental policies are coordinated. In practice, this is what has been happening since the 1972 Stockholm Conference and considerable progress has been made.

Acid rain may have been the first issue demanding international agreement. The phenomenon had been known for more than a century, and the monitoring of atmospheric acidity over northern and western Europe began in the 1950s (ALLABY, 1989, pp. 106–107), but it was in the 1970s that it was observed over Scandinavia and north-eastern North America. North-western Europe, eastern North America, and eastern China are still the regions most seriously affected but, as Figure 6.14 illustrates, several other parts of the world are potentially at risk. Because the acid derives in part from industrial atmospheric emissions that often travel long distances carrying sulphur dioxide and nitrogen oxides, its effects can be reduced only if nations collaborate. The fact of such transport was established by a case study Sweden presented to the Stockholm Conference in 1972, which led to larger study programmes, from 1972 to 1977 sponsored by the OECD, and starting in 1978 sponsored by the Economic Council for Europe (ECE). The ECE Convention on Long-Range Transboundary Air Pollution (*www.unece.org/env/lrtap*) was signed by 34 countries in 1979, a conference on the matter was held in 1982 in Stockholm, and in 1985 21 of the countries that signed the ECE Convention agreed the '30 per cent Protocol'. This committed them to reducing sulphur emissions by at least 30 per cent of their 1983 levels by 1993 at the latest. A further Protocol to the Convention, signed by 27 countries in 1988, called for emissions of nitrogen oxides to be no higher than their 1987 levels by the end of 1994 (TOLBA AND EL-KHOLY, 1992, p. 25).

The Convention has proved highly successful. Between 1980 and 1999 European emissions of sulphur were reduced by half, those of nitrogen oxides by 16 per cent, of volatile organic compounds by 20 per cent, and of ammonia by 18 per cent. The reductions have not been uniform, however. Some countries achieved much more than others. A further protocol was added to the Convention in November 1999. This dealt with acidification, ground-level ozone, and the eutrophication of surface waters.

Acid rain, more correctly called 'acid deposition' because airborne acid can travel as mist, fog, snow, and in dry air, as well as in rain, affects trees, soils, and surface water. The processes by which it does so are complex and often indirect. It has often been stated, for example, that airborne sulphate dissolves in cloud droplets to form sulphuric acid. While this is true, the resulting acid is often too dilute to harm plants directly. The damage it causes arises from chemical reactions in certain soils. The extent of the effect on forests is also difficult to determine and has sometimes been overstated (ALLABY, 1999, pp. 164–169). Research into the many facets of the problem has continued over many years and the full account of its nature and extent has at last been assembled (MACKENZIE AND EL-ASHRY, 1989).

Signing an agreement does not make the problem vanish, of course. The agreement must be ratified through national parliaments or assemblies and then implemented. This may prove more difficult than was predicted at the time of signing, either technically or because of fluctuations in the economic fortunes of the signatory states or the political fortunes of their governments. Even if the agreement is implemented, it may prove insufficient. Nevertheless, no improvement is possible without international agreement, so an agreement's conclusion and signature is an essential first step toward amelioration, even though acid rain continues to cause concern.

A similar, but in some ways even more intractable, problem affects what UNEP calls 'regional seas'. These are seas which are landlocked, or almost so, such as the Mediterranean and Red Seas, and seas bordering continents that are vulnerable to pollutants reaching them from the land. The first to be debated was the Mediterranean and it provides an excellent, if extreme, illustration of the difficulties involved.

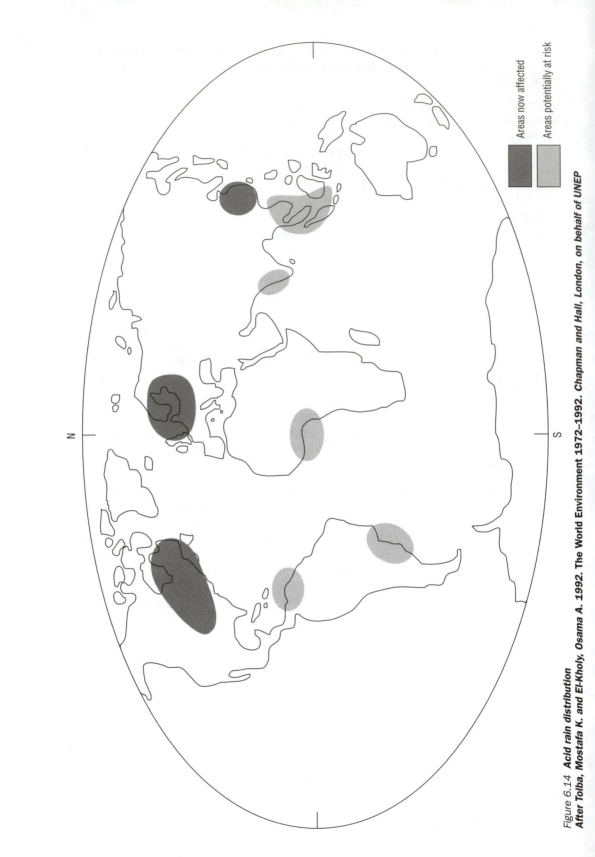

Figure 6.14 *Acid rain distribution*
After Tolba, Mostafa K. and El-Kholy, Osama A. 1992. The World Environment 1972–1992. Chapman and Hall, London, on behalf of UNEP

Areas now affected

Areas potentially at risk

N

S

As Figure 6.15 shows, the Mediterranean and Aegean are bordered by Spain, Gibraltar, France, Monaco, Italy, Croatia, Yugoslavia, Greece, Macedonia, Turkey, Cyprus, Syria, Lebanon, Israel, Egypt, Libya, Malta, Tunisia, Algeria, and Morocco. The countries to the north are industrialized, those to the south not, the cultures are Christian (both Eastern and Western), Jewish, and Islamic, and relations were less than friendly between Greece and Turkey at the time of the negotiations. Oil tankers sailing between the Gulf and Europe and North America pass through the Suez Canal and cross the Mediterranean, and many of them used to wash out their tanks in the Mediterranean as they returned empty to collect a fresh cargo. In addition, the Mediterranean system of inter-linked basins receives industrial discharges carried by the Ebro, Rhône, Po, Danube, Dnieper, Don, and Nile, as well as many smaller rivers.

Despite the apparent hopelessness of the task (at least Yugoslavia was still united and at peace at the time), meetings of the nations bordering the Mediterranean were held under UNEP auspices. These led to the adoption, at Barcelona in February 1976, of the Convention for the Protection of the Mediterranean Sea Against Pollution (the Barcelona Convention). It dealt with oil pollu-tion and a range of industrial discharges, although it did not specify their sources. The Convention included an Action Plan setting out specific means for achieving its objectives.

This was the first agreement in what became the Regional Seas Programme. It has produced agreements to reduce pollution in such enclosed marine areas as the Red Sea and the waters south of Kuwait, and the Black Sea (bordered by Bulgaria, Romania, Ukraine, Russia, Georgia, and Turkey) was brought into the area covered by the Barcelona Convention (see Figure 6.15). In all, seven conventions have been agreed since the Barcelona Convention, covering Kuwait (1978), West and Central Africa (1981), the South-East Pacific (1981), the Red Sea and Gulf of Aden (1982), the Wider Caribbean (1983), Eastern Africa (1985), and the South Pacific (1986). The next to be agreed will cover the South Asian Seas, East Asian Seas, and North-West Pacific. Most of these agreements also involve the conservation and management of resources, such as fish stocks (TOLBA AND EL-KHOLY, 1992, pp. 775–777). Figure 6.16 shows the areas included in the Regional Seas Programme.

Countries bordering them have also concluded agreements independently of UNEP to protect particular seas. The Helsinki Convention, signed in 1980, covers the Baltic. In the North Sea, water enters through the Strait of Dover and is carried by the tidal system around the coasts of bordering countries, collecting river discharges as it goes, so coastlines are especially vulnerable, the German Bight being the most seriously affected area. Pollution of the North Sea is covered by the Oslo Convention, signed in 1972 and dealing with discharges from ships at sea, and the Paris Convention of 1974, dealing with pollution from land sources (CLARK, 1992, pp. 126–147). Rivers that cross or mark international borders have also been the subjects of agreements to reduce pollution.

As well as brokering international agreements, UNEP is responsible for coordinating the moni-toring networks without which protection of the global environment would falter for want of data. These data are obtained from surface stations distributed throughout the world.

Their observations are augmented by satellite data. SPOT (Système Probatoire d'Observation de la Terre), launched in 1986, has a monochrome resolution of 10 m and provides information for farmers, geologists, and land-use planners. ERS–1 monitors ice patterns and surface temper-atures. ERS-2 does the same, but also monitors ozone levels. JERS–1 gathers a wide range of data and Radarsat, launched in 1995, measures the Earth's surface. Several more satellites are planned. ADEOS, to be launched in 1996, will study atmospheric chemistry and collect land and sea data. Meteor 3M–1, to be launched in 1998, will study atmospheric aerosols and chemical compounds. In 1999, ADEOS II will begin studying surface wind speeds and directions over the oceans. The

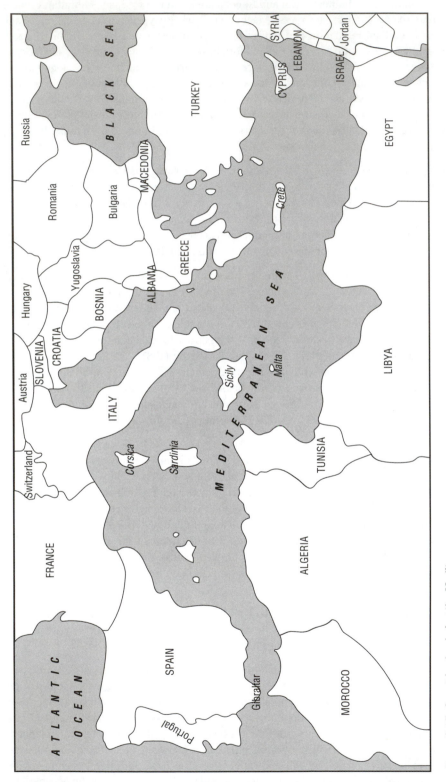

Figure 6.15 *Countries bordering the Mediterranean*

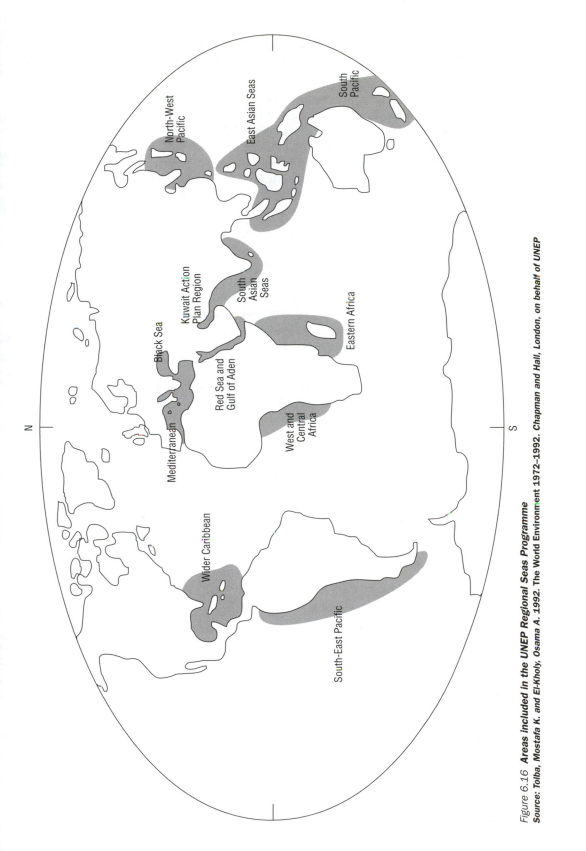

N

S

North-West
Pacific

East Asian Seas

South
Pacific

Kuwait Action
Plan Region

South
Asian
Seas

Eastern Africa

Black Sea

Red Sea and
Gulf of Aden

West and
Central
Africa

Mediterranean

Wider Caribbean

South-East Pacific

Figure 6.16 **Areas included in the UNEP Regional Seas Programme**
Source: Tolba, Mostafa K. and El-Kholy, Osama A. 1992. The World Environment 1972–1992. Chapman and Hall, London, on behalf of UNEP

United States has a continuing programme of eight satellite launches between 1998 and 2003 and launch of the space station in 2001, which will considerably increase the flow of environmental data returned to Earth (LAWLER, 1995).

Data are processed and distributed through several systems. The Global Observing System (GOS) specializes in meteorological data. It is coordinated by the World Meteorological Organization (WMO) and operated by WMO member states. The Global Environmental Monitoring System (GEMS), established by UNEP, covers 142 countries and handles data pertaining to atmosphere, climate, pollution, and renewable resources. GEMS has several component systems. The Human Exposure Assessment Locations Programme (HEALS), for example, established in 1984 and run by the World Health Organization (WHO) in collaboration with UNEP, monitors human exposure to a number of pollutants (TOLBA AND EL-KHOLY, 1992, pp. 612–618).

Transnational pollution is not a recent phenomenon. Acid rain, first reported more than a century ago, must already have been causing damage for many decades and in those days of intensive heavy industry deriving its energy from coal-burning, it would be surprising if acid pollutants had not been transported from Britain across the North Sea by the generally westward movement of air masses, or from the industrial cities of the north-eastern United States and eastern Canada to adjacent rural areas. Similarly, industrial and domestic wastes were discharged into rivers and from there into the sea long before the means existed to measure the scale on which pollutants from one country were affecting the shores of others.

What has changed is not so much the character of transnational pollution as our increasing ability to measure and track it, and to observe and quantify its effects. It is through advances in technological sensitivity and scientific understanding that we are able to determine when and where pollution originates and the degree of its severity. These same advances allow us to devise ways of minimizing the harm it causes.

With the exception of greenhouses gases, the discharge of which may lead to changes in the global climate, most pollution affects local areas or regions. Local, regional or global, however, it can be ameliorated only through international cooperation. Recognition of this fact has compelled governments, often reluctantly, to modify their concepts of national sovereignty by admitting a wider duty to impose constraints on domestic activities for the benefit of citizens and environments beyond their borders. In this century we have learned that truly we all inhabit a single planet. As Barbara Ward and René Dubos wrote in their introduction to *Only One Earth*, the background book to the 1972 Stockholm Conference: 'As we enter the global phase of human evolution it becomes obvious that each man has two countries, his own and Planet Earth.'

The perceived need for international collaboration to protect and enhance the environment we all share offers the hope that such collaboration may expand into other areas of concern and that we may learn at last how to live in closer harmony with one another as well as with our planet.

End of chapter summary

Concern for the natural environment is far from new. Attempts to conserve resources such as forests and to reduce air pollution date from medieval times. In the course of this century, however, those attempts have broadened and, perhaps for the first time in history, they promise to be effective.

The most important change in our attitude to the environment comes from our recognition of its global extent. We have learned to appreciate that events happening here and now can produce environmental consequences far removed in place and time. This is novel. In previous centuries thoughtful people often expressed concern about the environment, but it was always the local environment they sought to improve, because they lacked the technological means to compile a picture that encompassed an entire country, region, or continent simultaneously, far less the entire world. With comprehensive satellite monitoring and electronic means of communication we now have that ability. Every point on the surface of our planet is under constant surveillance. It has given us the concept of 'Spaceship Earth'.

Our newly acquired broad view is accompanied by a deeper technical understanding of the way the environment works. This understanding is very far from complete, but it is improving rapidly and already it allows practical steps to be taken to protect species, habitats, landscapes, and human health that have a good chance of succeeding. Zoos, once menageries for the amusement of the rich and then potential breeding stations for the production of new farm animals, are now dedicated to the conservation of rare and endangered species. Agricultural improvements have greatly increased crop yields in Europe and North America, so less land is needed for farming. This makes it practicable to set land aside for non-agricultural purposes, such as recreation and wildlife conservation. So-called 'industrial' farming may remove much of the natural flora and fauna from the farmed area, but it also frees large areas elsewhere.

It is not only our comprehension of the environment that has assumed a global dimension. So have the economies and political structures of nations. International institutions now exist to resolve a wide range of issues by negotiation and agreement. Once the institutional framework had been set in place, environmental matters affecting more than one nation were fitted into it. The creation of the United Nations, first mooted in 1942 and finally achieved in 1945, made it possible to form the United Nations Environment Programme (UNEP) in 1973. UNEP and other UN agencies have been instrumental in nurturing many of the international environmental treaties and conventions that have been implemented over the last few decades.

Despite the fears of some environmentalist scaremongers, we have good reason to be optimistic about the future. It is simply not true that the world is irredeemably damaged. The well documented fact that in almost every corner of the world people are healthier and living longer than they did a generation ago gives the lie to the idea that the human environment is becoming more hostile. There are environmental problems, areas that have been severely damaged, and valuable habitats that are being lost and their species threatened with extinction. But as our knowledge expands and informs our political will to protect, restore, and reconstruct, we may hope and expect that in the years to come the condition of the global environment will improve.

End of chapter points for discussion

How did zoos and botanic gardens originate?
What is the best way to control farm pests?
Do world conservation strategies have any effect?
What is the precautionary principle and what are its advantages and disadvantages?

See also

Human populations and demographic change (section 54)
Genetic engineering (section 55)
Zoos, nature reserves, wilderness (section 57)

Further reading

The DDT Story. Kenneth Mellanby. 1992. British Crop Protection Council, Farnham. Short and simple to read, this book tells the full history of the world's most notorious insecticide and is written by a scientist closely involved in monitoring its environmental effects.

North–South: A Programme for Survival. The Independent Commission on International Development Issues. 1980. Pan Books, London. A detailed analysis of the economic and environmental situation facing the world.

Our Common Future. The World Commission on Environment and Development. 1987. Oxford University Press, Oxford. The famous Brundtland Report, which provided background material for the 1992 Rio Conference.

Silent Spring. Rachel Carson. 1963. Hamish Hamilton, London. Still in print, this is the book that first aroused public concern about the condition of the environment.

References

Abramov, Valery. 1990. 'Man vs Mosquito: The arms race is on', in *Developing World Health*, Grosvenor Press International, London, pp. 34–36.

Allaby, Michael. 1989. *Green Facts.* Hamlyn, London.
——. 1990. *Into Harmony with the Planet.* Bloomsbury Publishing, London.
——. 1999. *Ecosystem: Temperate Forests.* Fitzroy Dearborn, London.

Begon, Michael, Harper, John L., and Townsend, Colin R. 1990. *Ecology,* 2nd edn. Blackwell Science, Oxford.

BMA. 1991. *Hazardous Waste and Human Health.* Oxford University Press, Oxford.

Bragg, James R., Prince, Roger C., Harner, E. James, and Atlas, Ronald M. 1994. 'Effectiveness of bioremediation for the *Exxon Valdez* spill', *Nature*, 368, 413–418.

Brewer, Richard. 1988. *The Science of Ecology,* 2nd edn. Saunders College Publishing, Forth Worth, Texas.

Brooks, Robert, Anderson, Chris, Stewart, Robert, and Robinson, Brett. 1999. 'Phytomining: growing a crop of a metal', *Biologist*, 46, 5, 201–205.

Carson, Rachel. 1963. *Silent Spring.* Hamish Hamilton, London.

Clark, R.B. 1992. *Marine Pollution*, 3rd edn. Oxford University Press, Oxford.

Clement, Keith. 1995. 'Investing in Europe: government support for environmental technology', *Greener Management International*, January 1995. pp. 41–51. Greenleaf Publishing, Sheffield.

Cockburn, Andrew. 1991. *An Introduction to Evolutionary Ecology.* Blackwell Scientific Publications, Oxford.

Conway, G.R., Gilbert, D.G.R., and Pretty, J.N. 1988. 'Pesticides in the UK: Science, policy and the public', in Harding, D.J.L. (ed.) *Britain Since 'Silent Spring'.* Institute of Biology, London. pp. 5–32.

Culotta, Elizabeth. 1995. 'Bringing back the Everglades', *Science*, 268, 1688–1690.
——. 1995a. 'Ecologists flock to snowbird for varied banquet of findings', *Science*, 269, 1045–1046.

Dizon, Andrew E., Lockyer, Christina, Perrin, William F., Demaster, Douglas P., and Sisson, Joyce. 1992. 'Rethinking the stock concept: a phylogeographic approach', *Conservation Biology*, 6, 1, 24–36. Blackwell Scientific Publications, Oxford.

Ehrlich, Paul R. and Ehrlich Anne H. 1970, 1972. *Population, Resources, Environment: Issues in Human Ecology.* W.H. Freeman and Co., San Francisco.

Fincham, J.R.S. 1995. 'The killing fields' (book review), *Nature*, 377, 297.

Foster, Karen Polinger. 1999. 'The earliest zoos and gardens', *Scientific American*, July 1999, pp. 48–55.

Goldsmith, Edward, Allen, Robert, Allaby, Michael, Davoll, John, and Lawrence, Sam. 1972. 'A Blueprint for Survival', *The Ecologist*, 2, 1.

Green, M.B. 1976. *Pesticides, Boon or Bane?* Elek Books, London.

Holmes, Bob. 1999. 'A dark cloud over Asia', *New Scientist*, 13 February, p. 25.

Hooper, Paul D. and Gibbs, David. 1995. 'Cleaner technology: a means to an end, or an end to a means?', *Greener Management International*, January 1995. pp. 28–40. Greenleaf Publishing, Sheffield.

Independent Commission on International Development Issues. 1980. *North-South: A Programme for Survival.* Pan Books, London.

IUCN. 1980. *World Conservation Strategy: Living Resource Conservation for Sustainable Development.* IUCN, Gland.

Lachance, L.E. 1974. 'Status of the sterile-insect release method in the world', in *The Sterile-Insect Technique and its Field Applications*, International Atomic Energy Agency, Vienna, pp. 55–62.

Lawler, Andrew. 1995. 'NASA mission gets down to Earth', *Science*, 269, 1208–1210.

Mace, Georgina M. 1995. 'Classification of threatened species and its role in conservation planning', in Lawton, John H. and May, Robert M. (eds). *Extinction Rates.* Oxford University Press, Oxford. pp. 198–200.

MacKenzie, James J. and El-Ashry, Mohamed. 1989. *Air Pollution's Toll on Forests and Crops.* Yale University Press, New Haven, Conn.

Mason, C.F. 1991. *Biology of Freshwater Pollution,* 2nd edn, Longman Scientific and Technical, Harlow.

Meadows, Donella H., Meadows, Dennis L., Randers, Jorgen, and Behrens, William W. III. 1972. *The Limits to Growth.* Earth Island, London.

Mellanby, Kenneth. 1992. *The DDT Story.* British Crop Protection Council, Farnham.
—. 1992. *Waste and Pollution: The Problem for Britain.* HarperCollins, London.

Moffat, Anne Simon. 1995. 'Plants proving their worth in toxic metal cleanup', *Science*, 269, 302–303.

Nowicki, Maciej. 1992. *Environment in Poland: Issues and Solutions.* Ministry of Environmental Protection, Natural Resources, and Forestry, Warsaw.

Patterson, Colin. 1978. *Evolution.* Routledge and Kegan Paul, with the British Museum (Natural History), London.

Payne, C.C. 1988. 'Prospects for biological control', in Harding, D.J.L. (ed.) *Britain Since 'Silent Spring'.* Institute of Biology, London, pp. 103–116.

Potier, Michel. 1995. 'China charges for pollution', *OECD Observer*, February/March 1995. OECD, Paris. pp. 18–22.

Raven, Peter H., Berg, Linda R., and Johnson, George B. 1993. *Environment.* Saunders College Publishing, Forth Worth, Texas.

Stonehouse, Bernard. 1981. *Saving the Animals.* Weidenfeld and Nicolson, London.

Taubes, Gary. 1995. 'Epidemiology faces its limits', *Science*, 269, 164–169.

Tolba, Mostafa K. and El-Kholy, Osama A. 1992. *The World Environment 1972–1992.* Chapman and Hall, London, on behalf of UNEP.

Tudge, Colin. 1993. *The Engineer in the Garden.* Jonathan Cape, London.

Tyson, D. 1988. 'Future trends in the control of pests, weeds and diseases in the UK – with regard to environmental impact', in Harding, D.J.L. (ed.) *Britain Since 'Silent Spring'.* Institute of Biology, London, pp. 91–102.

von Felbert, Dirk. 1995. 'Trade, environment and aid', *OECD Observer*, August/September. OECD, Paris. pp. 6–10.

Ward, Barbara and Dubos, Rene. 1972. *Only One Earth: The care and maintenance of a small planet.* André Deutsch, London.

World Commission on Environment and Development. 1987. *Our Common Future.* Oxford University Press, Oxford.

World Wildlife Fund UK, Nature Conservancy Council, Countryside Commission, Countryside Commission for Scotland, The Royal Society of Arts, and Council for Environmental Conservation. 1983. *The Conservation and Development Programme for the UK: A response to the World Conservation Strategy.* 2 vols. Kogan Page, London.

End of book summary

The concept of the environmental sciences is still evolving. Like the life and Earth sciences, it comprises elements taken from many disciplines. It differs from the other broad scientific groupings in its political and social implications and this means it attracts much more attention from the public.

Public concern about the condition of the natural environment has stimulated the development of a large, sophisticated, and politically powerful environmentalist movement. Its leaders have their own ideas about the way the environment should be managed and the social and economic reforms that, in their view, are necessary if this is to be achieved. Their ideology is essentially moral and not necessarily dependent upon a scientific appraisal of the issues that engage them.

This book deals with environmental science, not with environmental philosophy or ideology. Although the two are distinct, they remain linked, however. An understanding of the science will inform your opinions and help you to distinguish between fact and fancy, to allot priorities, and to judge the likely effectiveness of particular courses of action.

You may not count yourself an environmentalist, but if you are to enter any profession that requires you to alter any aspect of the natural environment in any way at all you will need a basic grounding in environmental science. It will be your responsibility either to estimate the likely environmental consequences of the change you propose or to evaluate critically the estimation made by someone else. To do that you will need to understand the science.

Basics of Environmental Science provides no more than an introduction to a truly vast topic. Its purpose is to give you a broad idea of what environmental science entails and to whet your appetite so you may be encouraged to pursue your studies further.

In the course of the six chapters into which the book is divided you will have encountered the environmental dimension of most of the disciplines that comprise the environmental sciences. You will also have been introduced to many of the key concepts. Although you need not read them in order, the chapters follow a logical sequence, from Earth sciences to physical resources, from biosphere to biological resources, and finally to environmental management.

Scientists do not and cannot provide certainty. Science is based on scepticism and a proper response to any scientific assertion is to challenge it. In this ruthless selection process only the most robust explanations survive. Consequently, scientific discoveries and ideas are necessarily somewhat provisional. They may be modified or replaced in the light of new knowledge. This is the way science works, but it is profoundly unsatisfactory to those seeking reassurance or absolute answers that will remain true throughout eternity. Not surprisingly, therefore, many scientific issues are controversial.

You have been introduced to a few of the controversies. Is global warming happening? If so, is it due to our emissions of greenhouse gases or to some other cause, such as changes in the solar output? If land is set aside for conservation, should it be in the form of a single block or divided into many smaller blocks linked by pathways? Given that not all areas of ecological value can be protected from development, how shall we choose which to protect and which to abandon? Is modern intensive farming good or bad for wildlife?

As an environmental scientist you will join the search for answers to these questions and many others like them. Even if you never find an answer, the search will reveal new knowledge that will add to our understanding of the world in which we live. The search is valuable, even if it proves fruitless. That is science.

Good luck.

Glossary

adaptation The process by which species adapt evolutionarily (i.e. acquire adaptedness).

adaptedness The condition of a species that is adapted to the conditions under which it lives.

adiabatic (*adj.*) Without exchanging energy with the surrounding medium. Applied to a change in air temperature occurring solely as a consequence of the air rising (and cooling) or descending (and warming).

aeolian Wind-blown.

aeroplankton Spores and other microscopic organic particles and organisms that drift in the air because they weigh so little that they fall only slowly and are likely to be carried aloft again by rising air currents before settling on a surface.

aerosol A colloid, in which particles are dispersed in a gas, usually air. Because of their small size (0.01–10 μm), airborne aerosols fall very slowly.

aggressive mimicry The resemblance of a predatory or parasitic species to its prey or host.

albedo A measure of the amount of radiation a non-luminous body reflects.

Allen's rule A rule, proposed by J.A. Allen, stating that projecting parts of the body of mammals (ears, muzzle, and tail) tend to be larger in animals living in warm climates than in closely related animals living in cold climates.

allogenic (*adj.*) Applied to an ecological succession driven by abiotic environmental changes.

alluvial Formed from sediment deposited by a river.

amino acid One of the constituent units of a protein, comprising an amino group (NH_2) and carboxyl group (COOH) both attached to the same carbon atom. Of more than 80 naturally occurring amino acids, about 20 are commonly found in proteins.

anion A negatively charged ion.

aphelion The point at which an orbiting body is furthest from the body it orbits.

aquifer A body of permeable material below ground through which water flows.

autogenic (*adj.*) Applied to an ecological succession driven by environmental changes induced by vegetation.

autotroph An organism that can synthesize the organic compounds it needs for nourishment from simple, inorganic molecules.

Batesian mimicry The close resemblance between a species that is palatable to predators and an unrelated species that is unpalatable, leading predators to avoid both. It was first observed by H.W. Bates.

Bergmann's rule A rule, proposed by C. Bergmann, stating that animals living in cold regions are larger than closely related animals living in warm regions.

biodiversity An abbreviation of 'biological diversity' and usually taken to mean the total number of species presently living on Earth.

biogeochemical cycles The transport of elements between rock, water, air, and living organisms in an approximately circular fashion that periodically returns them to each stage.

biological control The control of pest populations without using chemical pesticides, most commonly by stimulating parasites and predators of the pest.

biological oxygen demand (BOD, biochemical oxygen demand) A measure of the pollution of water by organic matter (e.g. sewage), calculated by measuring the weight (in milligrams) of oxygen utilized, and so removed from a sample of water, by microorganisms during 5 days at a constant temperature of 20 °C.

biomagnification (bioaccumulation) The accumulation in the body of repeated small doses of an ingested substance until it reaches a sufficient concentration to produce physiological effects.

biomass The total mass of all living organisms within a defined area, or at a particular trophic level within that area.

black body A body that absorbs all the radiation falling upon it and reradiates it at a wavelength determined by its temperature.

'black smoker' A hydrothermal vent on the ocean floor emitting fluids containing iron, manganese, and copper. These tend to be black and resemble smoke. Vent fluids containing zinc and arsenic are usually white, and known as 'white smokers'.

'blue baby' syndrome *See* methaemoglobinaemia.

carbon ⁻14 (^{14}C) A radioactive isotope of carbon with a half-life of 5730 ± 30 years.

carbonation A chemical reaction between a compound and carbonic acid (H_2CO_3) that forms soluble bicarbonates.

carrying capacity The maximum population a specified environment can support over a prolonged period without degrading the environment.

cation A positively charged ion.

chromosome A thread of DNA and organic compounds found in the nuclei of all living plant and animal cells each species has a typical number (e.g. 46 in humans). Chromosomes consist of two paired strands. When cells divide, the chromosome strands separate, each daughter cell receiving one strand (called a chromatid). In body (somatic) cells the complementary chromatid is then synthesized.

climax The stable, enduring plant community that is the final result of a succession of communities (sere). The succession begins with the colonization of bare ground or water that sustains no plant life.

clone (*n.*) Two or more individuals that are genetically identical, or any one of those individuals. (*v.*) To produce a clone.

commensalism A relationship between two organisms of different species in which one (the commensal) benefits and the other (the host) is unaffected.

dewpoint The temperature at which atmospheric water vapour condenses or water evaporates.

disseminule A seed, spore, or fertilized egg that is broadcast (i.e. disseminated) randomly by an organism.

Dryas Two periods of cold climate, known as the Older Dryas and Younger Dryas, identified by the widespread occurrence of pollen from *Dryas octopetala* (mountain avens), an alpine plant. The Younger Dryas is dated at about 11 000–10 000 years ago.

ecliptic The great circle apparently traced by the Sun in relation to the 'fixed' stars in the course of a year. The plane of the ecliptic is the plane of the Earth–Moon orbit about the Sun. If the visible universe is imagined as a sphere (the 'celestial sphere') with the Earth at its centre, the ecliptic is at an angle to the equator of this sphere.

endozoochory The carrying of plant seeds by an animal inside its body.

epizoochory The carrying of plant seeds by an animal on the outside of its body (e.g. in its coat or feathers).

euphotic zone The upper region of a lake or sea within which light intensity is sufficient for photosynthesis.

eutrophic Over-enriched (with nutrients).

evolution The development of new species from pre-existing species, with no implication of a mechanism by which this occurs.

fluvial Pertaining to a river.

gamete An ova or sperm cell; these are the cells which fuse at fertilization and develop to form the new individual.

gene The basic unit of heredity, comprising a segment of DNA (or RNA in some viruses) of varying length that codes for a particular function or several related functions and occupies a fixed position on a chromosome.

gene bank An establishment where tissues, seeds, or genetic material are stored for purposes of conservation.

gene pool The total number of genes possessed by all those members of a sexually reproducing species or population that are capable of reproduction (i.e. in the reproductive phase of their lives).

genome The entire complement of genetic information carried by an individual in a single set of chromosomes.

genotype The genetic constitution of an organism.

geomorphology The scientific study of landforms and their relationship to underlying geologic structures.

Gloger's rule A rule, proposed by C.W.L. Gloger, stating that animals living in warm climates are more darkly coloured than closely related animals living in cold climates.

half-life The time taken for half an amount of a radioactive substance to decay.

heterotroph An organism that obtains the organic substances it needs for nourishment by consuming other organisms.

homoeostasis The tendency of a system to maintain its own equilibrium by resisting or adjusting to change.

homoiotherm An organism in which the internal body temperature varies only within narrow limits. Regulation may be achieved by internal (endothermic) or behavioural (ectothermic) mechanisms or a combination of both (as in humans, who sweat to cool themselves and shiver to warm themselves, but also wear variable amounts of clothing and light fires in winter).

humidity The amount of water vapour held in the air. The most commonly used measure is relative humidity, the amount of water vapour present in air as a proportion of the amount needed to saturate the air at the same temperature, and expressed as a percentage.

hydrosere A sere (ecological succession) that begins in water.

igneous (*adj.*) Applied to rocks formed directly by the cooling of material from the mantle.

inversion (temperature inversion) The situation in which air temperature remains constant, or rises, with increasing altitude within a layer of the atmosphere. Warm, rising air is trapped by an inversion, because it enters a region where the surrounding air is at the same temperature and density as itself. Thus inversions tend to trap pollutants carried upward in warm air.

ion An atom or molecule that has gained or lost one or more electrons and so possesses a net electric charge.

isostasy The flexing of the Earth's crust in response to local changes in its mass. If material is deposited at the surface, the crust beneath that material is depressed; surface erosion causes the crust to rise.

isotope An atom with a mass differing from other atoms of the same element because its nucleus contains more or fewer neutrons. The nucleus contains the same number of protons as other atoms of the element, so the atom is chemically identical.

lapse rate The rate at which air temperature changes with altitude. This depends on local conditions, especially humidity.

Little Ice Age A period when average temperatures were lower than they are today throughout middle latitudes in the northern hemisphere. It lasted from about 1550 to 1860, with an especially severe episode in the early 1660s.

maximum sustainable yield (MSY) The crop that is equal to the number of individuals or biomass recruited to a population during the period (usually 1 year) for which it is calculated. Theoretically, this is the crop that can be harvested without depleting the population.

metamorphism The alteration of rock by exposure to changed temperature, pressure, or chemical environment, or some combination of these, most commonly leading to the dissolution and recrystallization of its mineral constituents.

methaemoglobinaemia ('blue baby' syndrome) A rare condition in young babies that occurs when nitrates ingested from food or water are converted to nitrites in their very acidic gastric environment (gastric acidity falls to levels at which this ceases to occur in infants older than about 6 months). The nitrites react with blood haemoglobin to form methaemoglobin, which forms a compound with oxygen that is more stable than oxyhaemoglobin and, therefore, gives up oxygen less readily to tissues requiring it. This produces blue skin coloration, due to oxygen deficiency in the blood, and consequent oxygen starvation of tissues. In extreme cases the condition can prove fatal.

microhabitat A defined part of a habitat within which a particular species or community is most likely to occur.

migration The mass movement of organisms from one place to another.

mitosis The process by which body (somatic) cells divide, each dividing call yielding two daughter cells. Although continuous, for convenience the process is described as comprising four distinct phases. (1) At prophase, the chromosomes contract, becoming visible under a microscope. Each separates into two parallel strands (called chromatids), and the nuclear membrane (enclosing the cell nucleus) disappears. (2) At metaphase, the pairs of chromatids move to a structure formed from protein fibres (called the spindle) and align themselves across its centre (called the equator). (3) At anaphase, the chromatid pairs separate and move to the opposite ends (called poles) of the spindle. (4) at telophase, the cell constricts across its centre, the spindle appratus disappears, the nuclear membrane re-forms, and the chromosomes extend to become long, fine filaments, not clearly visible under a microscope. Each daughter cell now contains a complete set of chromosomes. Between mitotic divisions (called interphase) each chromosome duplicates itself.

morph A particular form of an organism.

Müllerian mimicry The resemblance of one dangerous or poisonous species to another, leading predators to avoid all species of that appearance. It was first described by F. Müller.

mutation A change in the structure or amount of genetic material. Most mutations involve alterations within individual genes, but some comprise major restructuring of chromosomes or changes in the number of chromosomes in the cell nucleus. Mutations are inherited only if they occur in gametes or cells destined to become gametes.

mutualism A relationship between two organisms of different species from which both derive benefit.

nucleic acid Genetic material, occurring as DNA or RNA, that comprises strings of nucleotides.

nucleotide An organic molecule that forms the basic unit in nucleic acids and comprises a pentose sugar (ribose or deoxyribose), a phosphate group, and a purine or pyrimidine base.

oligotrophic Deficient in nutrients.

optimum sustainable yield (OSY) The crop that can be harvested without depleting the population. It is commonly calculated as half the carrying capacity.

oxygen-isotope analysis Examing the ratio of two isotopes of oxygen, $^{18}O : {}^{16}O$, in a sample and inferring from the result information regarding the source or temperature of water or the water from which the sample (e.g. calcium carbonate or silica) was deposited. 'Light' water ($H_2{}^{16}O$) evaporates more readily than 'heavy' water ($H_2{}^{18}O$).

ozone layer A region of the stratosphere, at 20–25 km, where the concentration of ozone (O_3) is higher than elsewhere.

parasitism A relationship between two organisms of different species in which one (the parasite) lives inside or on the surface of the other (the host) and obtains food from it. Parasitism usually injures the host, although in some cases the effect is very slight.

pedogenesis The natural processes by which soil forms.

pedology The scientific study of soils.

perihelion The point at which an orbiting body is closest to the body it orbits.

permafrost Ground that is permanently frozen below the surface. It forms where temperatures below freezing have persisted through two consecutive winters and the intervening summer.

perturbation A disturbance to a system (e.g. the atmosphere, or an ecosystem) caused by something external to that system.

phenotype The observable appearance or characteristics of an organism.

phytoplankton Plankton comprising plants, most of them single-celled algae.

plankton Small organisms, including the juvenile stages of many larger organisms, that drift near the surface of water.

plate tectonics The theory that describes the Earth's crust as comprising a number of rigid sections which move in relation to one another. This movement causes ocean basins to open and close and continents to change their position and orientation.

poikilotherm (exotherm) An organism (e.g. a fish) in which body temperature varies according to that of its environment.

polymorphism The existence of different forms among the members of a species. These may be externally visible (e.g. the number of spots on the wings of ladybirds) or invisible (e.g. blood groups in humans) and may be transitory or permanent.

polyploidy The possession of one or more sets of chromosomes in addition to the two sets ordinarily found. It results from the replication of complete chromosome sets without a subsequent division of the cell nucleus.

purine One of the types of base found in nucleic acids. Adenine and guanine are purines and link only to pyramidines.

pyrimidine One of the types of base found in nucleic acids. Cytosine and thymine are the bases in DNA; in RNA uracil takes the place of thymine. Pyrimidines link only to purines.

radiocarbon dating A dating method based on the ratio of carbon-12 (^{12}C) to carbon-14 (^{14}C) present in a sample. Both isotopes occur in the atmosphere and are absorbed by living organisms. During life, therefore, the $^{12}C : ^{14}C$ ratio in cells is identical to that in the atmosphere. After death, ^{14}C decays and the ratio changes. By measuring the ratio in a sample and relating it to the known half-life of ^{14}C, samples up to about 70 000 years old can be dated. The ^{14}C content of the atmosphere is not constant, but corrections are made for fluctuations over the last 8000 years by reference to the ^{14}C content of tree rings in certain long-living species (e.g. bristlecone pine, *Pinus longaeva*).

radioisotope An isotope that is radioactive and can therefore be used to identify the location or track the movement of the element of which it is an isotope.

radiometric dating Calculation of the time that has elapsed since a rock or organic deposit formed by measuring the ratio of a radioisotope to its decay products or to a non-radioactive isotope and using the known half-life of the radioisotope to determine the time taken for an initial amount to decay to the concentration found in the sample.

residence time (removal time) The length of time a molecule or particle remains in a particular part of the environment (e.g. the air).

sequence (*n.*) The order in which bases occur in nucleic acid or amino acids in a protein. (*v.*) To identify and record that order.

seral stage One of the plant communities forming part of a sere (or succession).

sere The succession of distinct plant communities that follow one another and lead from the community of primary colonizers to the climax community.

serotinous Occurring late, as in plants that flower late in the year or events that take place late in the day.

solar constant The total amount of solar radiation in free space at the mean distance of the Earth from the Sun. It is approximately 1.36 W m^{-2}.

solarization The inhibition of photosynthesis that occurs at very high light intensity.

somatic mutation A mutation that affects a cell in the body (i.e. not a gamete) and that is transmitted only to those cells derived by mitosis from the cell carrying the mutation.

species diversity The number of species present in a particular area or community.

symbiont An organism that lives in symbiotic association with another.

symbiosis A situation in which two organisms of different species live together and are closely associated. The term generally implies some benefit to both, but is sometimes used to include parasitism.

synzoochory The collection of plant seeds by an animal and the storing of them for future consumption.

tectonic (*adj.*) Applied to a rock structure formed directly by movements within the crust (e.g. by faulting or folding).

tectonics The scientific study of the major structural features of the Earth's crust.

till Material that has been transported by a glacier.

trophic Pertaining to nourishment.

varve A layer of silt and sand deposited on the bed of a lake. Coarse, pale material is deposited in summer, dark, fine-grained material in winter, so each varve comprises one pale and one dark band. The sediment can be dated by counting the varves. Varves commonly form near the edges of glaciers.

weathering The breaking down of rocks into smaller particles by physical and chemical processes.

'white smoker' *See* **black smoke**.

xerophyte A plant adapted to arid conditions.

xerosere A sere (ecological succession) that begins in arid conditions.

zooplankton Plankton consisting of animals.

Bibliography

Abelson, Philip H. 1995. 'Renewable liquid fuels', *Science*, 268, 995.

Abelson, Philip H. and Hines, Pamela J. 1999. 'The plant revolution', *Science*, 285, 367–368.

Abramov, Valery. 1990. 'Man vs mosquito: the arms race is on', in *Developing World Health*, Grosvenor Press International, London. pp. 34–36.

Adams, Peter. 1977. *Moon, Mars and Meteorites*. HMSO for the British Geological Survey.

Alderson, Lawrence. 1978. *The Chance to Survive*. Cameron and Tayleur in association with David and Charles, Newton Abbot.

Allaby, Ailsa and Michael (eds.) 1999. *The Concise Oxford Dictionary of Earth Sciences*. 2nd edition. Oxford University Press, Oxford.

Allaby, Michael. 1977. *World Food Resources, Actual and Potential*. Applied Science Publishers, London.

——.1981. *A Year in the Life of a Field*. David and Charles, Newton Abbot.

——.1986. *The Woodland Trust Book of British Woodlands*. David and Charles, Newton Abbot.

——.1987. 'Environment' entry in *Britannica Book of the Year 1987*. Encyclopaedia Britannica, Chicago.

——.1989. *Green Facts*. Hamlyn, London.

——.(ed.). 1989. *Thinking Green: An Anthology of Essential Ecological Writing*. Barrie & Jenkins, London.

——.1990. *Into Harmony with the Planet*. Bloomsbury Publishing, London.

——.1992. *Air: The Nature of Atmosphere and the Climate*. Facts on File, New York.

——.1992. *Water*. Facts on File, New York.

——.1993. *Earth*. Facts on File, New York.

——.1993. *Fire*. Facts on File, New York.

——.1995. *Facing the Future*. Bloomsbury, London.

——.1998. *Dangerous Weather: Floods*. Facts on File, Inc., New York.

——.1998. *Dangerous Weather: Droughts*. Facts on File, Inc., New York.

——.1999. *Ecosystems: Temperate Forests*. Fitzroy Dearborn, London.

Angel, Heather, Duffey, Eric, Miles, John, Ogilvie, M.A., Simms, Eric, and Teagle, W.G. 1981. *The Natural History of Britain and Ireland*. Michael Joseph, London.

Appleby, A. John. 1999 'The Electrochemical Engine for Vehicles', in *Scientific American*, July 1999, pp. 58–63.

Avery, Dennis. 1995. 'Saving the Planet with Pesticides: Increasing Food Supplies While Preserving the Earth's Biodiversity', in Bailey, Ronald (ed.) *The True State of the Planet*, The Free Press, New York.

Balling, Robert C., Jr. 1995. 'Global Warming: Messy Models, Decent Data and Pointless Policy', in Bailey, Ronald (ed.) *The True State of the Planet*, The Free Press, New York, pp 83–107.

Baron, Stanley. 1972. *The Desert Locust*. Eyre Methuen, London.

Barry, Roger G. and Chorley, Richard J. 1982. *Atmosphere, Weather and Climate*. Methuen, London.

Begon, Michael, Harper, John L., and Townsend, Colin R. 1990. *Ecology: Individuals, Populations and Communities*, 2nd edn. Blackwell Science, Oxford.

Bender, Barbara. 1975. *Farming in Prehistory*. John Baker, London.

Berry, R.J. 1977. *Inheritance and Natural History*. Collins New Naturalist, London.

BMA. 1991. *Hazardous Waste and Human Health*. Oxford University Press, Oxford.

Boehmer-Christiansen, Sonja A. 1994. 'A scientific agenda for climate policy?', *Nature*, 372, 400–402.

Bolin, Bert. 'Politics of climate change', *Nature*, 374, 208.

Bowler, Peter J. 1992. *The Fontana History of the Environmental Sciences*. Fontana Press (HarperCollins), London.

Bragg, James R., Prince, Roger C., Harner, E. James, and Atlas, Ronald M. 1994. 'Effectiveness of bioremediation for the *Exxon Valdez* spill', *Nature*, 368, 413–418.

Braidwood, Robert J. 1960. 'The Agricultural Revolution', *Scientific American*, May.

Brewer, Richard. 1988. *The Science of Ecology*, 2nd edn. Saunders College Publishing, Fort Worth, Texas.

Brooks, Robert, Anderson, Chris, Stewart, Robert, and Robinson, Brett. 1999. 'Phytomining: growing a crop of a metal', *Biologist*, 46, 5, 201–205.

Burbank, Douglas W. 1992. 'Causes of recent Himalayan uplift deduced from deposited patterns in the Ganges basin', *Nature*, 357, 680–682.

Burgess, G.H.O. 1967. *The Curious World of Frank Buckland*. John Baker, London.

Burton, Maurice and Burton, Robert. 1977. *Inside the Animal World*. Macmillan, London.

Calder, Nigel. 1999. 'The Carbon Dioxide Thermometer and the Cause of Global Warming', *Energy and Environment*, 10, 1, 1–18.

Campbell, Joseph. 1962. *The Masks of God: Oriental Mythology*. Arkana (Penguin Books), London.

Cane, Mark A., Eshel, Gidon, and Buckland, R.W. 1994. 'Forecasting Zimbabwean maize yield using eastern equatorial Pacific sea surface temperature', *Nature*, 370, 204–205.

Carson, Rachel. 1962. *Silent Spring*. Penguin Books, Harmondsworth.

Clark, John. 1977. *Coastal Ecosystem Management*. Wiley-Interscience, New York.

Clark, R.B. 1992. *Marine Pollution*, 3rd edn. Oxford University Press, Oxford.

Clement, Keith. 1995. 'Investing in Europe: government support for environmental technology', *Greener Management International*, January, pp. 41–51.

Clutton-Brock, Juliet. 1981. *Domesticated Animals from Early Times*. Heinemann and British Museum (Natural History), London.

Cockburn, Andrew. 1991. *An Introduction to Evolutionary Ecology*. Blackwell Scientific Publications, Oxford.

Colyer, Charles N. and Hammond, Cyril O. 1968. *Flies of the British Isles,* 2nd edn. Frederick Warne & Co., London.

Conway, G.R., Gilbert, D.G.R., and Pretty, J.N. 1988. 'Pesticides in the UK: science, policy and the public', in Harding, D.J.L. (ed.) *Britain since 'Silent Spring'*. Institute of Biology, London.

Cooke, G.W. 1972. *Fertilizing for Maximum Yield*. Crosby Lockwood, London.

Copley, Jon. 1999. 'Indestructible', in *New Scientist*, 23 Oct., pp. 45–46.

Corbet, G.B. and Southern, H.N. 1977. *The Handbook of British Mammals*, 2nd edn. Blackwell Scientific Publications, Oxford.

Cowling, Sharon A. 1999. 'Plants and Temperature-CO^2 Uncoupling', *Science*, 285, 1500–1501.

Crawford, R.M.M. 1995. 'Plant survival in the High Arctic', *Biologist*, 42, (3), 101–105. Institute of Biology, London.

Cruickshank, James G. 1972. *Soil Geography*. David and Charles, Newton Abbot.

Crystal, Ronald G. 1995. 'Transfer of genes to humans: early lessons and obstacles to success', *Science*, 270, 404–410.

Culotta, Elizabeth. 1995. 'Many suspects to blame in Madagascar extinctions', *Science*, 268, 1568–1569.

——.1995. 'Bringing back the Everglades', *Science*, 268, 1688–1690.

——.1995. 'Shades of rain-forest green', *Science*, 269, 31

——.1995. 'Ecologists flock to snowbird for varied banquet of findings', *Science*, 269, 1045–1046.

Dajoz, R. 1975. *Introduction to Ecology*, 3rd edn. Hodder & Stoughton, London.

Dalziel, Ian W.D. 1995. 'Earth before Pangaea'. *Scientific American*, January.

Darwin, Charles. 1859. *On the Origin of Species by Means of Natural Selection*.

Dawkins, Richard. 1978. *The Selfish Gene*. Granada Books, London.

De Baar, Hein J.W., de Jong, Jeroen T.M., Bakker, Dorothée C.E., Löscher, Bettina M., Veth, Cornelius, Bathmann, Uli, and Smetacek, Victor. 1995. 'Importance of iron for plankton blooms and carbon dioxide drawdown in the Southern Ocean', *Nature*, 373, 412–415.

DellaPenna, Dean. 1999. 'Nutritional genomics: manipulating plant micronutrients to improve human health', *Science*, 285, 375–379.

Dizon, Andrew E., Lockyer, Christina, Perrin, William F., Demaster, Douglas P., and Sisson, Joyce. 1992. 'Rethinking the stock concept: a phylogeographic approach', *Conservation Biology*, 6, (1), 24–36.

Donahue, Roy L., Miller, Raymond W., and Shickluna, John C. 1958. *Soils: An Introduction to Soils and Plant Growth*. Prentice-Hall, Englewood Cliffs, NJ.

Dowdeswell, W.H. 1984. *Ecology: Principles and Practice*. Heinemann Educational Books, Oxford.

Dunning, F.W., Mercer, I.W., Owen, M.P., Roberts, R.H., and Lambert, J.L.M. 1978. *Britain before Man*. HMSO for the Institute of Geological Sciences, London.

Ecker, Joseph R. 1995. 'The ethylene signal transduction pathway in plants', *Science*, 268, 667–675.

Eddy, John A. 1977. 'The case of the missing sunspots,' *Scientific American*, May, pp. 80–89.

Ehrlich, Paul R. and Ehrlich, Anne H. 1970, 1972. *Population, Resources, Environment: Issues in Human Ecology*. W.H. Freeman & Co., San Francisco.

Fincham, J.R.S. 1995. 'The killing fields' (book review), *Nature*, 377, 297.

Ford, E.B. 1981. *Taking Genetics into the Countryside*. Weidenfeld and Nicolson, London.

Foster, Karen Polinger. 1999. 'The earliest zoos and gardens', *Scientific American*, July 1999, pp. 48–55.

Foth, H.D. and Turk, L.M. 1972. *Fundamentals of Soil Science*. John Wiley & Sons, New York.

Freeman, Paul. 1973. 'Ceratopogonidae (biting midges, "sand-flies", "punkies")', in Smith, Kenneth G.V. (ed.) *Insects and Other Arthropods of Medical Importance*. British Museum (Natural History), London.

Fry, Norman. 1984. *The Field Description of Metamorphic Rocks*. Geological Society of London Handbook. Open University Press, Milton Keynes (published by John Wiley and Sons in the USA and Canada).

Gentry, A.W. and Sutcliffe, A.J. 1981. 'Pleistocene geography and mammal faunas', in Cocks, L.R.M. (ed.) *The Evolving Earth: Chance, Change and Challenge*, British Museum (Natural History) and Cambridge University Press, London.

Gibbons, Ann. 1995. 'When it comes to evolution, humans are in the slow class', *Science*, 267, 1907–1908.

Gill, Martin. 1999. 'Fisheries', in *Britannica Book of the Year 1999*, Encyclopedia Britannica, Chicago, p. 130.

Gimingham, C.H. 1975. *An Introduction to Heathland Ecology*. Oliver and Boyd, Edinburgh.

Gleick, James. 1988. *Chaos: Making a New Science*. Heinemann, London.

Glob, P.V. 1969. *The Bog People*. Faber & Faber, London.

Godwin, Sir Harry. 1975. *History of the British Flora: A Factual Basis for Phytogeography*, 2nd edn. Cambridge University Press, Cambridge.

Goldsmith, Edward, Allen, Robert, Allaby, Michael, Davoll, John, and Lawrence, Sam. 1972. 'A Blueprint for Survival', *The Ecologist*, 2, (1).

Goldstein, Jerome (ed.). 1978. *The Least Is Best Pesticide Strategy*. The JG Press, Emmaus, PA.

Goudie, Andrew. 1986. *The Human Impact on the Natural Environment*. Basil Blackwell, Oxford.

Graham, A.L. 1981. 'Plate tectonics', in Cocks, L.R.M. (ed.) *The Evolving Earth: Chance, Change and Challenge*. British Museum (Natural History) and Cambridge University Press, London.

Graham, Jeffrey B., Dudley, Robert, Agullar, Nancy M., and Gans, Carl. 1995. 'Implications of the late Palaeozoic oxygen pulse for physiology and evolution'. *Nature*, 375, 117–120.

Green, M.B. 1976. *Pesticides, Boon or Bane?* Elek Books, London.

Grime, J.P. 1979. *Plant Strategies and Vegetation Processes*. John Wiley & Sons, Chichester.

Grootes, P.M., Stuiver, M., White, J.W.C., Johnsen, S., and Jouzel, J. 1993. 'Comparison of oxygen isotope records from the GISP2 and GRIP Greenland ice cores', *Nature*, 366, 552–554.

Grove, Richard H. 1992. 'Origins of Western environmentalism', *Scientific American*, July, pp. 22–27.

Hambrey, M.J. and Harland, W.B. 1981. 'The evolution of climates', in Cocks, L.R.M. (ed.) *The Evolving Earth: Chance, Change and Challenge*. British Museum (Natural History) and Cambridge University Press, London.

Hammerstein, Peter and Hoekstra, Rolf F. 1995. 'Mutualism on the move', *Nature*, 376, 121–122.

Harman, Denham. 1992. 'Role of free radicals in aging and disease', in Fabris, Nicola, Harman, Denham, Knook, Dick L., Steinhagen-Thiessen, Elizabeth, and Zs.-Nagy, Imre (eds) *Physiopathological Processes of Aging*, Annals of the NY Academy of Sciences, Vol. 673. NY Academy of Sciences, New York.

Harris, David R. 1996. 'The origins and spread of agriculture and pastoralism in Eurasia: an overview', in Harris, David R. (ed.) *The Origins and Spread of Agriculture and Pastoralism in Eurasia*. Univ. Coll. London Press, pp. 552–573.

Harvey, John. G. 1976. *Atmosphere and Ocean: Our Fluid Environments*. Artemis Press, Horsham, Sussex.

Haub, Carl V. 1995. 'Demography', in *Britannica Book of the Year 1995*, Encyclopedia Britannica, Chicago, pp. 255–256.

——.1999. 'Demography', in *Britannica Book of the Year 1999*, Encyclopedia Britannica, Chicago, p. 300.

Helfferich, Carla. 1993. 'Sandwort, Seabirds, and Surtsey', Article 1132, Alaska Science Forum. (*www.gi.alaska.edu/ScienceForum/ASF11/1132.html*)

Hidore, John J. and Oliver, John E. 1993. *Climatology: An Atmospheric Science*. Macmillan, New York.

Hill, David K. 1995. 'Pacific warming unsettles ecosystems', *Science*, 267, 1911–1912.

Hillstead, A.F.C. 1945. *The Blackbird*. Faber & Faber, London.

Hodges, Carroll Ann. 1995. 'Mineral resources, environmental issues, and land use', *Science*, 268, 1305–1312.

Holm, Søren and Harris, John. 1999. 'Precautionary principle stifles discovery', letter to *Nature*, 400, 398.

Holmes, Arthur. 1965. *Principles of Physical Geology*, 2nd edn. Nelson, Walton-on-Thames.

Holmes, Bob. 1995. 'Fishing treaty wins unanimous approval', *New Scientist*, 12 August, p. 7.

Holmes, Bob. 1999. 'A dark cloud over Asia', *New Scientist*, 13 February, p. 25.

Hooper, Paul D. and Gibbs, David. 1995. 'Cleaner technology: a means to an end, or an end to a means?', *Greener Management International*, January, pp. 28–40.

Huck, U. William. 1995. 'Lemming migrations', in Macdonald, David (ed.) *The Encyclopedia of Mammals*, Andromeda Oxford, Abingdon, Oxon., pp. 656–657.

Hudson, Norman. 1971. *Soil Conservation*. B.T. Batsford, London.

Hunt, Charles B. 1972. *Geology of Soils: Their Evolution, Classification, and Uses*. W.H. Freeman & Co., San Francisco.

Hutchins, Ross E. 1966. *Insects*. Prentice-Hall, Englewood Cliffs, NJ.

Hutchison, R. 1981. 'The origin of the Earth', in Cocks, L.R.M. (ed.) *The Evolving Earth: Chance, Change and Challenge*. British Museum (Natural History) and Cambridge University Press, London.

Hyams, Edward. 1976. *Soil and Civilization*. John Murray, London.

Independent Commission on International Development Issues. 1980. *North–South: A Programme for Survival*. Pan Books, London.

Inhaber, Herbert and Saunders, Harry. 1994. 'Road to nowhere', *The Sciences*, November/December 1994. NY Academy of Sciences, New York.

Intergovernmental Panel on Climate Change. 1992. *1992 IPCC Supplement: Scientific Assessment of Climate Change*. WMO and UNEP.

International Union for the Conservation of Nature and Natural Resources (IUCN). 1980. *World Conservation Strategy: Living Resource Conservation for Sustainable Development*. IUCN, Gland.

Jacobs, G.A., Hurlburt, H.E., Kindle, J.C., Metzger, E.J, Mitchell, J.L., Teague, W.J., and Wallcraft, A.J. 1994. 'Decade-scale trans-Pacific propagation and warming effects of an El Niño anomaly', *Nature*, 370, 360–363. London.

Janzen, Daniel H. 1975. *Ecology of Plants in the Tropics*. Studies in Biology no. 58 (Institute of Biology). Edward Arnold, London.

Johnson, C.G. 1963. 'The aerial migration of insects', in Eisner, Thomas and Wilson, Edward O. (eds) *The Insects*. W.H. Freeman & Co., San Francisco.

Johnson, Nicholas and David, Andrew. 1982. 'A Mesolithic site on Trevose Head and contemporary geography', *Cornish Archaeology*, 21, 67–103.

Joseph, Lawrence E. 1990. *Gaia: The Growth of an Idea*. Arkana (Penguin Books), London.

Kempe, D.R.C. 1981. 'Deep ocean sediments', in Cocks, L.R.M. (ed.) *The Evolving Earth: Chance, Change and Challenge*. British Museum (Natural History) and Cambridge University Press, London.

Kendeigh, S. Charles. 1974. *Ecology with Special Reference to Animals and Man*. Prentice-Hall, Englewood Cliffs, NJ.

Kerr, Richard A. 1995. 'Sun's role in warming is discounted', *Science*, 268, 28–29.

Kupchella, Charles E. and Hyland, Margaret C. 1986. *Environmental Science,* 2nd edn. Allyn and Bacon, Needham Heights, MA.

Kvenvolden, Keith A. 1994. 'Natural gas hydrate occurrence and issues', in Sloan, E. Dendy Jr, Happel, John, and Hnatow, Miguel A. (eds) *Natural Gas Hydrates*, Annals of the NY Academy of Sciences vol. 715. NY Academy of Sciences, New York.

Lachance, L.E. 1974. 'Status of the sterile-insect release method in the world', in *The Sterile-Insect Technique and Its Field Applications*. International Atomic Energy Agency, Vienna. pp. 55–62.

Lamb, H.H. 1995. *Climate, History and the Modern World*. 2nd edn. Routledge, London.

Lavallée, H.-Claude. 1999. 'Wood products', in *Britannica Book of the Year 1999*, Encyclopedia Britannica, Chicago, p. 182.

Lave, Lester B., Hendrickson, Chris T., and McMichael, Francis Clay. 1995. 'Environmental implications of electric cars', *Science*, 268, 993–995.

Lawler, Andrew. 1995. 'NASA mission gets down to Earth', *Science*, 269, 1208–1210.

Lawton, John H. 1995. 'Population dynamic principles', in Lawton, John H. and May, Robert E. (eds) *Extinction Rates*. Oxford University Press, Oxford.

Leach, Gerald. 1975. *Energy and Food Production*. International Institute for Environment and Development, London.

L'hirondel, Jean-Louis. 1999. 'Are dietary nitrates a threat to human health?', in Morris, Julian and Bate, Roger (eds) *Fearing Food*, Butterworth Heinemann, Oxford. pp. 38–46.

Loder, Natasha. 1999. 'Journal under attack over controversial paper on GM food', *Nature*, 401, 731.

Long, Steven P. and Mason, Christopher F. 1983. *Saltmarsh Ecology*. Blackie, Glasgow and London.

Lovelock, James. 1979. *Gaia: A New Look at Life on Earth*. Oxford University Press, Oxford.

——.1988. *The Ages of Gaia*. Oxford University Press, Oxford.

MacKenzie, James J. and El-Ashry, Mohamed. 1989. *Air Pollution's Toll on Forests and Crops*. Yale University Press, New Haven, Conn.

McCormick, M. Patrick, Thomason, Larry W., and Trepte, Charles R. 1995. 'Atmospheric effects of the Mt Pinatubo eruption', *Nature*, 373, 399–404.

Mace, Georgina M. 1995. 'Classification of threatened species and its role in conservation planning', in Lawton, John H. and May, Robert M. (eds) *Extinction Rates*. Oxford University Press, Oxford.

Malthus, Thomas. 1798. *An Essay on the Principle of Population as It Affects the Future Improvement of Society*. Penguin edition 1985.

Marshall, Eliot. 1995. 'Gene therapy's growing pains', *Science*, 269, 1050–1055.

Marshall, N.B. 1979. *Developments in Deep-Sea Biology*. Blandford, Poole, Dorset.

Mason, C.F. 1991. *Biology of Freshwater Pollution*, 2nd edn. Longman Scientific and Technical, Harlow.

May, Robert M., Lawton, John H., and Stork, Nigel E. 1995. 'Assessing extinction rates', in Lawton, John H. and May, Robert E. (eds) *Extinction Rates*. Oxford University Press, Oxford.

Meadows, Donella H., Meadows, Dennis L., Randers, Jorgen, and Behrens, William W. III. 1972. *The Limits to Growth*. Earth Island, London.

Mellanby, Kenneth. 1992. *The DDT Story*. British Crop Protection Council, Farnham.

——.1992. *Waste and Pollution: The Problem for Britain*. HarperCollins, London.

Ministry of Agriculture, Fisheries and Food and Department of Agriculture and Fisheries for Scotland. 1968. *A Century of Agricultural Statistics: Great Britain 1866–1966*. HMSO, London.

Mlot, Christine. 1995. 'In Hawaii, taking inventory of a biological hot spot', *Science*, 269, 322–323.

Moffat, Anne Simon. 1995. 'Exploring transgenic plants as a new vaccine source', *Science*, 268, 658–660.

——.1995. 'Plants proving their worth in toxic metal cleanup', *Science*, 269, 302–303.

——.1999. 'Crop engineering goes south', *Science*, 285, 370–371.

——.1999a. 'Engineering plants to cope with metals', *Science*, 285, 369–370.

Moore, David M. (ed.) 1982. *Green Planet: The Story of Plant Life on Earth*. Cambridge University Press, Cambridge.

Moore, Peter D. 1995. 'Too much of a good thing', *Nature*, 374, 117–118, London.

Morell, Virginia. 1995. 'The earliest art becomes older — and more common', *Science*, 267, 1908–1909.

——.1995. 'Cowardly lions confound cooperation theory', *Science*. 269, 1216–1217.

Mumford, Lewis. 1961. *The City in History*. Pelican Books, London.

Myers, R.A., Barrowman, N.J., Hutchings, J.A., and Rosenberg, A.A. 1995. 'Population dynamics of exploited fish stocks at low population levels', *Science*, 269, 1106–1108.

Naeem, Shahid, Thompson, Lindsey J., Lawler, Sharon P., Lawton, John H., and Woodfin, Richard M. 1994. 'Declining biodiversity can alter the performance of ecosystems', *Nature*, 368, 734–737.

National Academy of Sciences. 1972. *Genetic Vulnerability of Major Crops*. NAS, Washington, DC.

National Environment Research Council. 1974. *The Clyde Estuary and Firth: An Assessment of Present Knowledge Compiled by Members of the Clyde Study Group*. NERC Publications Series C no. 11.

Nisbet, E.G. and Fowler, C.M.R. 1995. 'Is metal disposal toxic to deep oceans?', *Nature*, 375, 715.

Nowak, Martin A., May, Robert M., and Sigmund, Karl. 1995. 'The arithmetics of mutual help', *Scientific American*, June, pp. 50–55.

Nowak, Martin A. and Sigmund, Karl. 1992. 'Tit for tat in heterogeneous populations', *Nature*, 355, 250–252. .

——.1993. 'A strategy of win–stay, lose–shift that outperforms tit-for-tat in the Prisoner's Dilemma game', *Nature*, 364, 56–58.

Nowicki, Maciej. 1992. *Environment in Poland: Issues and Solutions*. Ministry of Environmental Protection, Natural Resources, and Forestry, Warsaw.

Odum, E.P. 1984. *Fundamentals of Ecology*. W.B. Saunders, Philadelphia.

Oldroyd, D.R. 1980. *Darwinian Impacts*. Open University Press, Milton Keynes.

Parkes, R. John. 1999. 'Oiling the wheels of controversy', book review in *Nature*, 401, 644.

Parkinson, Claire L. 1997. *Earth From Above: Using Color-Coded Satellite Images to Examine the Global Environment*. University Science Books, Sausalito, CA.

Partridge, Michael. 1973. *Farm Tools through the Ages*. Osprey Publishing, Reading.

Patterson, Colin. 1978. *Evolution*. Routledge & Kegan Paul in association with the British Museum (Natural History), London.

Payne, C.C. 1988. 'Prospects for biological control', in Harding, D.J.L. (ed.) *Britain since 'Silent Spring'*. Institute of Biology, London.

Pearce, Fred. 1999. 'Counting down', *New Scientist*, 2 October 1999, pp. 20–21.

Pellew, Robin. 1995. 'Biodiversity conservation — why all the fuss?', *RSA Journal*, CXLIII (5461), 54.

Pennington, Winifred. 1974. *The History of British Vegetation*, 2nd edn. Hodder and Stoughton, London.

Persson, Reidar. 1995. 'Myths and truths on global deforestation', *IRD Currents* (Uppsala), May, pp. 4–6.

——.1995. 'Tropical plantations — success or failure?', *IRD Currents* (Uppsala), May, pp. 7–9.

Phillipson, J. 1971. *Methods of Study in Quantitative Soil Ecology: Population, production and energy flow*. International Biological Programme, London, and Blackwell Scientific Publications, Oxford.

Piel, Gerard. 1992. *Only One World*. W.H. Freeman, New York.

Pimm, Stuart L. 1984. 'The complexity and stability of ecosystems', *Nature*, 307, 321–326.

Pimm, Stuart L., Russell, Gareth J., Gittleman, John L., and Brooks, Thomas M. 1995. 'The future of biodiversity', *Science*, 269, 347–350.

Plucknett, Donald L. and Winkelmann, Donald L. 1995. 'Technology for sustainable agriculture', *Scientific American*, September, p. 150.

Pollard, E., Hooper, M.D., and Moore, N.W. 1974. *Hedges*. Collins New Naturalist, London.

Post, Eric, Peterson, Rolf O., Stenseth, Nils Chr., and McLaren, Brian. 1999. 'Ecosystem consequences of wolf behavioural response to climate', *Nature*, 401, 905–907.

Potier, Michel. 1995. 'China charges for pollution', *OECD Observer*, February/March, pp. 18–22.

Preston-Mafham, Rod and Preston-Mafham, Ken. 1984. *Spiders of the World*. Blandford Press, Poole, Dorset.

Rackham, Oliver. 1976. *Trees and Woodland in the British Landscape*. J.M. Dent, London.

Raffensperger, Carolyn, Tickner, Joel, Schettler, Ted, and Jordan, Andrew. 1999. '… and can mean saying "yes" to innovation', letter to *Nature*, 401, 207.

Raven, Peter H., Berg, Linda R., and Johnson, George B. 1993. *Environment*. Saunders College Publishing, Orlando, FL.

Reinhold, Robert. 1992. 'The lingering US drought', in *Britannica Book of the Year 1992*. Encyclopaedia Britannica, Chicago.

Ridley, Mark. 1985. *The Problems of Evolution*. Oxford University Press (paperback), Oxford.

Roberts, Neil. 1989. *The Holocene: An Environmental History*. Basil Blackwell, Oxford.

Rose, Steven. 1995. 'The rise of neurogenetic determinism', *Nature*, 373, 380–382.

Royal Commission on Environmental Pollution. 1979. *Seventh Report: Agriculture and Pollution*. HMSO, London.

Rudwick, Martin J.S. 1976. *The Meaning of Fossils: Episodes in the History of Palaeontology*. University of Chicago Press, Chicago.

Sankaran, Mahesh and McNaughton, S.J. 1999. 'Determinants of biodiversity regulate compositional stability of communities', *Nature*, 401, 691–693.

Simmons, I.G., Dimbleby, G.W., and Grigson, Caroline. 1981 'The Mesolithic', in Simmons, Ian and Tooley, Michael (eds) *The Environment in British Prehistory*. Duckworth, London.

Simmons, I.G. and Tooley, M.J. (eds) 1981. *The Environment in British Prehistory*. Duckworth, London.

Simpson, Sarah. 1999. 'Making waves', *Scientific American*, August 1999.

Small, R.J. 1970. *The Study of Landforms*. Cambridge University Press.

Smith, George E. 1967. 'Fertilizer nutrients as contaminants in water supplies,' in Brady, Nyle C. (ed.) *Agriculture and the Quality of Our Environment*, American Association for the Advancement of Science, Washington, DC.

Staskawicz, Brian J., Ausubel, Frederick M., Baker, Barbara J., Ellis, Jeffrey G., and Jones, Jonathan D.G. 1995. 'Molecular genetics of plant disease resistance', *Science*, 268, 661–667.

Stock, Christina. 1995. 'Flying in the face of tradition', *Scientific American*, June, pp. 11–12.

Stone, Richard. 1995. 'Taking a new look at life through a functional lens', *Science*, 269, 316–317.

Stonehouse, Bernard. 1981. *Saving the Animals*. Weidenfeld and Nicolson, London.

Stratton, J.M. and Brown, Jack Houghton. 1978. *Agricultural Records AD 220–1977*, 2nd edn. John Baker, London.

Tansley, A.G. 1946. *Introduction to Plant Ecology*. George Allen and Unwin, London.

Taubes, Gary. 1995. 'Epidemiology faces its limits', *Science*, 269, 164–169.

Thomas, Keith. 1983. *Man and the Natural World*. Penguin Books, London.

Thompson, Dick. 1995. 'Soil — financial asset or environmental resource?', *RSA Journal*, CXLIII, (5461), 56–67.

Thompson, H.V. and Worden, A.N. 1956. *The Rabbit*. Collins New Naturalist, London.

Thomson, David J. 1995. 'The seasons, global temperature, and precession', *Science*, 268, 59–68.

Thorpe, Richard and Brown, Geoff. 1985. *The Field Description of Igneous Rocks*. Geological Society of London Handbook. Open University Press, Milton Keynes (published by John Wiley & Sons in the USA and Canada).

Tilman, David and Downing, John A. 1994. 'Biodiversity and stability in grasslands', *Nature*, 367, 363–365.

Tivy, Joy. 1993. *Biogeography: A Study of Plants in the Ecosphere*, 3rd edn. Longman Scientific and Technical, Harlow.

Tolba, Mostafa K. and El-Kholy, Osama A. 1992. *The World Environment 1972–1992*. Chapman & Hall, London, on behalf of UNEP.

Townshend, J.R.G., Tucker, C.J., and Goward, S.N. 1993. 'Global vegetation mapping', in Gurney, R.J., Foster, J.L., and Parkinson, C.L. (eds) *Atlas of Satellite Observations Related to Global Change*. Cambridge University Press, Cambridge.

Tranter, Neil. 1973. *Population since the Industrial Revolution: The Case of England and Wales*. Croom Helm, London.

Trimble, Stanley W. 1999 'Decreased Rates of Alluvial Sediment Storage in the Coon Creek Basin, Wisconsin, 1975–93', *Science*, 285, 1244–1246.

Tucker, Maurice E. 1982. *The Field Description of Sedimentary Rocks*. Geological Society of London Handbook. Open University Press, Milton Keynes (published by John Wiley & Sons in the USA and Canada).

Tudge, Colin. 1993. *The Engineer in the Garden*. Jonathan Cape, London.

Tuomisto, Hanna, Ruokolainen, Kalle, Kalliola, Risto, Linna, Ari, Danjoy, Walter, and Rodriguez, Zoila. 1995. 'Dissecting Amazonian biodiversity', *Science*, 269, 63–66.

Tyson, D. 1988. 'Future trends in the control of pests, weeds and diseases in the UK — with regard to environmental impact', in Harding, D.J.L. (ed.) *Britain since 'Silent Spring'*. Institute of Biology, London.

Vaeck, Mark, Reynaerts, Arlette, Höfte, Herman, Jansens, Stefan, De Beuckeleer, Marc, Dean, Caroline, Zabeau, Marc, Van Montagu, Marc, and Leemans, Jan. 1987. 'Transgenic plants protected from insect attack', *Nature*, 328, 33–37.

Veldman, Meredith. 1994. *Fantasy, the Bomb, and the Greening of Britain*. Cambridge University Press, New York.

Von Felbert, Dirk. 1995. 'Trade, environment and aid', *OECD Observer*, August/September, pp. 6–10.

Wagner, Frederike, Bohncke, Sjoerd J.P., Dilcher, David L., Kürschner, Wolfram M., van Geel, Bas, and Visscher, Henk. 1999. 'Century-Scale Shifts in Early Holocene Atmospheric CO_2 Concentration', *Science*, 284, 1971–1973.

Ward, Barbara and Dubos, René. 1972. *Only One Earth: The Care and Maintenance of a Small Planet*. André Deutsch, London.

Webb, S. David. 1995. 'Biological implications of the middle Miocene Amazon seaway', *Science*, 269, 361–362.

Weissmann, Gerald, 1998, *Darwin's Audubon*. Plenum Press, New York.

Westbroek, Peter. 1992. *Life as a Geological Force*. W.W. Norton, New York.

Whitmore, T.C. 1975. *Tropical Rain Forests of the Far East*. Oxford University Press, Oxford.

Wilson, Edward O. 1992. *The Diversity of Life*. Penguin Books, London.

Wilson, Michael A., Hillman, John R., and Robinson, David J. 1999. 'Genetic modification in context and perspective', in Morris, Julian and Bate, Roger (eds.) *Fearing Food*. Butterworth-Heinemann, p. 67.

Windley, Brian F. 1984. *The Evolving Continents*. John Wiley & Sons, Chichester.

Winter, E.J. 1974. *Water, Soil and the Plant*. Macmillan, London.

World Commission on Environment and Development. 1987. *Our Common Future*. Oxford University Press, Oxford.

World Wildlife Fund UK, Nature Conservancy Council, Countryside Commission, Countryside Commission for Scotland, The Royal Society of Arts, and Council for Environmental Conservation. 1983. *The Conservation and Development Programme for the UK: A Response to the World Conservation Strategy*. 2 vols. Kogan Page, London.

Young, Arthur. 1808. *General Report on Enclosures*. Republished, 1971, by Augustus M. Kelley, New York.

Young, J.Z. 1981. *The Life of Vertebrates*, 3rd edn. Clarendon Press, Oxford.

Zahn, Rainer. 1994. 'Linking ice-core records to ocean circulation', *Nature*, 371, 289.

Index

disseminules 214, 216
distillation 101
diversity 183–188
DNA 205, 251–252, 266
Dobson units 53
Dokuchaev, Vasily Vasilievich 110
dolomite 26, 27
Douglas, Mary 11
Downing, John A. 186–187
drainage 33–34, 104, 144
Du Rietz, Gustaf Einar 164
dunes 112
Dust Bowl 111, 115

Earth: age 19; black body temperature
    44–45; flows 32; formation/structure
    19–23; orbit 37, 54–56, 76–78
Earth Summit *see* Rio Summit
earthquakes 23, 32, 104
ecological energetics 163
ecology 9–10, 13, 274–277
Economic Council for Europe (ECE) 289
ecosystems 163–168, 183–188
ecotones 139, 167
Ecotron 187
Eddy, John A. 78–79
edge effect 186
Ehrlich, Anne 277
Ehrlich, Paul 277
El Niño-Southern Oscillation Event (ENSO)
    49, 59–61, 64, 80, 81, 230
electric vehicles 127
electrolysis 101, 102, 132
Elton, Charles 160
Endangered Species Act 1973 262
endozoochory 214
energy budgeting 242–243
environmental impact assessment 2
environmental quality 7–8
environmentalism 2–3, 9, 10, 16, 50, 274
epidemiology 8
epizoochory 214
erosion 30, 32–33, 36, 107, 119–123, 241
estuaries 34–36
ethylene 254, 255
euphotic zone 138
European Union (EU) 8, 94
eutrophication 95, 96–99, 151
evapotranspiration 83

Everglades restoration 274–275
evolution 12, 13, 171, 194, 200–214, 224
evolutionarily stable strategies (ESSs) 207,
    208
exfoliation 107
exotic species 222
extinctions 72, 224, 226–227, 232, 261–262,
    264–265, 267
extremophiles 146–147
Exxon Valdez spill 276

feedback 43, 188–190, 191–192
feeding relationships 155–159
feldspars 28, 29
Ferrel, William 57
fertilizers 96, 115, 117, 119, 149, 151, 193
fire 172–173, 174, 176
fisheries 227–233
fitness 211, 214
flocculation 35
Flöhn, H. 81
flood plains 113, 115, 144
flora 141; regions 83–85; and soil 116
Florida, Everglades restoration 274–275
flowslide 32
Food and Agriculture Organization (FAO)
    241, 243–244, 278–279
food chains 155–157, 160, 269
food webs 157–159
foraging strategies 208–209
Forestry Commission 15, 235–237
forests 233–239; clearance 8, 14, 15, 44,
    234–235, 238–239; community 237;
    tropical 83, 143–144, 146, 184–185, 238
fossil fuels 123–127, 150, 151, 242, 285
fossils 12, 27, 73–74, 123, 224
Friends of the Earth 278
frontal systems 66–68
Fry, G.L.A. 184
fuel cells 127
fuels 41, 44; alcohol 41, 127; biomass 41, 42,
    127, 154, 276; fossil 123–127, 150, 151,
    242, 285

Gaia 6–7, 191
game theory 206–209
gasohol 41
Gause, Georgyi Frantsevich 173
gelifluction 113

gene banks 268
gene pool 262, 263–264, 266
gene therapy 252, 253
General Agreement on Tariffs and Trade
    (GATT) 288
general circulation models (GCMs) 49–50
genes 204–205, 209, 251–256
genetic determinism 209
genetic drift 205
genetic engineering 250–256
genetically modified (GM) foods 253,
    254–255
genomic libraries 266, 268
genotypes 176, 205, 211
geomorphology 32
geothermal energy 21
geothermal gradient 20–21
glaciers 69–73, 111–112, 113, 170
Gleason, Henry Allan 171
gleying 109, 117
Global Environmental Monitoring System
    (GEMS) 294
Global Observing System (GOS) 294
global warming potentials 46
Gloger, C. W. L. 212
Gloger's rule 212
granite 28, 29, 109, 130
gravitational water 106
grazing 175
Green Revolution 241
greenhouse effect 40, 41, 44–50, 152
greenhouse gases 6, 46–48, 50, 79, 81, 294
Greenpeace 194, 195, 253
ground water 91, 92, 96
Grove, William R. 127
Gulf Stream 63
gyres 61

habitats 183, 218–222, 262; fragmentation
    261–262, 263–264, 267; quality 8
Hadley cells 57, 145
Hadley, George 56, 57
Haeckel, Ernst 13
Halley, Edmond 56
Hawk Dove game 207
heat capacity 44
Helsinki Convention 291
hematite 28
herbicides 254–255

heterotrophs 156
Himalayas 23, 24, 25
homeostasis 188, 189, 190, 192
Human Exposure Assessment Locations
    Programme (HEALS) 294
Humboldt, Alexander von 12
humidity 65
hydration 27, 29, 107
hydrologic cycle 90–95
hydrolysis 27, 107
hydroseres 169–170
hydrosphere 5
hydrothermal vents 138
hydrothermal weathering 28

Ice Ages 34, 63–64, 69, 71–73, 78, 79–80,
    111–112, 189
ice sheets 30–32, 63–64, 75, 79, 80, 142, 170
iceberg mining 100
igneous rocks 4, 24, 25, 27, 30, 107
inbreeding depression 264, 266
index cycle 57–59
inheritance 203–204
insecticides 9, 156–157, 202, 269–270
instability 180, 182, 183
integrated pest management (IPM) 273
intensification 241–242
International Council for Science (ICSU) 49
International Panel on Climate Change
    (IPCC) 49, 50
International Union for Conservation,
    Nature and Natural Resources (IUCN)
    227, 262, 268, 278
interstades 71
intertropical convergence zone (ITCZ) 57,
    61
introduced species 216–218, 222
Ionian School 11
irrigation 103–104, 106–107
islands 178–179, 267–268
Isle Royale 155–156
Iterated Prisoner's Dilemma 207–208

jet streams 57, 58, 59
jump dispersal 216

K-selection 182–183
kaolin 28–29, 130
kaolinite 28–29, 117

oil 125; spills 275–276
oligotrophic lakes 97, 98
Olney, Warren 15
optimal foraging 208–209
optimum sustainable yield (OSY) 229–230
Organization for Economic Cooperation and
    Development (OECD) 284, 289
orogeny 24–25
Oslo Convention 291
osmosis 99, 102–103
outbreeding 264
oxidation 27, 107
oxygen 15, 53, 152–153; isotopes 80
ozone layer 53–54

Paine, Robert 222
Pangaea 21, 124
Panthalassa 21
parasitism 166, 190
Paris Convention 291
Parker, G.A. 207
Pavlov 207–208
peat 123, 124
pedestal erosion 120
pedogenesis 109, 110, 115, 117
Pellow, Robin 224
Penck, Walther 33
peneplain 33
peppered moths 202–203
perihelion 55, 56, 76
permafrost 30, 31, 112–113, 116, 126
pest control 250–251, 269–274
pesticides 8, 269–272, 273–274
petroleum 124
phenotypes 176, 205
pheromones 272–273
phosphorus 150, 151
photorespiration 154
photosynthesis 40, 103, 138, 151–155,
    176–177
photovoltaic cells 41
phylogenetic species concept 223
phytochoria 85
phytomining 277
phytoremediation 276
phytosociology 164, 165
Pinatubo, Mount 80
plate tectonics 12, 21–23, 24, 37
Pliny the Elder 11

ploughs 240–241
podzolization 116
polar regions 100, 142, 145
polar stratospheric clouds 53, 54
pollarding 238
pollution 8, 15, 196, 203, 256; air 13–14, 203,
    288, 289; control 281–287; mining 131,
    132, 134; oil spills 275–276; residence time
    4; transnational 288–294; water 282
polyclimax 173
polymerase chain reaction (PCR) 252
population 245–250; size 180–182, 190–191,
    221
precautionary principle 50
precipitation efficiency 83
Price, G.R. 207
primitive solar nebula (PSN) 19–20
Prisoner's Dilemma 206–208
Pusztai, Arpad 253
pyramid of biomass 161–162
pyramid of energy 162–163
pyramid of numbers 160–161

quadrats 165

r-selection 182–183
Rackham, Oliver 14
radiocarbon dating 75
radioisotopes 4
radiolabelling 4
radiometric dating 75–76
rainforest, definition 8
raised beaches 34
Ramsar Convention 145
range: expansion 216; fragmentation
    261–262, 263–264, 267
Rapoport, Anatol 207
Rayleigh scattering 38, 40
recognition species concept 223
Red Data Books 262
Red Queen effect 201
reduction 107
regional seas 289, 291, 293
Regional Seas Programme 291, 293
reintroductions 264, 265–266
renewable energy 40, 41, 127
renewable resources 94–95, 249
reproductive strategies 180–183
reservoirs: cycles 147; irrigation 104